Bridging Science and Policy Implication
for Managing Climate Extremes

World Scientific Series on Asia-Pacific Weather and Climate

This series will continue the collaboration with the Working Group on Tropical Meteorology Research, a component of the World Weather Research Programme, and other international weather and climate programs in publication projects.

The Editorial Board welcomes book or program/project proposals relevant to weather and climate or related fields, including interdisciplinary topics. Interested scientists please contact the Editor-in-Chief Professor Chih-Pei Chang (cpchang@nps.edu).

WORLD SCIENTIFIC SERIES ON ASIA-PACIFIC WEATHER AND CLIMATE

Vol. 10

Bridging Science and Policy Implication for Managing Climate Extremes

edited by

Hong-Sang Jung
APEC Climate Center, South Korea

Bin Wang
University of Hawaii, USA

APCC
APEC CLIMATE CENTER

World Scientific

Published by

World Scientific Publishing Co. Pte. Ltd.

5 Toh Tuck Link, Singapore 596224

USA office: 27 Warren Street, Suite 401-402, Hackensack, NJ 07601

UK office: 57 Shelton Street, Covent Garden, London WC2H 9HE

and

APEC Climate Center

12 Centum 7-ro, Haeundae-gu, Busan 48058, Korea

British Library Cataloguing-in-Publication Data

A catalogue record for this book is available from the British Library.

World Scientific Series on Asia-Pacific Weather and Climate — Vol. 10
**BRIDGING SCIENCE AND POLICY IMPLICATION FOR MANAGING
CLIMATE EXTREMES**

ISBN 978-981-3235-65-6

Preface

Over the past four decades, the number of climate-related disasters has been increasing globally. Scientific consensus based on the IPCC fifth report suggested that global warming will bring more intense and frequent extreme climate events. The Asia-Pacific, as the most disaster-prone region in the world, has already borne more than 50 percent of the global death toll from natural disasters since 1970, with a total of 6.3 billion people affected — nearly 7 times that of the rest of the world.

Increasing intensity and frequency of these extreme climate events spell major challenges ahead. These events hinder the achievement of sustainable economic growth and prosperity by disrupting supply chains, impeding production, destroying infrastructure, and necessitating high-cost rebuilding and recovery. Because of this, the reduction of disaster risk has been vaulted to the forefront of international agendas, including that of the Asia-Pacific Economic Cooperation (APEC) and the APEC Climate Center. Indeed, the Sendai Framework for Disaster Risk Reduction, the most encompassing international accord to date, was signed in 2015 to renew coordinated efforts. To mitigate the climate extreme risks and possible losses, it is essential to maximize the utilization of scientific outputs and to share best practices in disaster risk management. It is in this interest that the tenth volume of the World Scientific Series on Asia-Pacific Weather and Climate was developed.

Each year, to promote cutting edge research and help bridge the gaps between science and policy, the APEC Climate Center hosts the APEC Climate Symposium (APCS). APCS focused on "Regional Cooperation on Drought Prediction Science to Support Disaster Preparedness and Management" in Indonesia in 2013, "Managing Climate Extremes and Hydrologic Disasters" in China in 2014, and "From Science to Action: The Use of Weather and Climate Information for Efficient Disaster Risk Management" in the Philippines in 2015. This book is based on peer-reviewed manuscripts derived from a selection of the most significant research presented during APCS in 2013–2015. The aim of this book is to provide a critical understanding of climate extreme prediction and services and its application studies with a focus on climate extremes such as typhoons, droughts, and floods. Like other volumes in the World Scientific Series on Asia-Pacific Weather and Climate, it is expected that this book will serve as an important reference for researchers, forecasters, and policy makers across the region.

On behalf of the APEC Climate Center, I would like to express our deep appreciation to the series editor-in-chief Professor C.P. Chang, to all authors, reviewers and editors, to the APCS chairs, panelists and presenters, and to all the participants of the symposia for their valuable inputs and collaboration.

Hong-Sang Jung
Executive Director
APEC Climate Center

Contents

Bridging Science and Policy Implication for Managing Climate Extremes

Chapter 1

Assessing Seasonal Climate Forecasts Over Africa to Support Decision-Making

Niko Wanders* and Eric F. Wood[†]

Department of Physical Geography,
Utrecht University, Utrecht, The Netherlands
n.wanders@uu.nl
[†]*Department of Civil and Environmental Engineering,*
Princeton University, Princeton, NJ 08544, USA

Recent drought events like in the 2011 Horn of Africa and the ongoing drought in California have an enormous impact on nature and society. Reliable seasonal weather outlooks are critical for drought management and other applications like, crop modelling, flood forecasting and planning of reservoir operation, and would help reduce the potential economic damage from extremes as well as help optimize crop yields during more normal weather years from improved agricultural management. However, most seasonal forecasts are limited by low spatial and temporal model resolutions. The newly released North American Multi-Model Ensemble phase 2 (NMME-2), provides subseasonal forecast that increase the temporal resolution from monthly to daily, enabling subseasonal forecasting for end-user applications that rely on a daily temporal resolution. In this study we give an overview of the current status of the NMME subseasonal forecasts ensembles, their skill over the African continent and the forecast skill of the ensembles for applications related to agriculture and hydrology. We show that the NMME-2 subseasonal forecasts are significantly skilful for both precipitation and 2 m air temperature for large parts of Africa. The precipitation forecasts are skilful up to a lead time of two months, while temperature anomaly forecasts have a significant skill beyond the three months lead for most of Africa. Potential applications that would benefit from the new NMME-2 ensemble were studied in more detail for West Africa. We show that the models have a significant skill in forecasting the onset of the annual rain season in West Africa, and thereby the start of the growing season. Additionally, the models have a significant forecast skill to predict the onset and peak of the high flow season for most parts of West Africa. The low uncertainty in the forecasts compared to the observed anomalies indicates that local stakeholders will benefit from the high temporal resolution that the NMME-2 provides. Results encourage future research into the potential of the new subseasonal NMME-2 forecast ensemble to forecast more specific end-user applications and climate services, which require skilful high temporal resolution forecasts.

1. Introduction

Floods and drought events occur in all regions of the world with large societal impact (Kundzewicz and Kaczmarek 2000). Recent drought events like in the Horn of Africa (United Nations 2011; Sida *et al.* 2012) and the on-going drought in California (Howitt *et al.* 2014) have an enormous impact on nature and society. The 2011/2012 drought over the Horn of Africa was an example where food aid and

other international help were not deployed until it was too late and the drought was already established. This led to an increased number of fatalities and economical damage (e.g. crop and livestock losses). With adequate planning supported by long-term meteorological forecasts and hydrological modelling, the severe impact of this drought could have been reduced. Preventive measures could have been taken and valuable resources could have been reserved for the extreme drought conditions. These drought

Bridging Science and Policy Implication for Managing Climate Extremes
Edited by Hong-Sang Jung and Bin Wang

events clearly show the need for an early warning system that allows accurate monitoring and forecasting of drought conditions and other natural disasters at a continental to global scale. This also calls for improved high resolution (both in space and time) long-term meteorological forecasts.

For short-range meteorological forecasts (up to 2 weeks in advance) weather model forecasts are available at high spatial and temporal resolution from a number of centers (Yan and van der Dool 2011; Magnusson and Källén 2013). However, to increase the time that is needed to optimally use weather information for a range of end-user applications, it is important to extend the forecast range beyond the two-week period. Applications that will benefit from information from these extended forecasts include, amongst others, crop modelling, flood and drought forecasting, and planning of reservoir operations. Seasonal forecast models produce weather outlooks that range from 14 days to one year, thereby, bridging the gap between weather and climate models. They are available at a coarser resolution and lower temporal resolution (typically monthly timescale and 1° spatial resolution) compared to the high resolution weather models. Studies by Yuan *et al.* (2013) and the DEWFORA project (Dutra *et al.* 2014) showed that although they have a lower resolution, the seasonal forecasts still have the potential to detect anomalies in precipitation, surface air temperature and soil moisture over Africa up to several months in advance. However, it remains unknown how this seasonal forecast signal can impact the forecasts of the crop growing season or other hydrological variables. This is due to the fact that the monthly resolution is too coarse for many applications (e.g., onset of the growing season and start of river flows). To overcome this existing gap between the monthly scale data and climate services that rely on higher temporal resolution forecasts (i.e. daily), the North American Multi-Model Ensemble (NMME) provides subseasonal forecasts. These new higher

temporal resolution data could open new potential applications for end users that want to use subseasonal forecasts in their decision support systems and for their climate services.

In this study we provide an overview of the current status of the NMME subseasonal forecasts models, their skill over the African continent and the forecast skill of the ensemble set for applications related to agriculture and hydrology. First, we look at the subseasonal forecast skill of the models and validate the forecasts of precipitation and air temperature over Africa. Thereafter, we study the ability of the models to forecast the onset of the growing season over Western Africa. Finally, the skill in reproducing the river discharge over Western Africa is discussed.

2. Material and Methods

2.1. *Seasonal forecast models*

In this study we use the forecasts from the North American Multi-Model Ensemble phase 2 (NMME-2) project that provides subseasonal forecasts for a 31-year period at a daily temporal resolution (Kirtman *et al.* 2014). NMME-2 provides the first multi-model ensemble that produces seasonal forecast with a daily temporal resolution. Due to the improved temporal resolution this newly available forecast ensemble can help to enhance the added value of seasonal forecasts for climate service.

We assess the subseasonal forecast skill of the NMME-2 ensemble for daily precipitation and 2 m air temperature forecasts over Africa using the output from four readily available NMME-2 models for over the period 1982–2012 (Kirtman *et al.* 2014). The models used in the analysis are: CanCM3 (Merryfield *et al.* 2013), CanCM4 (Merryfield *et al.* 2013), FLOR-B01 (Vecci *et al.* 2014) and CCSM4 (Hurrell *et al.* 2013), where for CCSM part of the forecasts is not available (Table 1). Two additional models are part of the NMME-2 ensemble, however, at the time of writing this paper their precipitation forecasts

Table 1. Number of ensemble member of the individual NMME-2 models, including data availability and applied reprocessing.

	CanCM3	CanCM4	FLOR-B01	CCSM4
Temperature data availability	100%	100%	100%	71.4%
Precipitation data availability	100%	100%	100%	76.5%
Resampling	None	None	Bilinear	Bilinear
Ensemble members	10	10	12	10

were not available. The added value of these incomplete models would be hampered by the absence of important meteorological variables like precipitation.

2.2. Reference datasets

To validate the subseasonal forecasts over Africa made by the NMME-II ensemble, we use the Princeton Global Forcing (PGF, Sheffield *et al.* 2006) dataset. This dataset covers the NMME-2 re-forecast period 1982–2012 and is available with a daily temporal resolution at a 0.25° spatial resolution at the global scale, and was aggregated up to 1° to match the spatial resolution of NMME-2. Details on the dataset can be found in the respective publication.

2.3. Hydrological modelling

To evaluate the ability to use the NMME-2 for a hydrological decision support system over Africa, we used the hydrological model VIC (Liang *et al.* 1994; 1996). This model allows us to propagate the precipitation and temperature forecasts and make forecasts of hydrological variables (e.g. soil moisture, river discharge and baseflow). VIC is widely used in many hydrological studies (e.g. Sheffield *et al.* 2014; Wanders *et al.* 2015) and has been calibrated globally against observed discharge from the Global Runoff Data Centre (GRDC) to ensure accurate hydrological simulations. Using the daily temperature, VIC is able to resolve the surface energy balance, thereby providing additional valuable information on important

variables like the surface temperature and evaporation flux.

To evaluate the performance of the subseasonal forecasts using VIC, we created a baseline where VIC has been forced by observational data from the Princeton Global Forcing dataset. This reference simulation is used both for the verification of the forecasts as well as the initialization of the forecast. To ensure that the hydroclimatology of the forecast and the reference match, the NMME-2 forecast meteorological variables are bias corrected against the PGF. The bias correction adjusts the rainfall intensities and number of rain days based on matching the cumulative density functions (CDF-matching) of the forecasts and the observational data. For daily 2 m air temperature, the time series were bias-corrected against the PGF using CDF matching.

2.4. Canonical event analysis

We follow the approach developed by Roundy *et al.* 2015 to perform a canonical event analysis over a range of temporal and spatial scales. Canonical events are specifically defined spatially and temporally aggregated forecasts, with the finest event being the model resolution (1° spatially and daily) and alternative events being aggregated forecasts — for example 3° spatially and seasonally averaged forecasts. Model forecast skill can vary with canonical events. When a seasonal forecast model shows skill over a range of canonical events (i.e. a range of temporal and spatial scales), it shows signs of robust forecast skill and is therefore more

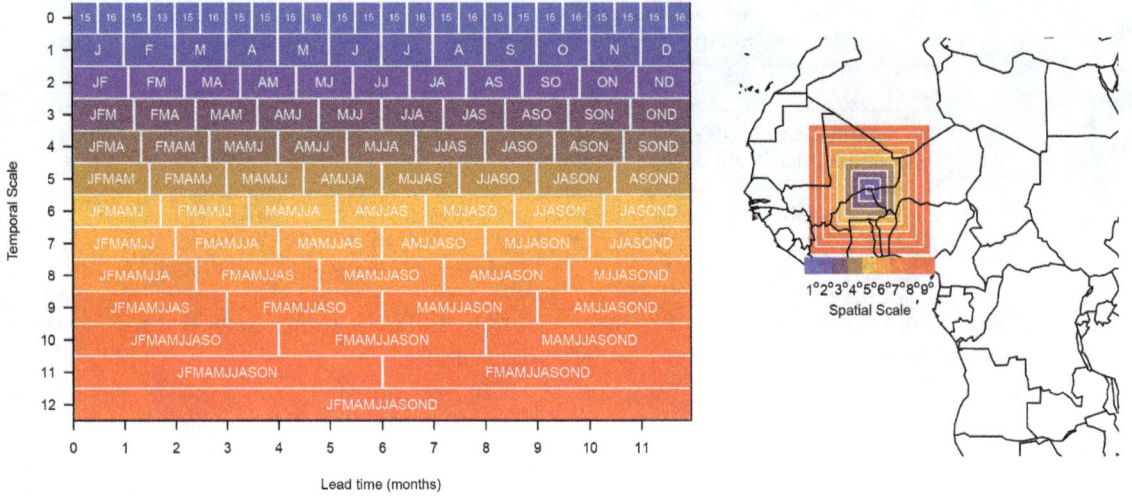

Fig. 1.   Temporal (*left*) aggregation setup of a forecast initialized in January, and the spatial aggregation (*right*) of the same January forecast.

reliable when applied in real-life forecasting situations. Forecast skill for canonical events is computed using the Spearman ranked correlation between the seasonal forecast and observed data. When at a location the correlation is statistically significant ($p < 0.05$) and positive ($R > 0.0$) we find that the model has skill for that particular temporal and spatial scale at that location. This statistically significant event is used to calculate the probabilistic predictability metric (PPM) that is given by:

$$\text{PPM} = E_S/E_t \qquad (1)$$

where E_s is the number of events that show skill and E_t is the total number of events. Following Eq. (1), it can be shown that a PPM of 0 indicates no skill for any temporal or spatial scale at that location for that initialization month, while a PPM of 1 indicates skill across all scales.

The temporal and spatial scales under consideration in this work are provided in Fig. 1, where we show the scales for a January forecast. We show that the January forecast can be used to obtain a monthly average of a meteorological variable for the coming month. However, it can also be used to get the same monthly average of an 11 month lead period. Similarly for a two

monthly temporal aggregation we can obtain the 0 to 10 month lead forecasts, while the 12 month forecast can only be obtained at a lead time of 0 months (Fig. 1). The same January forecast can also be used to produce forecasts for different temporal aggregation periods, ranging from subseasonal two week periods to 12 month periods. In total this provides us with 102 unique temporal timescales for the January forecast and 8 unique spatial scales. This leads to a combined total of 816 unique combinations of temporal and spatial scale. When the model shows a PPM of 1 this would indicate that the correlation is significantly positive for all these 816 events.

2.5. *Detection of start of agricultural and high flow season*

The start of the growing season in the sub-Saharan regions of Africa is largely dominated by large scale annual rain events coming from the South around April. Here, we used the normalized cumulative density function of precipitation to estimate the onset of the growing season. By using a normalized CDF the systematic over- or underestimation of the precipitation by the seasonal forecasts model is accounted for and

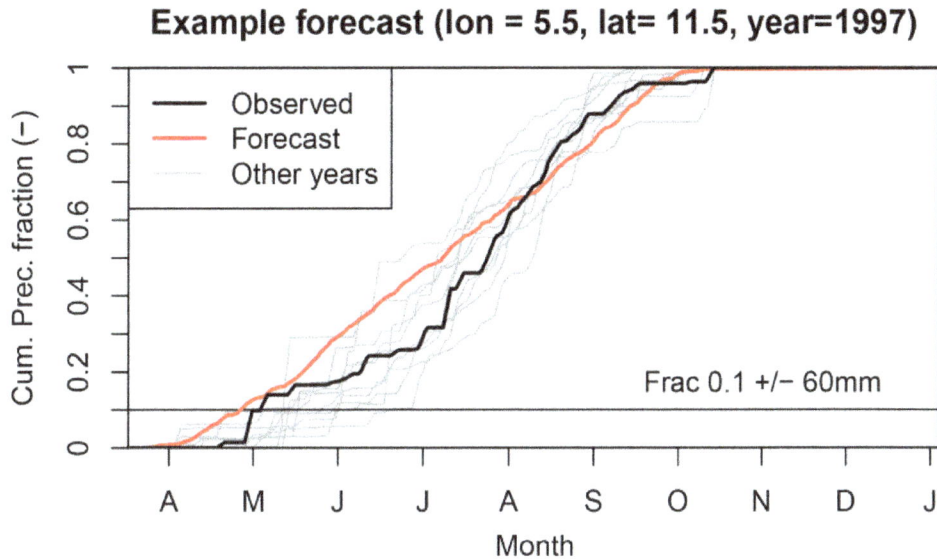

Fig. 2. Detection of the growing season as applied in this study. The normalized cumulative precipitation is shown for a seasonal model forecast from CanCM3 (red), the observed precipitation (black) and the spread over different years. The onset of the growing season is defined as the day that the precipitation exceeds the 0.1 threshold for the first time.

the percentiles can be directly compared to the observations. For all forecasts, the day that the exceedance of the 10th percentile is forecasted is selected as the day that the growing season starts and the absolute difference between this day and the reference (observed day) is calculated (Fig. 2). When the mean absolute difference between the forecasts and observations is smaller than the randomly selected year and the actual year. To assess the skill of the model to forecast the onset of the growing season, we only used the forecast issued on the 1st of March. This forecast was selected based on local information from The Centre Regional de Formation et d'Application en Agrométéorologie et Hydrologie Opérationnelle in Niger (AGRHYMET) that the onset of the growing season starts after the 15th of March and they use a seasonal forecast from the 1st of March in their operational system. Earlier forecasts lack forecast skill so the agricultural planners at AGRHYMET do not require information before the 1st of March. To ensure that the assessment of the model

forecasts skill is in agreement with the demands from local end users and farmers, we used the same forecasts as they would in their operational system.

2.6. Detection of start of agricultural and high flow season

The start of the growing season in the sub-Saharan regions of Africa is largely dominated by large scale annual rain events coming from the South around April. Here, we used the normalized cumulative density function of precipitation to estimate the onset of the growing season. By using a normalized CDF the systematic over- or underestimation of the precipitation by the seasonal forecasts model is accounted for and the percentiles can be directly compared to the observations. For all forecasts, the day that the exceedance of the 10th percentile is forecasted is selected as the day that the growing season starts and the absolute difference between this day and the reference

Example forecast (lon = 5.5, lat= 11.5, year=1997)

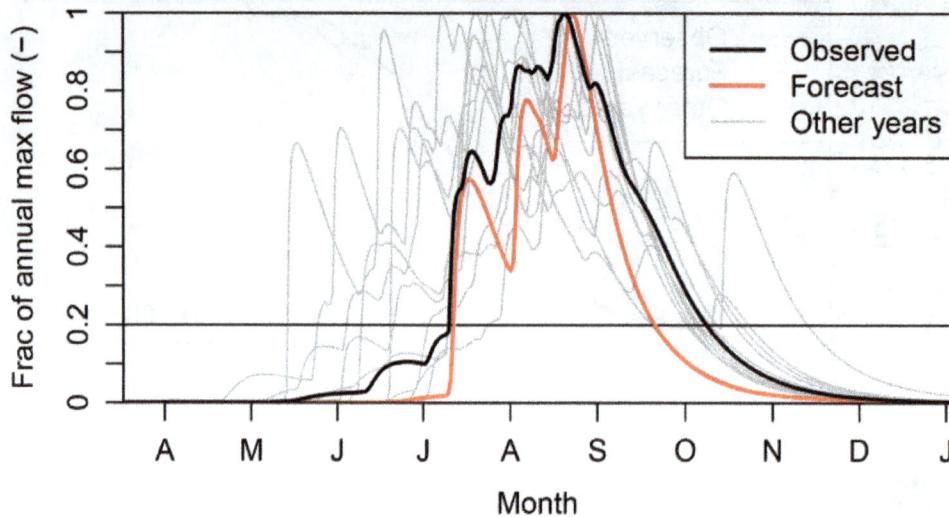

Fig. 3.   Detection of the high flow season as applied in this study. The start of the high flow season is defined as the first exceedance of the 20th percentile threshold. The peak of the high flow season is the day that the forecast hits the maximum annual discharge.

(observed day) is calculated (Fig. 2). When the mean absolute difference between the forecasts and observations is smaller than the interannual spread between the observed days, the forecast is deemed skilful. This implies that the model has a significant forecast skill when the differences between model forecast and observations are smaller than the differences between a random selected year and the actual year. To assess the skill of the model to forecast the onset of the growing season, we only used the forecast issued on the 1st of March. This forecast was selected based on local information from The Centre Regional de Formation et d'Application en Agrométéorologie et Hydrologie Opérationnelle in Niger (AGRHYMET) that the onset of the growing season starts after the 15th of March and they use a seasonal forecast from the 1st of March in their operational system. Earlier forecasts lack forecast skill so the agricultural planners at AGRHYMET do not require information before the 1st of March. To ensure that the assessment of the model

forecasts skill is in agreement with the demands from local end users and farmers, we used the same forecasts as they would in their operational system.

To detect the start of the high flow season, we used the hydrological model VIC to provide baseflow estimates. We define two metrics to quantify the skill of the model to accurately forecast the start and peak of the high flow season. First, the high flow season is normalized where the peak discharge is used to normalize the discharge (Fig. 3). The start of the high flow season is detected by looking at the first exceedance of the 20th percentile of the annual maximum flow. The mean absolute difference between the forecast and observed start of the high flow season is used as a metric for the model performance. Similar to the detection of the growing season, the offset is compared to the interannual spread between the observed onsets to identify if the model has a lower signal to noise ratio than would be obtained with using historical information. The start of the high flow season is important to determine for end-user applications that depend on the presence of open water (e.g., irrigation), while the peak flow is important for the replenishment of water resources for the upcoming dry season in this region (e.g., reservoir operations).

3. Results

3.1. *Meteorological forecast skill over Africa*

To analyze the skill of each seasonal forecast model, we first computed the Spearman ranked correlation for all forecast months separately. The obtained correlations have been used to calculate the PPM (Eq. (1)) for all initialization months, and all temporal and spatial scales. Significant positive correlations have been identified as events that show skill (E_s). The anomaly correlation for precipitation in the first week is high for the CanCM and CCSM models (Fig. 4). The forecast skill fades quickly after the first weeks, leading to low or negligible skill after a one month lead. In general, the precipitation is difficult to predict due to its high natural variability — something that is confirmed for the subseasonal forecasts. The forecast skill of FLOR is additionally hampered by the poor initial conditions that result in a poor overall forecast skill of the subseasonal precipitation. The FLOR model uses the AMIP climatological runs to initialize the land surface states in the model (Gates *et al.* 1998), which limits the forecasts skill for the first weeks. The forecast skill in these weeks largely dependent on the initial conditions of the land surface and are therefore, strongly hampered by the initialization procedure currently used by FLOR, hence the lower forecast skill.

In general, higher anomaly correlations are found for temperature forecasts compared to precipitation forecasts (Fig. 5). The decrease in the predictability metric as a function of increased lead time is not as strong as found for precipitation forecasts. The predictability over

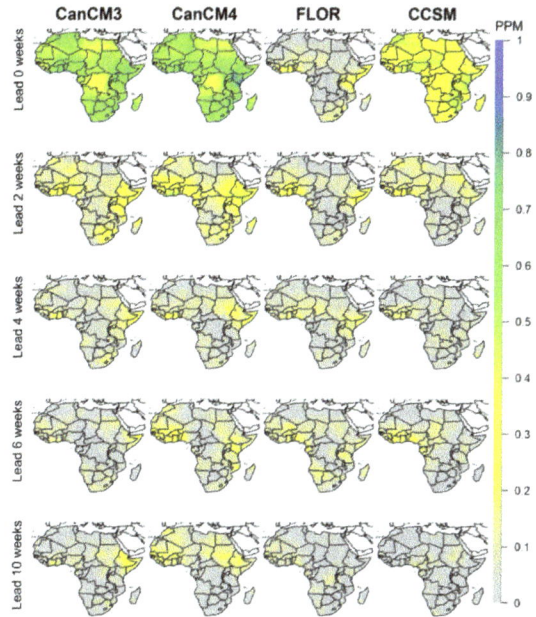

Fig. 4. Skill in precipitation forecasts for different leads and subseasonal forecast models from the NMME-2 ensemble. A PPM of 1 indicates all forecasts are skilful, for the given lead time and location.

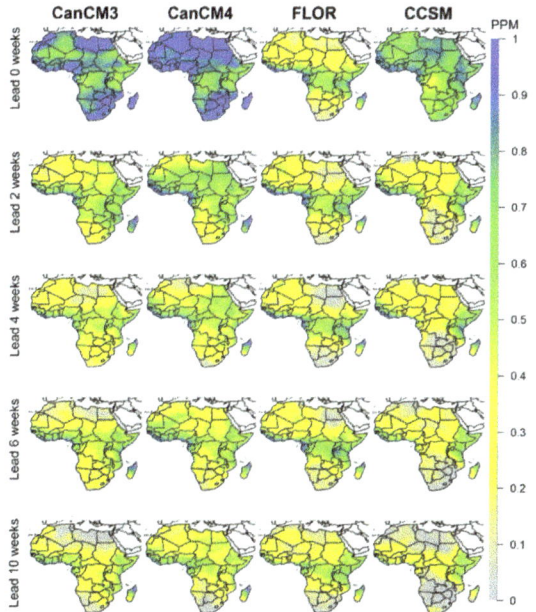

Fig. 5. Skill in temperature forecasts for different leads and subseasonal forecast models from the NMME-2 ensemble. A PPM of 1 indicates all forecasts are skilful, for the given lead time and location.

Africa (especially Equatorial Africa) is very persistent and high, with around 80–90% (PPM between 0.8 and 0.9) of the forecast scenarios showing significant skill. In contrast to the precipitation forecast, significant skill can be found beyond 3 month lead time. In addition, we see that the FLOR model outperforms the CanCM models for the longer lead times. The model still suffers from the AMIP climatological initialization method, however, the skill for the longer leads is in general higher than can be found for the other models.

For all seasonal forecast models, the seasonal precipitation forecast skill reduces with increased lead times (Fig. 6). Lead times beyond a 2 month lead time have no, to negligible prediction skill regarding precipitation anomalies at small temporal aggregation periods. This feature is not unique for Africa. Larger parts of the world are characterized by a low predictability, which quickly disappears with increasing lead time of one or two months. In general, a lower prediction skill for the higher resolution temporal anomalies is found in CanCM and CCSM models, while FLOR shows a high skill at the longer leads at longer temporal scales.

3.2. *Forecast crop growing season*

The growing season in Sub-Saharan Africa is limited by the availability of rain water that falls in the wet season between April and October because of the predominance of rain-fed agriculture. For an optimal yield, it is important that the planting of the fields be done just with the onset of the seasonal rains. If planting is too early, then the seedlings may suffer drought before the real rain season onset, planted too late and yields tend to be significantly lower. We looked at the skill of the seasonal forecast models to reproduce the onset of the rain season

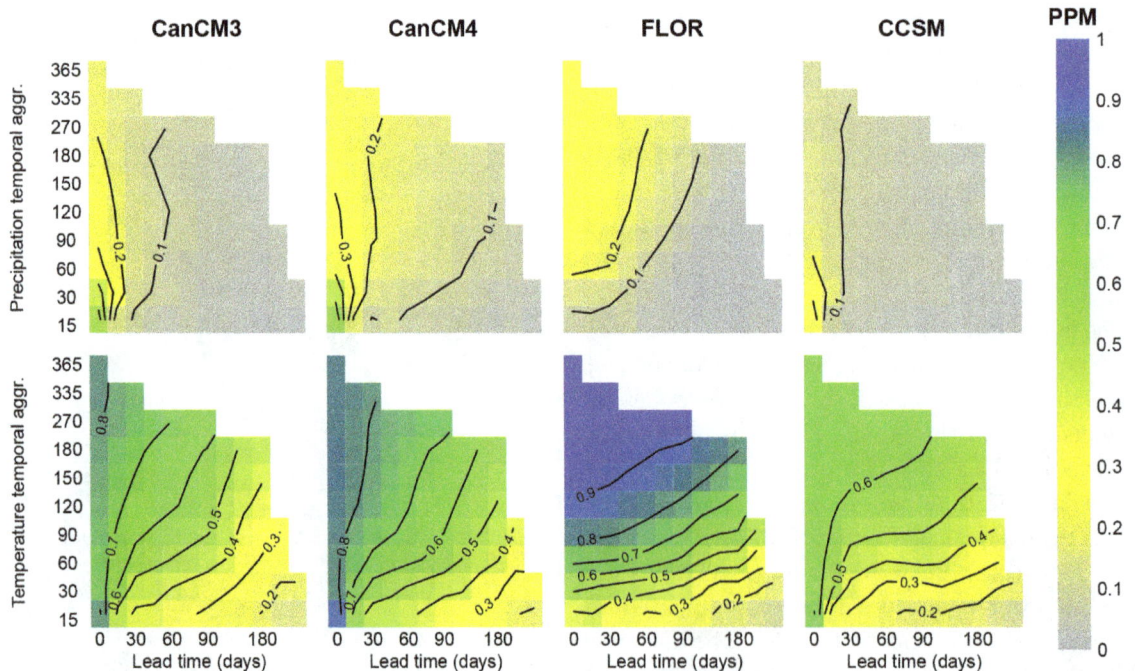

Fig. 6.   Skill of African subseasonal forecast across different temporal scales and lead times. A PPM of 1 indicates all forecasts are skilful in the given combination of temporal scale and lead time.

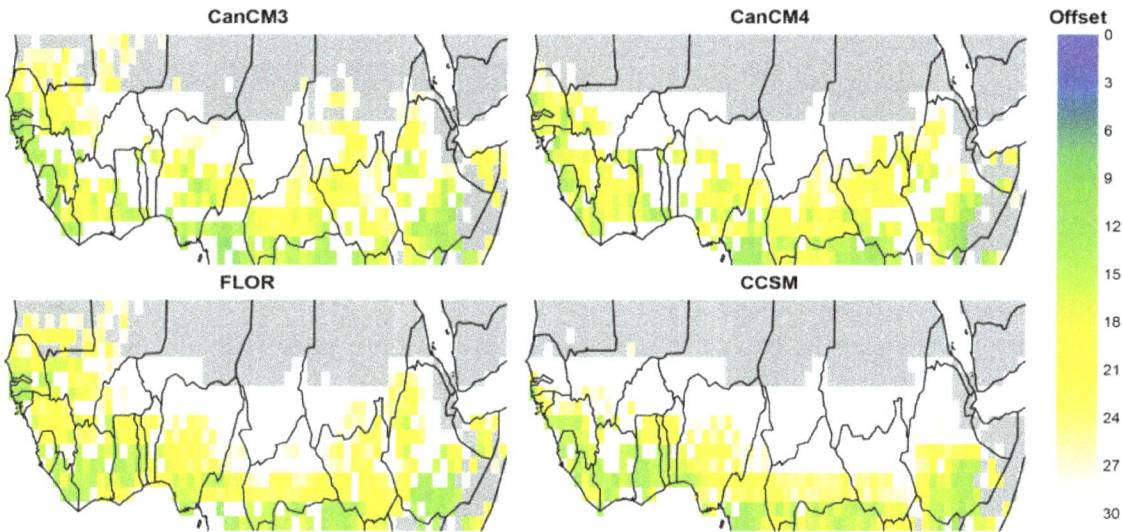

Fig. 7. Forecasts skill for the start of the growing season. Grey regions indicate where the signal to noise ratio is higher than the interannual spread in the start of the growing season. White areas indicate an offset of 30 days or more.

with the forecast issued on the 1st of March, roughly 8 weeks ahead (depending on the geographical location) of the start of the growing season (Fig. 7). We found that the forecast skill of the model is high for the coastal regions of West Africa, where the precipitation onset is early in the year and more abundant. The offset in the forecast increases in the Sahel, where rain is less predictable and the spread in the observations is larger. The models all show skill in the wetter parts of Chad, Niger and Mali, while the northern parts of these countries (with an annual precipitation of less than 200 mm) show a low predictability of the rain onset, partly due to the higher interannual variability. Small differences are found between models, indicating that the choice of model is of less importance.

3.3. Forecast high flow season

Accurate forecast of the start of the high flow season could help in water resources management at the end of a dry season. When the onset of the high flow season is accurately forecasted one could optimize the utilisation of the remaining water in reservoirs, lakes and the groundwater. Here we show that the forecasting of the start of the high flow season can be done with an accuracy of around 1 month (Fig. 8). The observed spread and onset of the start of the high flow season is larger than was observed for the start of the growing season (e.g. comparing Fig. 2 and Fig. 3). This wider initial spread makes it more difficult to produce accurate forecasts of the exact onset date. Additionally, the onset of the high flow season is later than that of the growing season. As a result, the 1st of March forecast has to accurately forecast the start of the high flow season 2–4 months ahead. It is shown that large discrepancies exist between the models, where CCSM lacks the required skill to provide forecasts with a higher accuracy than a 60 days offset. Of the remaining models, CanCM4 and FLOR show the highest skill in most regions and have similar spatial patterns. For some regions in the Sahara desert, an offset of zero days is found; this is caused by the

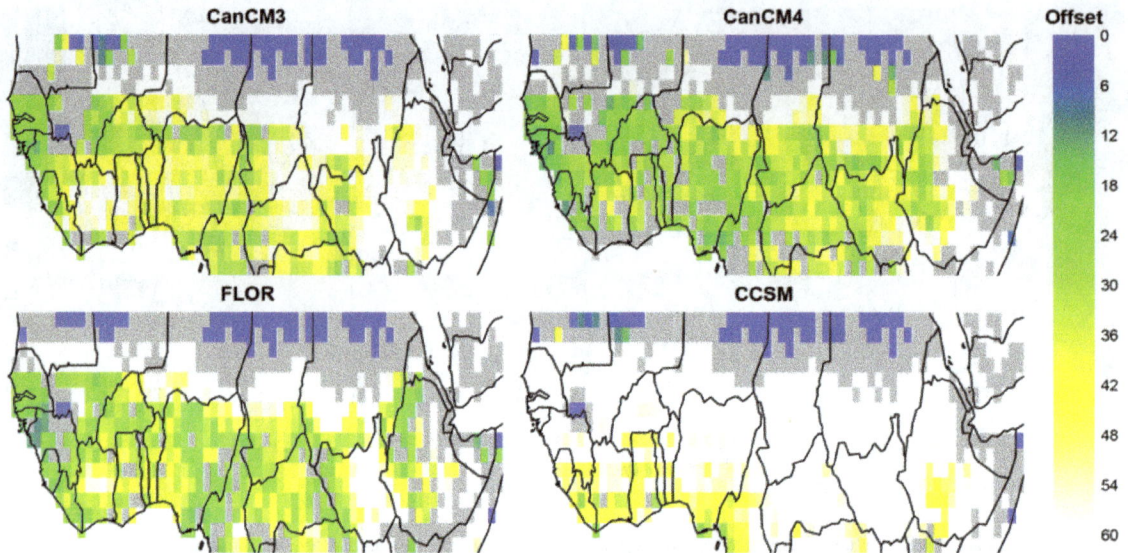

Fig. 8. Forecast skill for the start of the high flow season. Grey regions indicate where the signal to noise ratio is higher than the interannual spread in the start of the high flow season. White areas indicate an average offset of 60 days or more.

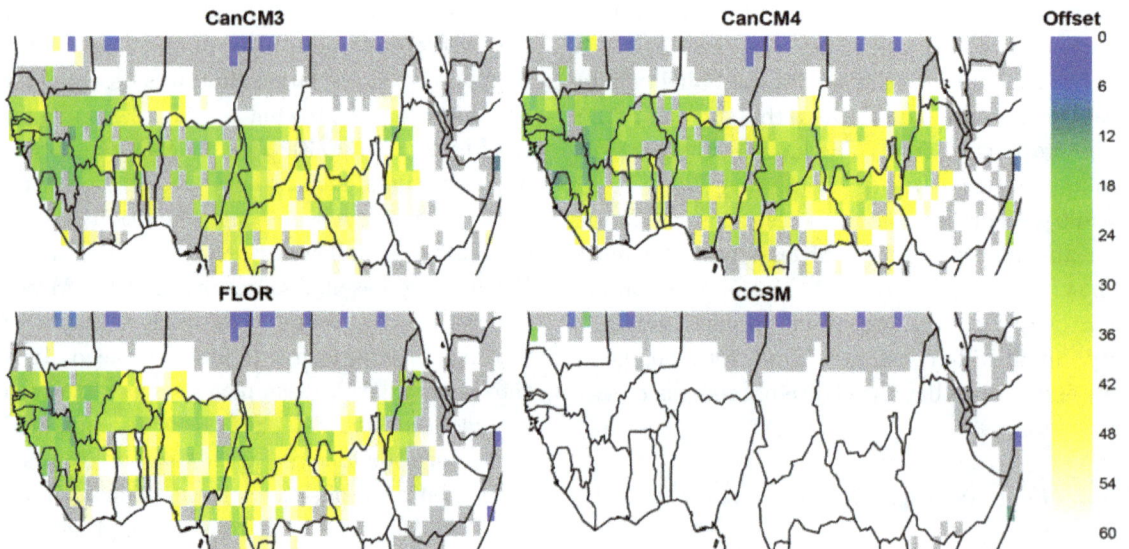

Fig. 9. Forecast skill for the peak of the high flow season. Grey regions indicate where the signal to noise ratio is higher than the interannual spread in the peak of the high flow season. White areas indicate an average offset of 60 days or more.

complete absence of river discharge in the model forecast and observations.

In addition to the start of the high flow season, it is important for water resource management to have forecast knowledge of the peak and end of the high flow season. We show that the ability of the models to accurately forecast the peak of the high flow season is almost identical to their skill in forecasting the start of the high flow season. In general the peak discharge is found 4–6 months after the 1st of March forecast initialization and hence is more difficult to accurately forecast. Nonetheless, the models show a similar offset in the forecast of the peak flow timing as they do for the onset of the high flow season. The highest skill is found in the transition regions between the wet and dry climatic regions of West Africa. These regions also have a very distinct peak flow season, which reduces the false detection of this event in the model forecasts. A low skill is found for CCSM, which is hampered by lower data availability, leading to more uncertainty in the detection skill which results in a low forecast skill. The other models again show similar patterns in terms of forecast skill and the forecasts offset compared to the observations.

4. Discussion

4.1. *Canonical event analysis*

Most assessments of model forecast skill are done based on other metrics than with the canonical event analysis combined with the Probabilistic Predictability Metric (PPM). Anomaly correlations or Brier scores are methods often used to evaluate seasonal forecast. The downside of these methods is that they either provide information at set initialization months, for set lead times (e.g., using correlations) or require the definition of specific events (e.g., exceedance over threshold, Brier score). This is not a problem when scores are required for a fixed forecast scenario (e.g. April rain, forecasted in January), however, it does not

inform us on the general performance of the model. Therefore, we argue that unless these clear goals exist, it is beneficial to use the PPM as a score that informs the end user on the average model performance at a specific location or a decomposition of the performance as a function of temporal aggregation and lead time.

A question that remains is how the obtained forecasts skills in terms of PPM could be translated towards applications. From the description of the PPM, it can be derived that a PPM of 0.1 indicates significantly skilful forecast in 10% of the scenarios. This information may be more intuitive to end users than an anomaly correlation for a forecast. However, it does not inform the user on the actual strength of the skill in these 10% — this could be anything between a correlation of 0.35 (lower significance boundary) to 0.75 (highest correlation found). On the other hand, it has the advantage that it provides a linear skill metric that can be used to determine the average skill of the model over a specific location or season in a more comprehensive way.

From the PPM analysis (e.g., Fig. 6), it is shown that different forecast models should be used for different purposes. The CanCM and CCSM models clearly provide the best subseasonal short range, high temporal resolution forecast that can inform end users on events that are happening within the coming month. When more extensive planning is required (e.g. water resource management), FLOR seems to do a better job at predicting the long-term precipitation and temperature anomalies. The improved forecast skill of FLOR with longer leads could be strongly related to the ability to have a more accurate forecast of global sea surface temperatures (SST). Strong teleconnections have been found between SST anomalies and observed rainfall variability over the African continent (Rowell 2013). Strong teleconnections are found for the El Nino, Atlantic and Indian Ocean Dipole indices. Currently, FLOR is the only model in this ensemble that doesn't have advanced land initialization scheme embedded in their forecast system. This suggests that a potential improvement in the first months after forecast initialization could be expected when a more advanced land initialization is implemented. This is currently a topic that is under study and but no definitive results have been obtained so far.

The results with regard to differences in forecast skill in both short-term and long-term forecast for different models can help to improve the quality of decision support system and their role in providing valuable forecasts for climate services. Additionally, the PPM analysis can inform end users when not to use seasonal forecasts for their decision-making process. This prevents high-impact measures to be taken based on overall poor model forecasts quality or a poor forecast quality for the lead time and temporal scale of interest.

4.2. *Growing season*

To accurately detect the start of the growing season, multiple factors need to be taken into account. First of all, the start of the growing season is largely determined by the local climatology and whether the start of the growing season is limited by the amount of available energy (e.g., large parts of Europe) or the available water for crop growth (e.g., West Africa). For large parts of Africa the start of the growing season is largely dominated by precipitation patterns and amounts, and therefore has a more uncertain start than can be found for energy limited regions that are heavily dependent on incoming solar radiation. The added value of having accurate forecasts for the start of the rain season is significant since they dominate much of the agricultural activity in West Africa and the Horn of Africa (Barbé *et al.* 2002). These are regions that can clearly benefit from accurate subseasonal forecasts to optimize the potential crop yield.

In general, the detection of the growing season is hampered by multiple factors. Firstly, it is important to match the modelled start of the growing season with the perception of the farmers in these regions. Due to an existing collaboration with AGRHYMET, we could determine the guidelines used for the onset of the growing season in these regions in Africa. In general, AGRYHMET applies a threshold of 3 days of consecutive rain exceeding a total of 20 mm of accumulated precipitation. Since the seasonal forecast models are known to have too much drizzle (low rainfall intensities) in their forecast we transformed this 20 mm threshold to the 10th percentile of the annual rainfall. Using the observational dataset we validated this threshold and found that these two criteria are mostly fulfilled in a time period of one week. This ensured that we have a more objective threshold that could be applied for a large region. The second problem for detecting the growing season is related to the different crops that are planted in different regions. Every crop has specific demands and therefore depends on different planting dates with relation to the rainfall. This information is not available with the forecast and is therefore difficult to implement in the growing season detection methodology. We assumed here that the methodology supplied by AGHRYMET would be the best benchmark for the start of the growing season.

From the results it is clear that the detection of the growing season clearly benefits from the daily temporal resolution of the NMME-2 forecasts. By using daily resolution forecasts, we can now actually address these more detailed questions related to local climate services. It is shown that we can use these models to forecast the onset of the growing season roughly two months in advance. This could help local farmers and authorities to prepare for the coming growing season and make sure resources are in place in time. Additionally, when a very late onset of the growing season is forecasted, farmers can switch to crops that have lower water demand or produce their yield in a shorter growing season (Brumbelow and Georgakakos 2001).

This is only possible due to the higher temporal resolution of the data. This leads to a major improvement in the usability of seasonal forecasts in agricultural planning.

4.3. *High flow season*

The high flow season in most parts of Africa is highly seasonal and strongly linked with the precipitation season (e.g. Fig. 3). In West Africa this very strong seasonality creates a high need for water resource planning and management, since available resources are low during large parts of the year. The added value of subseasonal forecasts could be substantial if the duration and magnitude of the flow season can be accurately forecasted. The timing of the peak discharge is important for water resources management, where the peak flow determines potential to replenish reservoirs after long dry periods, the timing of the peak discharge is important. A large number of reservoirs depends on a distinct wet season to refill their storage for the upcoming drier periods. Additionally, the remaining water in the reservoirs can be optimally utilized when the onset of the high flow season and peak discharge is known. We show here that the seasonal forecast models are capable of accurately forecasting the timing of the annual discharge peak in West Africa and could therefore, be a valuable tool for water resource planning in early spring.

A valuable next step would be to perform a more detailed assessment of the added value of subseasonal forecasts for hydrological modelling and end users that require detailed hydrological subseasonal forecasts. These results provide an initial result on the potential of the new subseasonal forecast models for application-based climate services.

5. Conclusions

In this study, we show that subseasonal forecasting from the NMME-2 ensembles can substantially improve the climate services that depend on high temporal resolution forecasts. We show that the forecast skill for both precipitation and 2 m air temperature is significantly skilful for many regions in Africa. Most notably, there is a high skill for both subseasonal precipitation and temperature over West Africa and the Horn of Africa. In both regions, we find relatively high population densities in combination with highly seasonal rainfall and a high dependency of the agriculture on these seasonal rainfall events (Barbé *et al.* 2002). The skill in the precipitation forecast fades quickly with increased forecast lead, while the forecast skill for temperature anomalies remains high for lead times beyond 3 months. It is shown that the skill of the models varies across the temporal aggregation window and lead times. For example, the FLOR model shows a relatively low skill on the subseasonal scale for short leads. However, this model shows a stronger forecast skill for longer lead times and larger temporal aggregation windows. This indicates that models should be used accordingly for applications that rely on forecast at a longer lead or larger temporal resolution.

To study the impact of the NMME-2 ensemble for application driven scientific questions, we looked at two important end-user applications in West Africa. This region is of particular interest since we have information on the end-user requirements. The weather patterns have a large impact on the agricultural system and the seasonal forecast models show significant skill in this region. Results show that the models have significant skill in forecasting the onset of the annual rain season and thereby the start of the agricultural growing season. This is a promising result, where the increased temporal resolution of the seasonal climate models, from a monthly to a daily temporal resolution, helps meet end user requirements.

In addition to the detection of the growing season, we analyzed the skill of the NMME-2 ensemble to forecast important hydrological events. First, we looked at the predictability of the onset of the high flow season. This is important moment for agriculture and water resource management, since resources are needed for irrigation and replenishing the lost water resources over the dry winter period. In West Africa there is a clear seasonal pattern, going from none to negligible river discharge to a high flow season 1 to 2 months after the first rainfall events. The models have a significant forecast skill to predict the onset of the high flow season for most parts of West Africa. The average onset is within the sub-monthly domain, indicating that the there is an additional gain by using the subseasonal forecast. Similar results were found for the onset of the peak flow season, where the uncertainty in the forecast is often lower than 30 days for many regions in West Africa. This indicates that local stakeholders will benefit from the NMME-2 ensemble for their decision support systems.

Overall, it is concluded that with the introduction of NMME-2, progress has been made going from seasonal to subseasonal forecasting. We argue that additional progress can be made, by extending the existing the ensemble with more models and by looking more in-depth at specific climate services.

Acknowledgments

This research was supported by the NOAA Climate Program Office under grant NA15OAR431 0075 (Assessing Phase 2 NMME Forecasts for Improved Predictions of Drought and Water Management) to EFW, and NW was supported by a NWO Rubicon Fellowship 825.15.003. (Forecasting to Reduce Socio-Economic Effects of Droughts).

References

Brumbelow, K. and A. Georgakakos, 2001: Agricultural planning and irrigation management: The need for decision support. *The Climate Report*, **1**(4), 2–6.

Dutra, E., W. Pozzi, F. Wetterhall, F. Di Giuseppe, L. Magnusson, G. Naumann, P. Barbosa, J. Vogt, and F. Pappenberger, 2014: Global meteorological drought — Part 2: Seasonal forecasts. *Hydrol. Earth Syst. Sci. Discuss.*, **11**, 919–944. doi:10.5194/hessd-11-919-2014.

Fan, Y. and H. van den Dool, 2011: Bias correction and forecast skill of NCEP GFS ensemble week-1 and week-2 precipitation, 2-m surface air temperature, and soil moisture forecasts. *Weather Forecast.*, **26**, 355–370. doi:10.1175/WAF-D-10-05028.1.

Gates, W. L., J. Boyle, C. Covey, C. Dease, C. Doutriaux, R. Drach, M. Fiorino, P. Gleckler, J. Hnilo, S. Marlais, T. Phillips, G. Potter, B. Santer, K. Sperber, K. Taylor, and D. Williams, 1998: An overview of the results of the atmospheric model intercomparison project (AMIP I). *Bull. Am. Meteorol. Soc.*, **73**, 1962–1970.

Howitt, R., J. Medelln-Azuara, D. MacEwan, J. Lund, and D. Sumner, 2014: Economic analysis of the 2014 drought for California agriculture, Report of UC Davis, Center for Watershed Sciences.

Hurrell, J. W., M. M. Holland, P. R. Gent, S. Ghan, J. E. Kay, P. J. Kushner, J.-F. Lamarque, W. G. Large, D. Lawrence, K. Lindsay, W. H. Lipscomb, M. C. Long, N. Mahowald, D. R. Marsh, R. B. Neale, P. Rasch, S. Vavrus, M. Vertenstein, D. Bader, W. D. Collins, J. J. Hack, J. Kiehl, and S. Marshall, 2013: The community earth system model: A framework for collaborative research. *Bull. Am. Meteorol. Soc.*, **94**, 1339–1360. doi:10.1175/BAMS-D-12-00121.1.

Kirtman, B. P., Dughong Min, Johnna M. Infanti, James L. Kinter, III, Daniel A. Paolino, Qin Zhang, Huug van den Dool, Suranjana Saha, Malaquias Pena Mendez, Emily Becker, Peitao Peng, Patrick Tripp, Jin Huang, David G. DeWitt, Michael K. Tippett, Anthony G. Barnston, Shuhua Li, Anthony Rosati, Siegfried D. Schubert, Michele Rienecker, Max Suarez, Zhao E. Li, Jelena Marshak, Young-Kwon Lim, Joseph Tribbia, Kathleen Pegion, William J. Merryfield, Bertrand Denis, and Eric F. Wood, 2014: The North American multimodel ensemble: phase-1 seasonal-to-interannual prediction; phase-2 toward developing intraseasonal prediction. *Bull. Am. Meteorol. Soc.*, **95**, 585–601. doi:10.1175/BAMS-D-12-00050.1.

Kundzewicz, Z. W. and Z. Kaczmarek, 2000: Coping with hydrological extremes. *Water International*, **25**(1), 66–75. doi:10.1080/02508060008686798.

Luc Le Barbé, Thierry Lebel, and Dominique Tapsoba, 2002: Rainfall variability in West Africa during the years 1950–90. *J. Climate*, **15**, 187–202. doi:http://dx.doi.org/10.1175/1520-0442(2002)015<0187:RVIWAD>2.0.CO;2.

Liang, X., D. P. Lettenmaier, E. F. Wood, and S. J. Burges, 1994: A simple hydrologically based model of land surface water and energy fluxes for general circulation models. *J. Geophys. Res.*, **99**(D7), 14415–14428.

Liang, X., E. F. Wood, and D. P. Lettenmaier, 1996: Surface soil moisture parameterization of the VIC-2 L model: Evaluation and modification. *Global and Planetary Change*, **13**(14), pp. 195–206.

Magnusson, L. and E. Källén, 2013. Factors influencing skill improvements in the ECMWF forecasting system. *Mon. Weather Rev.*, **141**, 3142–3153. doi:10.1175/MWR-D-12-00318.1.

Merryfield, William J., Woo-Sung Lee, George J. Boer, Viatcheslav V. Kharin, John F. Scinocca, Gregory M. Flato, R. S. Ajayamohan, John C. Fyfe, Youmin Tang, and Saroja Polavarapu, 2013: The Canadian seasonal to interannual prediction system, Part I: Models and Initialization. *Mon. Weather Rev.*, **141**, 2910–2945.

Roundy, Joshua K., Xing Yuan, John Schaake, and Eric F. Wood, 2015: A framework for diagnosing seasonal prediction through canonical event analysis. *Mon. Weather Rev.*, **143**, 2404–2418. doi:http://dx.doi.org/10.1175/MWR-D-14-00190.1.

Rowell, D. P., 2013: Simulating SST Teleconnections to Africa: What is the state of the art?. *J. Climate*, **26**, 5397–5418. doi:10.1175/JCLI-D-12-00761.1.

Sheffield, J., G. Goteti, and E. F. Wood, 2006: Development of a 50-yr high-resolution global dataset of meteorological forcings for land surface modeling. *J. Climate*, **19**(13), 3088–3111.

Sheffield, J., E. F. Wood, N. Chaney, K. Guan, S. Sadri, X. Yuan, L. Olang, A. Amani, A. Ali, S. Demuth, and L. Ogallo, 2014: A drought monitoring and forecasting system for sub-Sahara African water resources and food security. *Bull. Am. Meteorol. Soc.*, **95**(6), 861–882.

Sida, L., B. Gray, and E. Asmare, 2012: Real-time evaluation of the humanitarian response to the horn of African drought crises. Tech. report, Inter-Agency Standing Committee.

United Nations, 2011: Humanitarian requirements for the horn of Africa drought 2011. Tech. report, Office for the Coordination of Humanitarian Affairs (OCHA), New York and Geneva.

Vecchi, G. A., T. Delworth, R. Gudgel, S. Kapnick, A. Rosati, A. T. Wittenberg, F. Zeng, W. Anderson, V. Balaji, K. Dixon, L. Jia, H.-S. Kim, L. Krishnamurthy, R. Msadek, W. F. Stern, S. D. Underwood, G. Villarini, X. Yang, and S. Zhang, 2014: On the seasonal forecasting of regional tropical cyclone activity. *J. Climate*, **27**, 7994–8016. doi:10.1175/JCLI-D-14-00158.1.

Wanders, N., M. Pan, and E. F. Wood, 2015: Correction of real-time satellite precipitation with multi-sensor satellite observations of land surface variables. *Remote Sens. Environ.*, **160**, 206–221. http://doi.org/http://doi.org/10.1016/j.rse.2015.01.016.

Yuan, X., E. F. Wood, N. W. Chaney, J. Sheffield, J. Kam, M. Liang, and K. Guan, 2013: Probabilistic seasonal forecasting of African drought by dynamical models. *J. Hydrometeorol.*, **14**, 1706–1720.

Chapter 2

Variability and Predictability of Climate Linked to Extreme Events

Swadhin Behera

Application Laboratory, JAMSTEC Yokohama, Japan
behera@jamstec.go.jp
http://www.jamstec.go.jp/res/ress/behera/

The climate variability in tropical oceans plays an important role not only in the regional weather and climate but also in the extreme events leading to socio-economic loses. Here, the roles of some of those modes of climate variability are investigated and reviewed. Particularly El Niño/Southern Oscillation (ENSO), Indian Ocean Dipole (IOD) and ENSO Modoki are explored for their roles in the extreme events ranging from blistering summers to overflowing rivers. While the role of ENSO in extreme events is extensively studied and recognized, the role of other climate modes is not so well understood. Here, it is shown that IOD and ENSO Modoki are responsible for some of the extreme events even more than ENSO.

1. Introduction

Many parts of the world, particularly the growing economies in the Asia-Pacific sector, is vulnerable to floods, droughts, and other extreme climate related disasters. Several extreme flood events are witnessed in recent years. Thailand floods in 2011 caused huge damages to the industrial areas of Bangkok (Tsai *et al.* 2015). In addition to the huge economic losses, the extreme event created a sudden scarcity of computer hard disks due to damages to that industry. Chennai, a busy industrial city in the southeastern part of India, experienced one of the worst floods of the century during November–December of 2015. While these extreme flood events consistently damaged parts of Asia, intermittent and successive droughts have also caused havoc in various regions: For example, several parts of Australia and the United States have been experiencing droughts for several years. These variations could possibly be linked to the changes in modes of climate variation in addition to the global warming. Furthermore, rapid urbanization, mis-management of civil constructions and piling of household and industrial wastes are severely affecting the rivers and regional hydrology and leading to climate hazards. Under these anthropogenic stresses, societal impacts of the extreme events are aggravated further.

The modes of climate variations that are seen to anchor some of the extreme events have been studied over the years. Climate variability, unlike the secular warming trend recently denoted as anthropogenic climate change, refers to a condition in which climate oscillates significantly around a normal state from seasons to decades due to various factors inherent to the earth system. In particular, the air-sea interactions in the tropical oceans give rise to several modes of climate variations, viz., the El Niño/Southern Oscillations (ENSO), the ENSO Modoki, the Indian Ocean Dipole (IOD), the subtropical Indian Ocean Dipole (SIOD) to name a few (e.g., the review in Yamagata *et al.* 2015).

Over the last few decades, considerable progresses have been made in the understanding of ENSO variability. It is also notable that the ENSO predictability has improved dramatically over the years. Nonetheless, for some of the recent events, for example for the El Niños of 2009 and 2012 climate predictions did not turn out to be so accurate to predict the triggerings and terminations of those events in addition to their exact amplitudes. While the efforts to improve our understanding of ENSO dynamics and to refine the ENSO predictability continue, the scope and dimensions of this research have expanded considerably over the last couple of decades to explore the role of other climate phenomena that directly or indirectly influence the periodicity, evolution and strength of ENSO besides the regional climate variations. The Indian Ocean Dipole (IOD) is found to be one such phenomenon.

The Indian Ocean is traditionally treated as a passive ocean basin with only variations associated with the ENSO-induced warming and cooling. This viewpoint has changed in the recent decades after the discovery of the IOD (Saji *et al.* 1999; Yamagata *et al.* 2004). The IOD is shown to influence the ENSO evolution and periodicity in addition to the weather and climate of several parts of the world. Hence, a better prediction of IOD is essential for improving the ENSO as well as regional climate variations. ENSO Modoki in the tropical Pacific (e.g., Ashok *et al.* 2007; Weng *et al.* 2008), different flavors of ENSO (Larkin and Harrison 2005; Yeh *et al.* 2009; Kug *et al.* 2009, Kao and Yu 2009), Atlantic Niño (Chang *et al.* 2006), subtropical dipole modes in southern Atlantic, Indian and Pacific Oceans (Behera and Yamagata 2001; Fauchereau *et al.* 2003; Suzuki *et al.* 2004; Hermes and Reason 2005; Morioka *et al.* 2010; 2012; 2013; 2014) are also important for their associated (with ENSO, IOD and other modes) as well as distinctive influences on the regional rainfall and temperature variability. In addition, several local ocean-atmosphere coupled phenomena are also recently identified near eastern boundaries. These coastal Niños and Niñas, e.g., Ningaloo Niño/Niña, California Niño/Niña, Dakar Niño/Niña, are shown to influence the coastal ecosystem as well as the

regional climate (Feng *et al.* 2011; Kataoka *et al.* 2014; Yuan and Yamagata 2014; Oettli *et al.* 2015; Yamagata *et al.* 2015).

The climate variation impact on the society is recognized through some of the societal implications. Ground hydrology and river stream flows are some of the indicators of climate impacts. In this chapter, the tropical climate impact on the stream flows of Citarum River (e.g., D'Arrigo *et al.* 2009; Sahu *et al.* 2012) and Pranaiaba River (Aceituno 1988; Marengo 1995; Sahu *et al.* 2014) are discussed. Stream flow of the Citarum River basin in Indonesia is mainly dependent on the seasonal monsoon rainfall from October to May. Some of the previous studies suggest that the rainfall pattern of Java has its strongest climatic connection with the ENSO (Aldrian and Susanto, 2003; Hendon 2003). It is found that the seasonal low-pressure and associated rainfall zone over Indonesia shifts eastward causing drought in most parts of the country during El Niño events. It is also recognized that cold sea surface temperature (SST) anomalies off Java-Sumatra associated with the positive phase of IOD suppress convection over western Indonesia (Behera *et al.* 1999; Behera and Yamagata 2003). On the other hand, the Parnaíba river discharge of northern Brazil, nearer to the eastern side of the Pacific Ocean, is seen to be influenced by the newly found El Niño Modoki (Sahu *et al.* 2014). These results are very promising for developing the stream flow prediction systems since the leading climate models are now routinely predicting those climate modes on long lead times (e.g., Luo *et al.* 2015).

It is widely recognized that the seasonal climate predictability is mostly rooted in the processes of the tropical upper ocean — the so-called ocean memory. The dynamical oceanic processes, such as the Rossby and Kelvin wave propagations and associated basin-wide ocean adjustment or subsurface temperature anomalies form the basis of this ocean memory. Those processes help trigger tropical climate anomalies which rapidly grow via active ocean-atmosphere interactions and hence provide key precursors for the dynamical climate prediction (e.g., Luo *et al.* 2011).

Sophisticated ocean-atmosphere coupled general circulation models (CGCMs) are now able to predict climate variations and their dynamically linked teleconnections all over the globe. Two general approaches are often adopted to assess climate predictability. One is for potential predictability assessment by assuming both model and initial conditions are perfect. While this method is useful for assessing the upper limit of climate predictability, the estimated predictability may be strongly biased to model errors. The other approach is for practical predictability assessment by measuring model ability in forecasting the observed climate from a realistic initial condition. A few decades ago, statistical models were routinely used for predictions of climate variations on seasonal to interannual time scales. Those are now replaced by sophisticated dynamical climate prediction systems. The advent in advanced computing system to churn out a huge amount of computations (now of the order of petaflop) in a very short time has helped in the rapid development of dynamical global prediction systems.

The skill scores of present generation dynamical models are comparable or better than that of the statistical models. Statistical models are built based on a linear or nonlinear relationship between a set of factors (i.e., predictor) and phenomena to be predicted (i.e., predictand). The relationships used in the model can be acquired from a physical association supplemented by dynamical understanding of those phenomena and processes, or pure statistical associations. However, those predictions are conducted by assuming that statistical associations built from historical observations will not necessarily remain intact in the future particularly under global warming scenarios. Hence, dynamical prediction using CGCMs has emerged as a better option. Moreover, the dynamical models

could capture the extreme events if the models resolve the necessary dynamics and physics.

2. Modes of Tropical Climate Variations Linked to Extreme Events

Traditionally, ENSO is well-recognized as a climate variability mode in the tropics. It is also shown as one of the most influential climate modes on earth; some of the extremes in regional rainfall and temperature variations are linked to ENSO variability. In recent years, a few other tropical climate modes are discovered. For example, the IOD is shown to be a dominant mode of climate variations in the Indian Ocean (Saji *et al.* 1999; 2003). In addition, a new flavor of the ENSO called ENSO Modoki (Ashok *et al.* 2007; Weng *et al.* 2008) is also identified in the Pacific. These newly identified climate modes also have quite significant influences on Indo-Pacific sector as well as other far off regions through the atmospheric and oceanic teleconnections.

The climate variation links to extreme events are discussed in several recent articles. The association of ENSO, IOD and ENSO Modoki with phenomena ranging from extreme summer monsoon rainfall to river stream flows is reviewed here.

2.1. *IOD and ENSO connections to Indian summer monsoon*

The ENSO connection to Indian summer monsoon rainfall is historically observed to be quite strong (e.g., Rasmusson and Carpenter 1983). However, the relationship is shown to have weakened in recent decades (Kumar *et al.* 1999). It has been revealed that the influence of the ENSO on the monsoon is complemented by the IOD (Behera *et al.* 1999) and that the frequent occurrence of the IOD in the last couple of decades has weakened the ENSO-monsoon relationship (Ashok *et al.* 2001; Behera *et al.*

2003). While investigating the unusual monsoon of 1994, Behera *et al.* (1999) observed that the moisture divergence in the southeast Indian Ocean, associated with the positive IOD (pIOD) of that year helped maintain the monsoon rainfall over India and several parts of Southeast Asia.

The significance of the eastern pole of the IOD as an important source of moisture during 1994 was depicted by the anomalies of moisture divergence from that region and convergence over Indian subcontinent. Ashok *et al.* (2001) further confirmed this relationship with a long record of data and found that the emergence of the IOD influence in the Indian Ocean has reduced the ENSO impacts on monsoon. This explains the observed normal monsoon rainfalls over India recently during some of the strong El Niño years such as 1997 and 2006 that co-occurred with pIOD. The Indian summer monsoon rainfall (ISMR) usually suffers from an ensuing El Niño in the absence of pIOD. Although this tripartite relationship appears robust, its past variation is obscure due to the lack of reliable instrumental observations. Nevertheless, based on available observations, it is noted that the relationship between the Southern Oscillation Index and ISMR has dramatically changed over the decades. The correlation between them was close to 1 during the first decade of the 20th century as compared to a negligible relationship during the last decade, frequented by pIODs, of that century (Fig. 1).

Some of the recent studies have also extended the IOD index to historical period through proxy information extracted from coral records (e.g., Abram *et al.* 2003; Kayane *et al.* 2006). The coral index further corroborates the changing relationship between the ENSO-monsoon and the IOD-monsoon relations. The relationship between the ENSO and Indian summer monsoon rainfall is weakened in the period frequently met by the IOD (e.g., in the 1990s). That is when the correlation between the coral IOD index and the ISMR is strong. On the other hand, during the early part of the 20th century, the ENSO-Indian

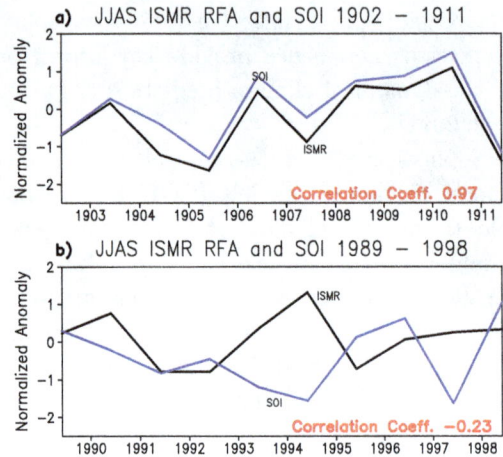

Fig. 1. The time series of Southern Oscillation Index (SOI) and Indian summer monsoon rainfall (ISMR) for (a) first and (b) last decades of the 20th century.

summer monsoon rainfall correlation was significantly high (e.g., Abram *et al.* 2008; Nakamura *et al.* 2009; 2011). The Indian Ocean was strongly influenced by the ENSO in that period, which has been marked by some of the worst droughts in India (Walker 1925). This ENSO domination during the early part of the 20th century has influenced the long-term weak correlation between the ISMR and the coral IOD index.

Nakamura *et al.* (2009), based on the coral records from the Lake Victoria in Kenya, suggested that the influence of the ENSO has decreased over the western Indian Ocean in recent decades. This resulted in frequent occurrences of the IOD and its influence on the Indian and East African climates. They also suggested that a mode shift in IOD variability related to the warming trend in the western Indian Ocean has raised the mean SST to a threshold value that encourages tropical convections. This has favored the atmospheric convection to shift to the west. The change in the mean condition is one of the triggers of frequent pIOD events and intense short rains in East Africa.

In an earlier study, Abram *et al.* (2008), based on the coral records from the eastern Indian Ocean, also suggested that the IOD season rainfall has progressively decreased in western Indonesia and increased in eastern Africa through the 20th century. Furthermore, based on a coral dipole mode index (by taking the difference between eastern and the western Indian Ocean coral records), they have suggested that there has been an exceptional increase in the frequency and strength of pIOD events during the 20th century associated with enhanced upwelling in the eastern pole of the IOD. Perhaps this is related to changes in the Walker circulation associated with the global warming as suggested by them. They also found that the IOD events were significantly associated with reduced monsoon rainfall before 1905 probably resulting indirectly from a strong inverse relationship between ENSO and the Asian monsoon.

2.2. *IOD and ENSO links to extreme hot summers*

Europe has experienced unprecedented heat waves during several recent summers. Many parts of Europe have seen some of the worst human and ecosystem losses on record during two recent unprecedented events of 2007 and 2010. The 2003 event alone was responsible for about 40,000 heat wave related human deaths across Europe as noted in several news reports. The heat waves of 2010 summer caused large-scale environmental damages in Eastern Europe though the total number of deaths was not as high as that of 2003. Some of the reports suggest that more than 600 wildfires broke out in Russia alone. It resulted in smog levels that were five to eight times higher than normal, leading to destruction of forest ecosystems, degradation of the air quality and widespread health issues.

The persistent warm days in summer over Europe are usually found to be associated with strong atmospheric blockings. These blockings are, in turn, related to amplification, breaking or dissipation of planetary-scale waves. In fact, being located at the intersection of subtropical and extratropical climate regimes,

the atmospheric condition is influenced by seasonal meridional migration of tropospheric baroclinic activities associated with the planetary scale waves that cause variations in the subtropical anticyclone. Regional migrations of this anticyclone and associated weather systems give rise to summer blockings over the region. This anticyclone together with a persistence deficit of soil moisture (e.g., Gracia–Herrera *et al.* 2010) from the beginning of the season is believed to be responsible for the extremely high surface temperatures in 2003 summer.

In a normal season, weather disturbances usually propagate eastward from the Atlantic on the southern flank of the European blocking highs. These systems alleviate local summer conditions in the Mediterranean region. However, the blocking high in 2003, located farther south of its normal position, barricaded the Atlantic disturbances and inhibited the formation of local weather systems over most parts of southwestern Europe. Analysis of the NCEP-NCAR reanalysis data revealed a seasonal blocking located over southwestern Europe during the summer of 2003.

In 2010, this situation was very different; the high was located over Eastern Europe (affecting Russia). Compared to the case in 2003, the southwestern part of Europe did not suffer much from the surface heat in 2010. In fact, several places in United Kingdom received torrential rainfall during that summer. The conditions on the eastern side of the Eurasian continent were also quite different during those extreme summers. Down southeast, over the Indian subcontinent, the 2010 summer monsoon was marked by a series of floods over Pakistan and over Northwest India. An intriguing aspect during this season was the strong westward shift of the West Pacific Subtropical High by nearly 15° longitudes relative to climatology. This period also witnessed the evolution of an intense La Niña in the Pacific Ocean (Mujumdar *et al.* 2012). To the east of the continent, abnormally cool summer was experienced in East Asia and Japan during summer in 2003 whereas it was exceptionally warm in 2010.

Besides the tropical teleconnection discussed in the following, the 2010 anomalies are linked to the Southern Annular Mode (SAM). The SAM exhibited a near record positive amplitude for austral spring 2010 as evidenced by zonally uniform strong high pressure anomalies in the Southern Hemisphere mid-latitudes and low pressure anomalies in the high latitudes. In 2010 eastern Australia experienced its wettest spring on record (Hendon *et al.* 2014). This SAM also impacts the summer monsoon rainfall variability over India (Amita *et al.* 2015) and over China (Nan and Li 2003).

The contrasting summer anomalies of those two extreme years were associated with different climate conditions. Atmosphere, on its own, has mostly the so-called short-term memory to form the base of some weather and climate phenomena. Therefore, atmospheric process like the blocking high alone cannot explain the sustained heat waves experienced during 2003 and 2010. Moreover, the persistence and anchoring of those highs, responsible for the overall consequences that these extremes carry to Europe, are generally associated with teleconnections arising from large-scale climate phenomena.

Based on the dipole pattern seen in the temperature anomalies of 2003 and 2010, further supported by EOF1 mode, Behera *et al.* (2012) have derived a Europe dipole temperature index by taking surface temperature anomaly (from NCEP reanalyses data) difference between area-averaged indices of Western (0–20E; 40–55N) and Eastern (30–60E; 45–60N) Europe. The derived event years from the time series show that three of the four extreme warm events of Western Europe were actually accompanied by pIOD events and all of the extremely hot summers of Eastern Europe are found to be related to La Niña events.

The IOD has a significant concurrent correlation with both the Western Europe index (0.45) and the Europe dipole index (0.51)

during the summer. However, the coefficient is weaker for the inverse correlation between IOD and Eastern Europe index (−0.34) though it remains significant. Interestingly, the correlation increases when the Eastern Europe index (−0.42) together with the Europe dipole index (0.41) leads the IOD index. Also, the concurrent correlation between Western Europe index and IOD index is 0.45 for the September–November season. This perhaps indicates a stronger influence of IOD on Europe during boreal fall.

The role of IOD is furthermore substantiated from the fact that ENSO, the other dominant mode of tropics, does not show significant correlation with the Western Europe index though it has a strong correlation with the Eastern Europe index as discussed in the following subsection. On the other hand, the extratropical North Atlantic Oscillation (NAO) has a moderate concurrent correlation with the Western Europe index (0.32) during summer. Since the correlation with NAO is lower than that with the IOD, it indicates a possible secondary role of North Atlantic in the development of the blocking highs as discussed in previous studies. Nevertheless, it may be noted that pIODs that either accompany El Niño events or are weaker during July–August do not support intense summers over Western Europe.

The monsoon-desert mechanism (Rodwell and Hoskins 1996) is suggested to explain the tropical teleconnection to Europe by the diabatic heating from the Indian region during the monsoon season. In a recent study, however, Behera *et al.* (2012) found that the enhanced rainfall over the Indian subcontinent and Western Pacific (Fig. 2) associated with IOD and La Niña during the July–August season of the composited years induce perturbations that are then projected on to the midlatitude waveguide. Teleconnection patterns associated with the positive rainfall anomalies of the composite of Western Europe hot summer are very well depicted in their composites of meridional wind anomalies at 300 hPa.

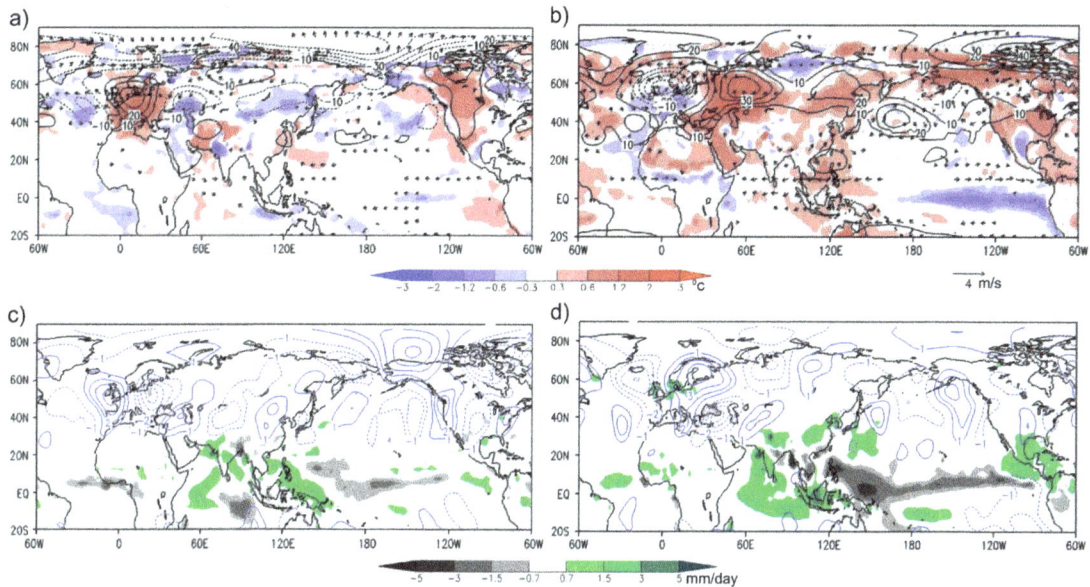

Fig. 2. (a) July–August composite anomalies of surface temperature (shaded), 850 hPa wind and geopotential height (contour) for the 4 extreme summers of Western Europe as given in Behera *et al.* (2012). (b) The corresponding composites of rainfall (shaded) and 300 hPa meridional wind anomalies. (b) and (d) same as (a) and (c) but for anomalies related to 4 extreme events of Eastern Europe. Values shown are above 85% level of statistical confidence from a 2-tailed t-test. Adapted from Behera *et al.* (2012).

The rainfall anomalies are responsible for developing the local baroclinic response to diabatic heating, which then disperses a circumglobal wavetrain of equivalent-barotropic nature on projecting the signal on to the midlatitude waveguide (Fig. 2). A double-jet structure is actually noticed in the zonal winds for the Western Europe composites (Fig. 2). This pattern usually tends to develop atmospheric blocking that stops the eastward propagation of cyclones and thereby supports the long-lasting warm anomalies. The second jet also disperses some of the Rossby wave energies to higher latitudes from the source regions in subtropics.

The process on the Eastern Europe side is somewhat opposite to that seen on the Western Europe side (Fig. 2). Composite SST anomalies related to Eastern Europe warm summers show the development of cold SST anomalies in the eastern Pacific associated with La Niña events during those composited years. Warm

SST anomalies develop on the northwest Pacific at this time and, associated with these warm anomalies, positive rainfall anomalies are seen in the adjacent landmasses of the Asian Continent. The signals associated with the diabatic heating of these above normal rainfall anomalies disperse to the midlatitude waveguide and a wave train emanates from East Asia to travel around the globe. As seen from the composites of 300 hPa meridional wind anomalies (Fig. 2), this wave train takes a somewhat different path from that of the composites of Western Europe and arrives over the affected region in Eastern Europe.

The role of tropical teleconnection to hot summers of East Asia is discussed in yet another study. By using the observational data during boreal summer of 1980–2010, Akihiko *et al.* (2014) investigated interannual variability of heatstroke related deaths in the Kanto region of Japan. Their results show that the death toll by heatstroke is highly correlated with the number of extremely hot days in which the daily maximum temperature exceeded 35°C. The death toll by heatstroke are seen to be linked with the tropical climate variability via the number of extremely hot days. In a detail investigation, they have found that extremely hot days are more frequently associated with the IOD than the ENSO. In a warmer climate, as expected in future, even a small temperature perturbation related to the tropical climate variability may lead to occurrence of the extremely hot days. At the moment, we are able to predict the associated climate variations at longer lead time. However, the extreme events themselves are not well-predicted beyond a couple of weeks' time. Therefore, it is urgent to develop good prediction systems to provide accurate predictions of the extremely hot days at long lead times.

2.3. *IOD and ENSO connections to extreme streamflows*

The ENSO and IOD influence could be seen in ground hydrology and stream flows besides global and regional climate variations. Here, the influence of those tropical climate modes on the stream flows of two rivers is discussed. The first is the Citarum River, which is the largest river basin in West Java and supplies 80% of the water demands to Jakarta alone. The river also plays a vital role for the economic development and livelihood of the people by supporting agriculture, fisheries, hydroelectric power generation, public water supply and industrial establishment of West Java Province and Jakarta City (Fares and Yudianto 2004).

Sahu *et al.* (2012) have found that the Citarum River extreme streamflows are affected by IOD and ENSO. Extreme stream-flow events of Citarum River in their study are derived from the daily stream-flows at the Nanjung gauge station. High and low stream-flow events are then classified based on the criteria of 1.5 and −1 standard deviations respectively, when such

high or low stream flows persisted for 5 or more days.

The extreme low-stream-flow events are found during pIOD years whereas extreme high-stream-flow events were associated with La Niña years (Sahu *et al.* 2012). A few of them were also associated with the negative phases of IOD (nIODs). The Indo-Pacific warm pool region is warmer than normal during La Niña and nIOD events. This facilitates low-level convergence and above normal rainfall over the Maritime Continent as seen in the anomalies of SST, wind and OLR (Fig. 3). Hence, this favorable climatic condition has caused some of the extreme rainfall events over the Citarum River catchment region and led to extreme stream flows. On the other hand, low-stream-flow events were associated with pIOD years when the Indian Ocean played a dominant role (Sahu *et al.* 2012). The cold anomalies in eastern Indian Ocean associated with the pIOD condition have caused the anomalous low-level divergence and reduced the normal seasonal rainfall over that region as

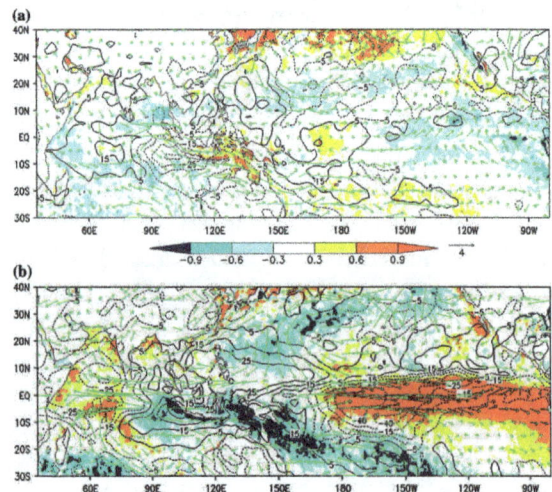

Fig. 3. Composite anomalies of SST (shaded), wind and OLR (contour) during boreal fall season for a) all extremely high stream-flow events, b) all extremely low-stream flow events of Citarum River. Unit for SST is °C, for wind is ms^{-1}, and for OLR is w/m^2. Values above 95% confidence level from a two-tailed student's t-test are shown. Adapted from Sahu *et al.* (2012).

noted from the positive OLR anomalies (Fig. 3). Since several of those pIOD events were accompanied by El Niño events, the El Niño signal in eastern Pacific is also noticed in the composite plot (Fig. 3). However, Sahu *et al.* (2012) have noted low stream-flow events were not always associated with El Niño though all of those were actually associated with pIODs. In addition, the correlation between seasonal stream flow and the eastern pole of DMI is 0.51 that is higher than the corresponding −0.39 correlation with Nino3 in boreal fall (Sahu *et al.* 2012). Therefore, pIODs are dominantly responsible for extreme stream-flow events of the Citarum River though some of those could be concurrent to co-occurring El Niños.

The relationships among the extreme stream flows, the ENSO and IOD apparently become nonlinear in boreal winter and spring. During the boreal winter season that coincides with the local monsoon season, the number of extreme events is higher and the durations of most of those events are longer as compared to that of the boreal fall. The rainfall in this season is controlled by several modes of climate variation such as intraseasonal disturbances, monsoon, IOD and ENSO. Therefore, the links to ENSO and IOD events are not as distinctive as those found during the previous season.

The Paranaíba River, which forms an important catchment of rainwater in northern Brazil, is on the other side of the Pacific and hence is expected to be influenced by the variability in the tropical eastern Pacific. The catchment area of the river is approximately 36,000 km^2. Major dams on its course are the Emborcação, Itumbiara and São Simão. Cachoeira Dourada near Itumbiara is one of the most important hydroelectric power stations in Brazil, providing energy to Goiânia and Brasilia. Sahu *et al.* (2014) studied the climate role in Paranaíba River variability based on the stream flow data from Fazenda Santa Maria gauge station in the upper Paranaíba River catchment and having a catchment area about 16,750 km^2. This upper

catchment is not artificially regulated, thus it is best suited for analysis to minimize anthropogenic influences on stream flow. Extremely low discharge events were cataloged based on a threshold of −1.5 standard deviation persisted for 7 days or more during December–February for the period 1978 to 2006 (Sahu *et al.* 2014).

Extremely-low discharge events of the Paranaíba River basin are found to be associated with the El Niño Modoki phenomenon (Sahu *et al.* 2014). 9 out of a total of 10 (90%) extremely low-discharge events are associated with El Niño Modoki during the austral summer season. Interestingly, unlike the common perception, not a single event is found to be associated with canonical El Niño. The anomalies of SST, wind and OLR composited for the duration (Fig. 4) of all those extremely low discharge events in the El Niño Modoki years clearly bring out the reasons of the low discharge events. The composite plot shows warm SST anomalies in the central Pacific flanked by colder SST on either side of the basin. Lower-tropospheric winds converge to this warmer SST anomalies and cause above normal rainfall west of the dateline. This gives rise to worldwide teleconnections as has been suggested in several

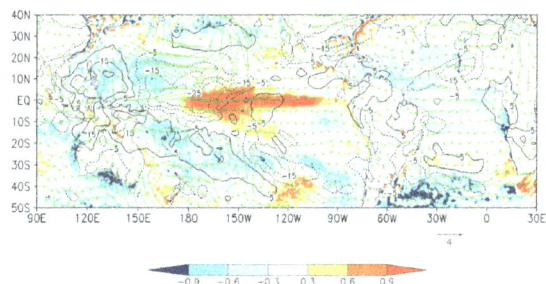

Fig. 4. Composite anomalies of SST (shaded), wind and OLR (contour) during austral summer season for all extremely high stream-flow events of Paranaiba River. Unit for SST is °C, for wind is ms^{-1}, and for OLR is w/m^2. Values above 95% confidence level from a 2-tailed student's t-test are shown. The red dot mark in the figure shows the location of the stream flow gauge station. Adapted from Sahu *et al.* (2014).

previous studies (e.g., Ashok *et al.* 2007; Weng *et al.* 2008). The low-level convergence in the central Pacific gives rise to subsidence over most parts of the Amazon and Paranaɪba catchments leading to reduced local rainfall and low river discharges.

3. Predictability

The seasonal climate variations are seen to influence the extreme events as is discussed in previous sections of this chapter. Hence, reliable predictions of those modes of climate variations are key to the success of catching the predictability of the associated extreme events. Of course, high-frequency climate variations and weather phenomena could also be responsible for some of the extreme events. However, those events have very low predictability on sub-seasonal to seasonal time scales. It is widely recognized that the tropical upper ocean has a major contribution to the seasonal and interannual climate predictability. The slowly varying ocean surface conditions form the basis of the so-called ocean memory: The dynamical oceanic processes, such as the Rossby and Kelvin waves (Behera *et al.* 2013) and associated basin-wide ocean adjustment or subsurface temperature anomalies, can trigger tropical climate anomalies, which rapidly grow via vigorous ocean–atmosphere interactions and hence provide key precursors for the climate prediction (e.g., Luo *et al.* 2011; Luo *et al.* 2015).

Over the decades, simple statistical models to comprehensive dynamical numerical models (Luo *et al.* 2003) are used to predict climate variations. Statistical models are cheap and often built based on statistical relationships among predictors and predictants. Statistical models, however, have drawbacks as the assumed relationship between predictor and predictant might change over time. Also, it is extremely difficult to find a perfect relation between a predictor and a predictant. On the other hand, due to advancement in computational environment,

numerical methods and model developments, CGCMs are becoming more reliable in recent years to predict global scale climate phenomena such as ENSO and IOD (Luo *et al.* 2015).

Here, the predictability of JAMSTEC CGCM based prediction system called SINTEX-F1 (Luo *et al.* 2003; Masson *et al.* 2005; Behera *et al.* 2006) is discussed. The atmospheric component (ECHAM4.6) of the SINTEX-F1 model has a resolution of 1.1° (T106) with 19 vertical levels. Its oceanic component (OPA8) has a relatively coarse resolution of a 2° Mercator horizontal mesh but with an equatorial enhancement of up to 0.5° in the meridional direction. It has 31 levels in the vertical of which 20 lie in the top 500-m with a resolution of 10-m from sea surface to 110-m depth. The ocean resolutions are enhanced to uniform 0.5° everywhere in a new version of the model called SINTEX-F2 (Doi *et al.* 2016).

Coupled models can be initialized with all available ocean-atmosphere observations through complex assimilation schemes. It is generally, however, complex and computationally expensive. So, SINTEX-F1 uses a simple but effective initialization approach by assimilating only observed SSTs, at the beginning of each monthly prediction update, in a coupled way (e.g., Luo *et al.* 2005).

The SINTEX-F1 prediction system has been used for real time forecast experiments with 27 members and demonstrated excellent performance in forecasting ENSO and IOD as well as associated climate anomalies over the globe (cf. www.jamstec.go.jp/frsgc/research/d1/iod/e/seasonal/outlook.html). In general, the model has high skills in predicting the tropical, and to a great extent, the extratropical surface temperatures up to a couple of seasons ahead (Fig. 5). The predictability for rainfall is not as widespread and skilful beyond a season or two (figure not shown). On the other hand, the predictability of surface temperature remains quite high in the tropical central and eastern Pacific (the regions of strong air-sea interactions) even

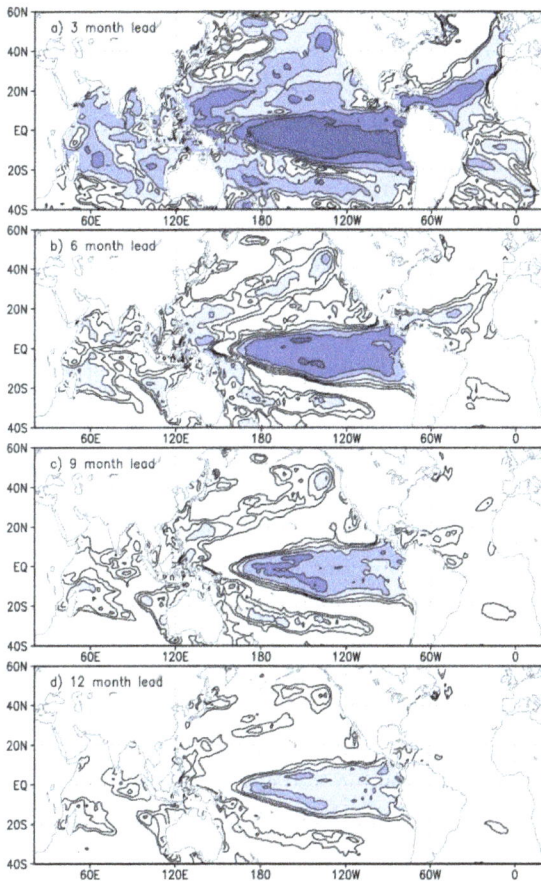

Fig. 5. SINTEX-F CGCM skill scores (anomaly correlation between prediction and observation) of global SST predictions at 3-, 6-, 9-, and 12-month lead times. Contour interval is 0.1 and regions with values above 0.6 are shaded. Adapted from Luo et al. (2005).

at a lead time of 12 months. However, the predictability in tropical Atlantic remains low on longer lead times. One primary reason for this is that most current coupled models fail to reproduce the mean zonal tilt of west warm-east cold structure along the equatorial Atlantic.

The SINTEX-F model can predict all interannual ENSO events over the past three decades up to 1-year lead as depicted in Luo et al. (2005, 2008a, 2015). Both the El Niño events in 1982/83, 1986/87, 1991/92, 1997/98, and 2002/03 and the La Niña events in 1984–86, 1988/89, 1995/96, 1999–2001, 2005/06, 2007–09, and 2010–12 are predicted up

to 12 months ahead. During the period 1982–2012, there are only two small false alarms: a weak warm event in 1990/91 and a weak cold one in 2003/04. Predictions of the El Niños in the 2000s appear to be less skilful compared to those in previous decades, a fact also found based on multi-model real time forecasts (Barnston et al. 2012; Wang et al. 2010). This is consistent with the change of El Niño characteristics in the 2000s: while in previous decades equatorial warm SST (>28°C), positive rainfall and warm subsurface temperature anomalies display an apparent eastward propagation in a coupled way in association with strong westerly anomalies in the central Pacific, the El Niño related warming and rainfall anomalies in the last decade tend to always stay in the western-central Pacific together with much weaker westerly anomalies in the central Pacific (Luo et al. 2012).

Spring-time barrier is often cited for low ENSO predictability during boreal spring associated with the phase locking of ENSO with seasonal cycle. However, the SINTEX-F1 model can successfully predict ENSO across the first spring barrier. This is related to the correct predictions of subsurface signals in the equatorial Pacific (Luo et al. 2005b), which provide the basis for the development of the up-coming ENSO event. On average, ENSO prediction skill for the period 1982–2012 reaches about 0.6 (0.4) at 14 (24) months lead with root mean square errors smaller than 0.8°C (see Luo et al. 2008a). Interestingly, several ENSO events over the past three decades could be predicted even at lead times of up to 1.5–2 years (Luo et al. 2008a). The strongest 1997/98 El Niño was found to be difficult to predict and missed by several coupled GCMs beyond 6 months lead time (e.g., Barnston et al. 1999). The SINTEX-F1 model, however, could predict this unprecedented event up to about 1.5-year lead with the forecasted amplitude reaching the 0.5°C, a criterion widely used for El Niño classification. The long-lasting La Niña events in 1984–86, 1999–2001, 2007–09,

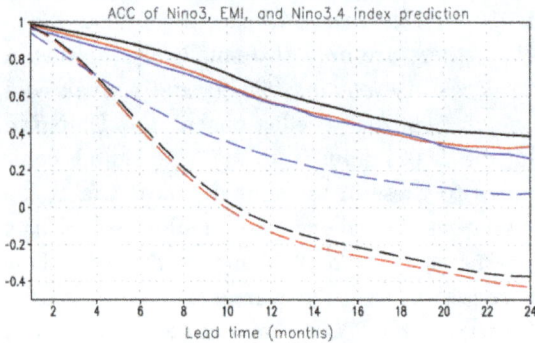

Fig. 6. Skill scores (anomaly correlations between observation and prediction) as a function of lead time of the Nino3 (red lines), ENSO Modoki index (blue lines), and Nino3.4 (black lines), based on persistence (dashed lines) and nine-member mean prediction of the SINTEX-F model for the period 1982–2012. Adapted from Luo *et al.* 2015.

and 2010–12 are predicted well up to 1.5–2 years lead in advance. Amazingly, the intermediate El Niño event in 2002/03 is also predicted well up to 2-year lead in advance; this suggests a link between the El Niño Modoki and the tropical Pacific decadal variability.

Unlike the ENSO-forced signals in the Indian Ocean that show high predictability, the coupled mode IOD has limited predictability (e.g., Wajsowicz 2005; Luo *et al.* 2005, 2008b; Wang *et al.* 2009). This is because air-sea coupling related to IOD is usually weak and more localized compared to ENSO. Several scales of phenomena, most prominently the intraseasonal oscillations affect the basin. A large asymmetry also exists between the surface warming and cooling intensity in the eastern Indian Ocean associated with negative and positive IOD events (Hong *et al.* 2008). The nIOD events do not appear to evolve into strong air-sea coupled processes in the Indian Ocean, and therefore their peak magnitudes are weak with low predictability in general (Luo *et al.* 2007).

While the SST predictions in the western and eastern Indian Ocean show some useful skills, prediction skill of the IOD index (i.e., SST anomaly gradient between the western and

eastern Indian Ocean) itself is quite low. The forecast skill drops at 3-month lead for February and May initial condition and at 6-month lead for August and November initial condition. As Wang *et al.* (2009) indicated, there is a July prediction barrier and a severe unrecoverable January prediction barrier for the IOD prediction. Luo *et al.* (2007) showed that the unique winter barrier for prediction of IOD and the eastern Indian Ocean SST is related to IOD's strong phase locking to the annual reversal of the monsoon. For the forecasts started from early May, while a July barrier exists, there is a robust bounce-back after July: This suggests that the mature phase of IOD in October–November is more predictable probably due to the predictability of the eastern Indian Ocean pole where the SST dominates the mature phase of IOD.

4. Summary

Climate variability in tropical oceans plays a significant role in the variability of regional weather and climate. Here, it is shown that the tropical modes of climate variations such as the ENSO, the ENSO Modoki and the IOD are associated with some of the extreme events ranging from hot summers to overflowing riverbanks. Some of the extreme summers of Europe and Asia are seen to be linked with IOD, the comparatively less investigated mode rooted in the Indian Ocean. The diabatic heating in the western Indian Ocean during pIOD years lead to circumglobal wave trains that affect the climate of East Asia and Europe. Some of the worst summers of Western Europe (e.g., like the 2003 event) happened when the circumglobal Rossby wave train emanating from the Indian region reinforced the blocking high over Europe. This in turn prolonged the hot spells over Europe, particularly in 2003, giving rise to record number of fatal heat strokes. Some of the heat stroke related deaths in northern Japan is also linked to IOD. On the other hand, the extreme summers of Eastern Europe are seen to be related with the La Niña transition phase in the tropical Pacific.

The impacts of these tropical climate modes are even registered in the extreme stream flows of some of the rivers in Asia and South America on the western and eastern sides of the Pacific Ocean. The extremely low stream flows of Citarum River in Indonesia are seen to be dominantly influenced by pIODs whereas the extremely high stream flows of the river are seen to be linked with La Niñas. On the eastern side of the Pacific Ocean, ENSO Modoki dominantly influences the extreme stream flows of the Parnaíba River in Brazil.

It is important to develop reliable prediction system for tropical climate modes considering their dominant roles in extreme events. Though it is also important to develop good prediction system for the extreme events, lack of their predictability beyond a couple of weeks time further portrays the need to focus on the predictability of their anchors — the tropical modes of climate variations. By understanding the scale interactions among those phenomena and developing predictability skills for those interactions would help us in extending the model's ability to predict extreme events. Furthermore, it would be possible to develop statistical-dynamical prediction system for extreme events through the representation of correct statistical relationship among extreme events and climate variations. Hopefully, the relationships depicted in some of the recent studies would lead us to develop such a skillful prediction system.

Acknowledgments

I acknowledge the contributions of many of my colleagues, especially Drs. Toshio Yamagata, Jing-Jia Luo and Netrananda Sahu, whose work I have extensively cited in this article I also thank two anonymous reviewers for their constructive comments.

References

Abram, N. J., M. K. Gagan, M. T. McCulloch, J. Chappell, and W. S. Hantoro, 2003: Coral reef death during the 1997 Indian Ocean Dipole linked to Indonesian wildfires. *Science*, **301**, 952–955. doi:10.1126/science.1083841.

Abram, N. J., M. K. Gagan, J. E. Cole, W. S. Hantro, and M. Mudelsee, 2008: Recent intensification of tropical climate variability in the Indian Ocean. *Nat. Geosci.*, **1**, 849–853. doi:10.1038/ngeo357.

Aceituno, P. 1988: On the functioning of the Southern oscillation in the South America sector. *Mon. Weather Rev.*, **116**, 505–524.

Akihiko, T., Y. Morioka, and S. K. Behera, 2014: Role of climate variability in the heatstroke death rates of Kanto region in Japan. *Sci. Rep.*, **4**, 5655. doi:10.1038/ srep05655.

Aldrian, E. and D. Susanto, 2003: Identification of three dominant rainfall regions within Indonesia and their relationship to sea surface temperature. *Int. J. Climatology*, **23**, 1435–1452.

Amita Prabhu, R. Kripalani, B. Preethi, and G. Pandithurai, 2015: Potential role of the February–March Southern Annular Mode on the Indian summer monsoon rainfall: a new perspective. *Clim. Dynam.*, doi:10.1007?s00382-015-2874-5.

Ashok, K., Z. Guan, and T. Yamagata, 2001: Impact of the Indian Ocean Dipole on the decadal relationship between the Indian monsoon rainfall and ENSO. *Geophys. Res. Lett.*, **28**, 4499–4502.

Ashok, K., S. K. Behera, S. A. Rao, H. Weng, and T. Yamagata, 2007: El Niño Modoki and its possible teleconnection. *J. Geophys. Res.*, **112**, C11007. doi:10.1029/2006JC003798.

Barnston, A. G., M. H. Glantz, and Y. He, 1999: Predictive skill of statistical and dynamical climate models in SST forecasts during the 1997–98 El Niño episode and the 1998 La Niña onset. *Bull. Am. Meteorol. Soc.*, **80**, 217–243.

Barnston, A. G., M. K. Tippett, M. L. L'Heureux, S. Li, and D. G. DeWitt, 2012: Skill of real-time seasonal ENSO model predictions during 2002–11. Is our capability increasing? *Bull. Am. Meteorol. Soc.*, **93**, 631–651.

Behera, S. K., R. Krishnan, and T. Yamagata, 1999: Unusual ocean-atmosphere conditions in the tropical Indian Ocean during 1994. *Geophys. Res. Lett.*, **26**, 3001–3004.

Behera, S. K. and T. Yamagata, 2003: Influence of the Indian Ocean Dipole on the Southern Oscillation. *J. Meteorol. Soc. Japan*, **81**, 169–177.

Behera, S. K., J. J. Luo, S. Masson, S. A. Rao, H. Sakuma, and T. Yamagata, 2006: A CGCM study on the interaction between IOD and ENSO. *J. Climate*, **19**, 1688–1705.

Behera, S. K, J. V. Ratnam, Y. Masumoto, and T. Yamagata, 2012: Origin of extreme summers in Europe — the Indo-Pacific connection. *Clim. Dynam.*, doi:10.1007/s00382-012-1524-8.

Behera, S. K., P. Brandt, and G. Reverdin, 2013: The tropical ocean circulation and dynamics, ocean circulation and climate — A 21st century perspective, Eds. Gerold Siedler, Stephen Griffies, and John Gould, John Church, pp. 385–412, International Geophysics Series Vol. 103, Elsevier.

Chang, P. *et al.*, 2006: Climate fluctuations of tropical coupled systems — The role of ocean dynamics. *J. Climate*, **19**, 5122–5174.

D'Arrigo, R., N. Abram, C. Ummenhofer, J. Palmer, and M. Mudelsee, 2009: Reconstructed streamflow for Citarum River, Java, Indonesia: linkages to tropical climate dynamics. *Clim. Dynam.* doi:10.1007/s00382-009-0717-2.

Doi, T., S. Behera, and T. Yamagata, 2016: Improved seasonal prediction by the SINTEX-F2 coupled climate model. *J. Geophys. Res.* (submitted).

Fares, Y. R. and D. Yudianto, 2004: Hydrological modelling of the upper Citarum catchment, West Java. *J. Environmental Hydrology*, **12**, Page 8.

Fauchereau, N., S. Trzaska, Y. Richard, P. Roucou, and P. Camberlin, 2003: Sea-surface temperature co-variability in the southern Atlantic and Indian Oceans and its connections with the atmospheric circulation in the Southern Hemisphere. *Int. J. Climatol.*, **23**, 663–677. doi:10.1002/joc.905.

Feng, M., M. J. McPhaden, S. Xie, and J. Hafner, 2013: La Niña forces unprecedented Leeuwin Current warming in 2011. *Sci. Rep.*, **3**, 1277. doi:10.1038/srep01277.

García-Herrera, R., J. Díaz, R. M. Trigo, J. Luterbacher, and E. M. Fischer, 2010: A review of the European summer heat wave of 2003. *Critical Reviews in Environmental Science and Technology*, **40**, 267–306.

Hendon, H. H., 2003: Indonesian rainfall variability: Impacts of ENSO and local air–sea interaction. *J. Climate*, **16**, 1775–1790.

Hendon, H. H., E.-P. Lim, J. M. Arblaster, and D. T. L. Anderson, 2014: Causes and predictability of the record wet spring over Australia in 2010. *Clim. Dynam.*, **42**, 1155–1174.

Hermes, J. C. and C. J. C. Reason, 2005: Ocean model diagnosis of interannual coevolving SST variability in the South Indian and South Atlantic Oceans. *J. Climate*, **18**, 2864–2882.

Hong, C.-C., T. Li, and J.-J. Luo, 2008: Asymmetry of the Indian Ocean dipole. Part II: Model diagnosis. *J. Climate*, **21**, 4849–4858.

Kao, H.-Y. and J.-Y. Yu, 2009: Contrasting eastern-Pacific and central-Pacific types of ENSO. *J. Climate*, **22**, 615–632.

Kataoka T., T. Tozuka, S. K. Behera, and T. Yamagata, 2014: On the Ningaloo Nino/Nina. *Clim. Dynam.* doi:10.1007/s00382-013-1961-z.

Kayanne, H., H. Iijima, N. Nakamura, T. R. McClanahan, S. K. Behera, and T. Yamagata, 2006: Indian Ocean Dipole index recorded in Kenyan coral annual bands. *Geophys. Res. Lett.*, **33**, L19709. doi:10.1029/2006GL027168.

Kug, J.-S., F.-F. Jin, and S.-I. An, 2009: Two types of El Niño events: cold tongue El Niño and warm pool El Niño. *J. Climate*, **22**, 1499–1515.

Kumar, K. K., B. Rajagopalan, and M. A. Cane, 1999: On the weakening relationship between the Indian Monsoon and ENSO. *Science*, **284**, 2156–2159.

Larkin, N. K. and D. E. Harrison, 2005: On the definition of El Niño and associated seasonal average U.S. weather anomalies. *Geophys. Res. Lett.*, **32**, L13705. doi:10.1029/2005GL022738.

Luo, J. J., S. Masson, S. K. Behera, P. Delecluse, S. Gualdi, A. Navarra, and T. Yamagata, 2003: South Pacific origin of the decadal ENSO-like variation as reproduced by a coupled GCM. *Geophy. Res. Lett.*, **30**, 2250. doi:10.1029/2003GL018649.

Luo, J.-J., S. Masson, S. Behera, S. Shingu, and T. Yamagata, 2005: Seasonal climate predictability with a high resolution coupled GCM. *J. Climate*, **18**, 4474–4497.

Luo, J.-J., S. Masson, S. Behera, and T. Yamagata, 2008a: Extended ENSO predictions using a fully coupled ocean–atmosphere model. *J. Climate*, **21**, 84–93.

Luo, J.-J., S. Behera, Y. Masumoto, H. Sakuma, and T. Yamagata, 2008b: Successful prediction of the consecutive IOD in 2006 and 2007. *Geophy. Res. Lett.*, **35**, L14S02. doi:10.1029/2007GL032793.

Luo, J.-J., S. Behera, Y. Masumoto, and T. Yamagata, 2011: Impact of global ocean surface warming on seasonal-to-interannual climate prediction. *J. Climate*, **24**, 1626–1646.

Luo, J.-J., W. Sasaki, and Y. Masumoto, 2012: Indian ocean warming modulates Pacific climate change. *PNAS*, **109**, 18701–18706, www.pnas.org/cgi/doi/10.1073/pnas.1210239109.

Luo, J.-J., J.-Y. Lee, C. Yuan, W. Sasaki, S. Masson, S. Behera, Y. Masumoto, and T. Yamagata, 2015: Current Status of Intraseasonal-seasonal-to-interannual prediction of the Indo-Pacific climate. Ch. 3, World Scientific Series on Asia-Pacific Weather and Climate, Vol. 7, eds. S. Behera and T. Yamagata.

Marengo, J. A. 1995: Variations and changes in South America streamflow. *Clim. Change*, **31**, 99–117.

Masson, S., J.-J. Luo, G. Madec, J. Vialard, F. Durand, S. Gualdi, E. Guilyardi, S. K. Behera, P. Delecluse, A. Navarra, and T. Yamagata, 2005: Impact of barrier layer on winter-spring variability of the South-Eastern Arabian Sea. *Geophy. Res. Lett.*, **32**, L07703. doi:10.1029/2004GL021980.

Morioka, Y., T. Tozuka, and T. Yamagata, 2010: Climate variability in the southern Indian Ocean as revealed by self-organizing maps. *Clim. Dynam.*, **35**, 1059–1072. doi:10.1007/s00382-010-0843-x.

Morioka, Y., T. Tozuka, S. Masson, P. Terray, J. J. Luo, and T. Yamagata, 2012: Subtropical dipole modes simulated in a coupled general circulation model. *J. Climate*, **25**, 4029–4047. doi:10.1175/JCLI-D-11-00396.1.

Morioka, Y., T. Tozuka, and T. Yamagata, 2013: How is the Indian Ocean subtropical dipole excited? *Clim. Dynam.*, **41**, 1955–1968. doi:10.1007/s00382-012-1584-9.

Morioka, Y., S. Masson, P. Terray, C. Prodhomme, S. K. Behera, and Y. Masumoto, 2014: Role of tropical SST variability on the formation of subtropical dipoles. *J. Climate*, **27**, 4486–4507.

Mujumdar, M., B. Preethi, T. P. Sabin, K. Ashok, S. Saeed, D. S. Pai, and R. Krishnan, 2012: The Asian summer monsoon response to the La Niña event of 2010. *Meteorol. Appl.*, **19**, 216–225.

Nakamura, N., H. Kayanne, H. Iijima, T. R. McClanahan, S. Behera, and T. Yamagata, 2009: The Indian Ocean Mode shift in the Indian Ocean climate under global warming stress. *Geophys. Res. Lett.*, **36**, L23708. doi:10.1029/2009GL040590.

Nan, S. and J. Li, 2003: The relation between summer precipitation in Yangzte River Valley and the previous Southern Hemisphere Annular Mode. *Geophys. Res. Lett.*, **30**(24), 2266. doi:10.1029/2003GL018381.

Rasmusson, E. M. and T. H. Carpenter, 1983: The relationship between eastern equatorial Pacific sea surface temperature and rainfall over India and Sri Lanka. *Mon. Weather Rev.*, **111**, 517–528.

Rodwell, M. J. and B. J. Hoskins 1996: Monsoons and the dynamics of deserts. *Q. J. Roy. Meteor. Soc.*, **122**, 1385–1404.

Sahu, N. S. Behera, Y. Yamashiki, K. Takara, and T. Yamagata, 2012: IOD and ENSO impacts on the extreme stream-flows of Citarum river in Indonesia. *Clim. Dynam.*, **39**(7–8), 1673–1680. doi:10.1007/s00382-011-1158-2.

Sahu, N., S. Behera, J. V. Ratnam, R. V. Da Silva, P. Parhi, W. Duan, K. Takara, R. B. Singh, and T. Yamagata, 2014: El Niño Modoki Connection to Extremely-low Streamflow of the Paranaíba River in Brazil. *Clim. Dynam.* **42**(5–6), 1509–1516.

Saji, N. H., B. N. Goswami, P. N. Vinayachandran, and T. Yamagata, 1999: A dipole mode in the tropical Indian Ocean. *Nature*, **401**, 360–363.

Saji, N. H. and T. Yamagata, 2003: Possible impacts of Indian Ocean dipole mode events on global climate. *Climate Res.*, **25**, 151–169.

Tsai, C. S. K. Behera, and T. Waseda, 2015: Indo-China Monsoon Indices, Scientific Reports 5, 8107. doi:10.1038/srep08107.

Walker, G. T. 1925: Correlations in seasonal variations of weather — a further study of world weather. *Mon. Weather Rev.*, **53**, 252–254.

Wang, B. and coauthors, 2009: Advance and prospects of seasonal prediction: assessment of the APCC/CliPAS 14-model ensemble retrospective seasonal prediction (1980–2004). *Clim. Dynam.*, **33**, 93–117.

Wang, W., M. Chen, and A. Kumar, 2010: An assessment of the CFS real-time seasonal forecasts. *Weather Forecast.*, **25**, 950–969.

Wajsowicz, R. C. 2005: Potential predictability of tropical Indian Ocean SST anomalies. *Geophys. Res. Lett.*, **32**, L24702. doi:10.1029/2005GL024169.

Weng, H., K. Ashok, S. K. Behera, S. A. Rao, and T. Yamagata, 2008: Impacts of recent El Niño Modoki on droughts/floods in the Pacific Rim during Boreal Summer. *Clim. Dynam.*, **29**, 113–129.

Yamagata, T., S. K. Behera, J-J. Luo, S. Masson, M. R. Jury, and S. A. Rao, 2004: The coupled ocean-atmosphere variability in the tropical Indian Ocean. Earth's climate: the ocean-atmosphere interaction. *Geophys. Monogr. Series*, **147**, AGU, pp. 189–211.

Yamagata, T., Y. Morioka, and S. Behera, 2015: Old and new faces of climate variations. Ch. 1, World Scientific Series on Asia-Pacific Weather and Climate, Vol. 7, Eds. S. Behera and T. Yamagata.

Yeh, Sang-Wook, J. Kug, B. Dewitte, M.-H. Kwon, P. Kirtman, and F. F. Jin, 2009: El Niño in a changing climate. *Nature*, **461**, 511–514. doi:10. 1038/nature08316.

Yuan, C. and T. Yamagata, 2014: California Niño/Niña. *Sci. Rep.*, **4**, 4801. doi:10.1038/srep 04801.

Chapter 3

Subseasonal Prediction of Extreme Weather Events

Bin Wang* and Ja-Yeon Moon

Department of Atmospheric Sciences, University of Hawaii at Manoa,
Honolulu, Hawaii, 96822, USA
*wangbin@hawaii.edu
http://iprc.soest.hawaii.edu/users/bwang/

Prediction of extreme weather events two-to-six weeks ahead has immense social-economic benefits for hazard prevention and risk management as well as economic planning. Physical basis for such subseasonal prediction is primarily rooted in the intrinsic predictability of large-scale circulation associated with the tropical intraseasonal oscillation (ISO). Such prediction requires not only accurate prediction of tropical ISO and its teleconnection, but also the knowledge of how ISO modulates extreme weather events in the tropics and extratropics. This chapter provides a review of current knowledge on the modulation of ISO on many types of extreme weathers in the global tropics and extratropics including flooding, heavy snow storms, cold surges, hurricane and tropical storms, wild fire, aerosols, tornado and hail days, etc. This type of knowledge affords useful prediction tools for subseasonal prediction of extreme events if provided ISO can be realistically forecasted. Recent progress on subseasonal forecasts of tropical storm activity made by world major operational centers is also briefly reviewed and issues are discussed.

1. Introduction

Extreme weather and climate events, such as flooding, drought, wild fire, hurricanes, heat waves and cold surges, have received increased attention in recent years due to frequently large fatality of human life and exponentially increasing loss associated with them (Karl and Easterling 1999). Prediction and attribution of extreme events is a challenging task and is currently undergoing considerable evolution (Herring *et al.* 2014).

Subseasonal prediction aims to forecast weekly mean anomalies beyond weather, mainly from week 2 to week 6, which bridges weather forecast and seasonal climate prediction. The physical basis for subseasonal prediction lies primarily in the predictability of the prominent Intraseasonal Oscillation (ISO) of tropical atmospheric circulations. In addition, there are other known and unknown sources for subseasonal variability from midlatitudes.

Tropical atmospheric motion exhibits a significant energy peak on intraseasonal (2–8 weeks) time scale, which is often referred to as tropical ISO. The ISO has considerable impacts not only in the tropics, but also in the middle and high latitudes through atmospheric teleconnection. The ISO is considered as a major source of global predictability on the subseasonal time scale (Waliser *et al.* 2011). In recent years, subseasonal prediction of extreme weather events has attracted broad attention of economists and social scientists. Such prediction requires both accurate ISO prediction and the ISO modulation of extreme weather events.

There have been increasing evidences of important ISO modulation of extreme events (Zhang 2013). The influence of the ISO has been demonstrated in monsoon systems, tropical storm activity, precipitation and temperature, as well as upper ocean variability in the tropics and high latitudes. Useful subseasonal predictions of extreme events are also possible

Bridging Science and Policy Implication for Managing Climate Extremes
Edited by Hong-Sang Jung and Bin Wang

in locations where a very strong ENSO (El Niño and Southern Oscillation) signal exists — in those locations ENSO can also cause a strong modulation of the subseasonal occurrence of extremes.

The main purpose of this chapter is to review and discuss the recent progress of the global and regional impacts of ISO on the extreme weather and climate events. In Sec. 2 we first briefly introduce what ISO is. Section 3 presents evidence of ISO modulation to various extreme weather events. Section 4 discusses current status on subseasonal prediction of tropical storms. The last section provides a summary.

2. The MJO and Boreal Summer ISO

Madden and Julian (1971) discovered a 40–50 day oscillation in tropospheric zonal winds at the equatorial central Pacific. This local 40–50 day oscillation is later linked to an eastward propagation of an equatorially trapped planetary-scale zonal circulation system (Madden and Julian 1972), which leads to an irregular 30–60 day oscillation in the tropics. This phenomenon is later named as Madden-Julian Oscillation (MJO). The MJO can be discerned from raw data without statistical manipulation (Zhang 2005). From dynamic standpoint, MJO can be defined as a tropical planetary-scale circulation system that couples a multi-scale convective complex and moves eastward slowly ($\sim$5 m/s) over the warm pool with a mixed Kelvin-Rossby wave structure (Rui and Wang 1990) and a baroclinic vertical structure in which planetary boundary layer low pressure and moisture convergence lead the major convective complex (Madden and Julian 1972; Wang 1988; Hendon and Salby 1994). The MJO disturbance amplifies over the Indo-Pacific warm pool and decays when passing through the dateline over cold eastern Pacific (Wang and Rui 1990). The MJO is primarily an atmospheric internal mode but vigorously exchanges surface entropy fluxes (Emanuel 1987) with the ocean mixed layer

especially via sea surface temperature (SST) variability (Krishnamurti et al. 1988).

The ISO is used as a more general term for the tropical atmospheric variability. MJO is the dominant mode of ISO that particularly prevails in boreal winter. During boreal summer the ISO behaves in a more complicated manner than the MJO (Wang et al. 2006). Boreal summer ISO (BSISO) has the following distinctive features from MJO: (a) a northward shift of the BSISO precipitation variability centers from the equatorial region to Asia-Pacific and North American summer monsoon regions (10–20°N); (b) a prominent northward propagation prevailing in the Indian, Western North Pacific (WNP)-East Asian and North American monsoon regions in addition to eastward propagation; (c) a Northwest-Southeastward tilted precipitation band prevailing as a leading mode in Asian-Pacific monsoon regions; (d) a considerable stationary convective seesaw component between the equatorial Indian Ocean and the WNP (Zhu and Wang 1993), and (e) two different periodicities of the BSISO: 30–60 days (e.g., Wang et al. 2006) and 10–20 days (e.g., Kikuchi and Wang 2010b). In view of these differences, for many applications, especially assessing the influences of ISO on tropical and extratropical extreme weather, it is instructive to consider tropical ISO as described by two modes, the MJO and the BSISO.

The evolutions of MJO and BSISO modes are presented in Figs. 1a and 1b, respectively. These typical evolution pictures were composited based on the first two leading empirical orthogonal function (EOF) modes of the OLR, 850 hPa wind, and 200 hPa wind anomalies. For the MJO, a Real-time Multivariate MJO (RMM) index developed by Wheeler and Hendon (2004) is the most widely used for monitoring and applications (e.g., Gottschalck et al. 2010). For the BSISO, two real time multivariate indices have been proposed to facilitate its detection, monitoring and prediction (Lee et al. 2013).

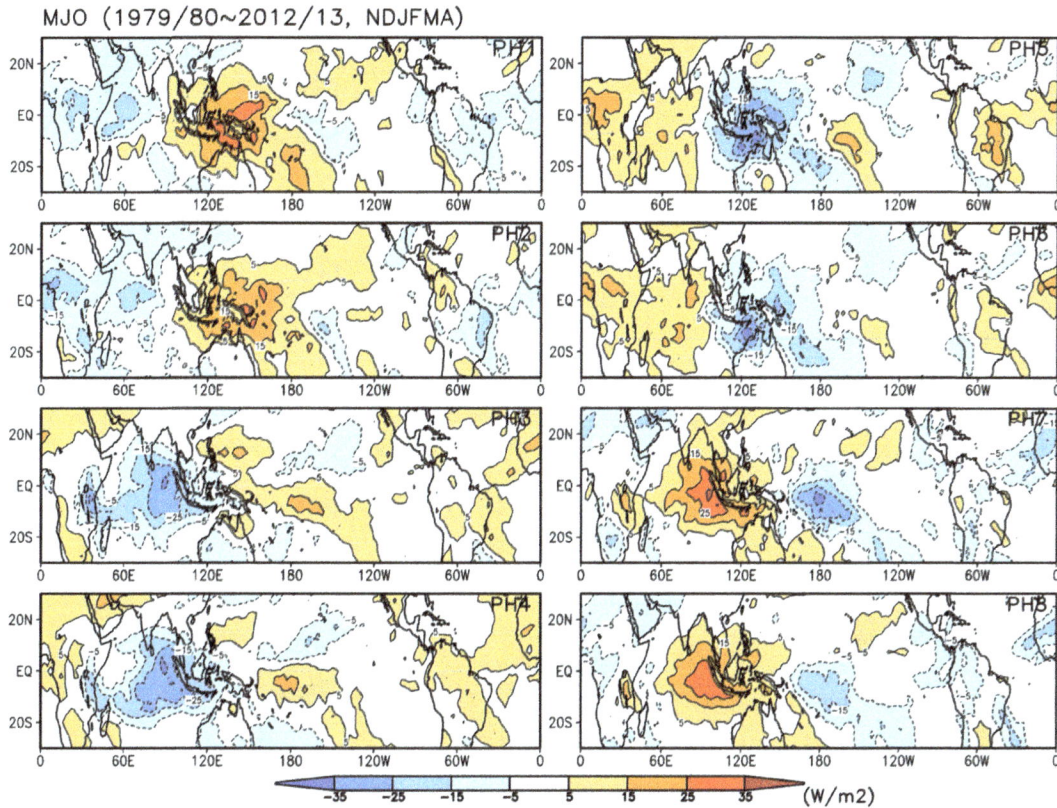

Fig. 1a.   The life cycle composite (phases 1–8) of Outgoing Longwave radiation (OLR) (shading) anomalies reconstructed based on PC1 and PC2 of Wheeler and Hendon (2004)'s MJO index (1979/80~2012/13 November through April). Negative OLR anomalies mean positive rainfall anomalies.

The aforementioned differences in MJO and BSISO evolution can be seen from comparison of Figs. 1a and 1b.

The wet and dry spells of the MJO and BSISO shown in Fig. 1 strongly influence tropical storms and extreme hydro-meteorological events, which are major driving forces of natural disasters, and thus impact socio-economic activities in the world's most densely populated region.

3. Intraseasonal Modulation of Extreme Weather Events

3.1. *Extreme rainfall (flood)*

During boreal winter, higher frequency of occurrences of extreme daily precipitation (exceeding 90th percentile of monthly gamma frequency distribution in intensity and spatial coverage) in Southeast Asia, Australia, Indonesia, and the western Pacific are found to occur in active phases of slow eastward propagating MJO. The convectively active (suppressed) phases of MJO can increase (decrease) the probability of extreme rain events over the land regions of Southeast Asia by about 30–50% (20–25%) during November–March season. Influence of MJO on localized rainfall extremes are also observed both in rainfall intensity and duration (Xavier *et al.* 2014). The greatest extreme weekly rainfall impact of the MJO occurs in northern Australia in (austral) summer associated with corresponding circulation anomalies (Wheeler *et al.* 2009). In the year 2002, 2007 and 2008 Jakarta

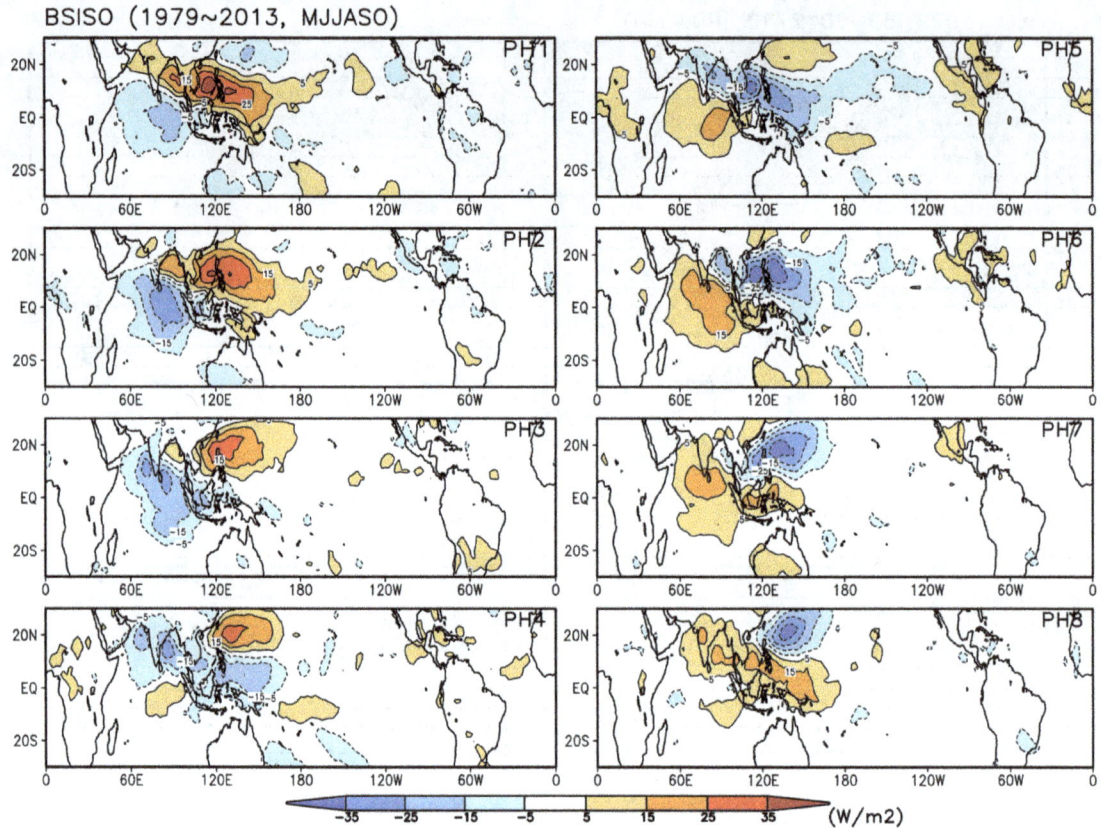

Fig. 1b.   The life cycle composite (phases 1–8) of OLR (shading) anomalies reconstructed based on PC1 and PC2 of Lee *et al.* (2013)'s BSISO1 index (1979∼2013 May through October).

experienced heaviest floods in early February. Those rainfalls took place in the early morning for a few consecutive days. The extreme early morning rainfall occurred due to combination of three dominant processes: the MJO wet phase, the cold surge and a cyclonic vortex in the southwest of Java (Aldrian 2008).

Over the eastern part of South America exhibits unambiguous and robust signals of increased frequency of extremes during active MJO situations (Jones *et al.* 2004). When MJO convective activity is suppressed over Indonesia and enhanced over the central Pacific, the 95th daily rainfall percentile increases over north-northeastern Brazil (Carvalho *et al.* 2004).

California receives most of its annual precipitation during boreal winter. The frequency of extreme precipitation events over California, US, are more common when tropical activity associated with the MJO is high as measured by MJO amplitude index. Higher number of extreme events preferably occur when the convective anomalies are located in the Indian Ocean (Jones 2000).

The variability of extreme precipitation over contiguous United States in winter was shown to be closely linked to the strength (amplitude) of the MJO in the tropics. When MJO is active (large amplitude), joint probabilities of the fractional area of contiguous US sectors are 2.0–2.5 higher than the probabilities during inactive days for both 75th and 90th percentile extreme precipitation (Jones and Carvalho 2012; Fig. 2). Over the Caribbean islands extreme rainfall

75th Percentile CREPS

90th Percentile CREPS

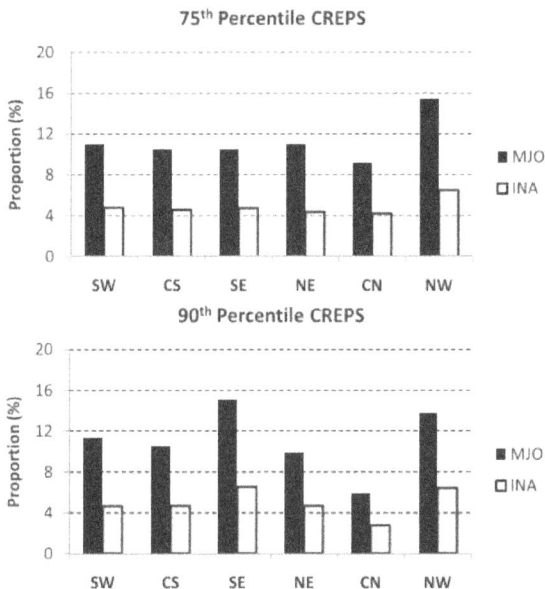

Fig. 2. Proportions of (top) 75th and (bottom) 90th percentile at each contiguous regions of extreme precipitation during active (black) and inactive (clear) MJO days. Horizontal axis indicates each sector in the contiguous United States. From Jones and Carvalho (2012).

events also show a strong relationship with the MJO phase, with extreme events being most common when MJO convection is strongly suppressed over the maritime continent and western Pacfic (Martin and Schumacher 2011).

During boreal summer the ISO has two prominent modes with broad band periods of 30–60 day (BSISO 1) and 10–30 day (BSISO2) (Lee et al. 2013). Both modes influence significantly the occurrence of extreme rainfall events over Yangtze River Valley and southern China. During phases 2–4 of the BSISO1 and phases 5–7 of the BSISO2, when the anomalous southwesterly moisture convergence is observed over southeast China, the probability of extreme rainfall events at 75th (90th) percentile increases by 30–50% (over 60%) relative to the non-BSISO periods (Hsu et al. 2016). Lee et al. (2017) have shown that compared to the climatological probability of extreme rainfall events, the BSISO-based probability provides advantageous information in narrowing down the area and

timing of high probability of extreme events (Fig. 3). The phases 4–6 averaged PA (phase-amplitude)-dependent probability during June and July over Indochina and southeast China is 1.5 times higher than climatological probability (upper panels of Figs. 3a and 3b). Over southeast China, the PA-dependent probabilities of extreme wet events in phases 4 reaches 30% in June while the climatological probability is only 10–20% (lower panel of Fig. 3a).

Eastern China experienced a series of severe floods during the summer of 1998. These floods are shown to be consistent with the propagation and activity of the 30–60 day intraseasonal oscillation over the western North Pacific, where the monsoon trough and the subtropical anti-cyclone appear as an anti-clockwise propagation with the enhanced and suppressed convective anomalies in a 30–60 day period (Zhu et al. 2003).

In South Asia, average daily precipitation, from 13 stations in Afghanistan for 1979–1985, decreases by 23% relative to the mean during the periods when the MJO enhances convection in the eastern Indian Ocean (positive phase) and increases by 21% when the MJO suppresses convection in the eastern Indian Ocean (negative phase). Individually, 3 out of the 13 stations showed more than a factor of 2 differences between the precipitations during the two phases. The distribution of extremes is also affected such that the 10 wettest days all occur during the negative MJO phase (Barlow et al. 2005).

3.2. Snow storms and cold surges

Intense wintertime snowfalls are major extreme weather events that tremendously influence the socio-economic environment and human activity. During the winter of 2009/2010, heavy snowstorms have resulted in tremendous amount of snowfall over the eastern part of the United States (US). This was attributed to the MJO with its unusual active convection over the

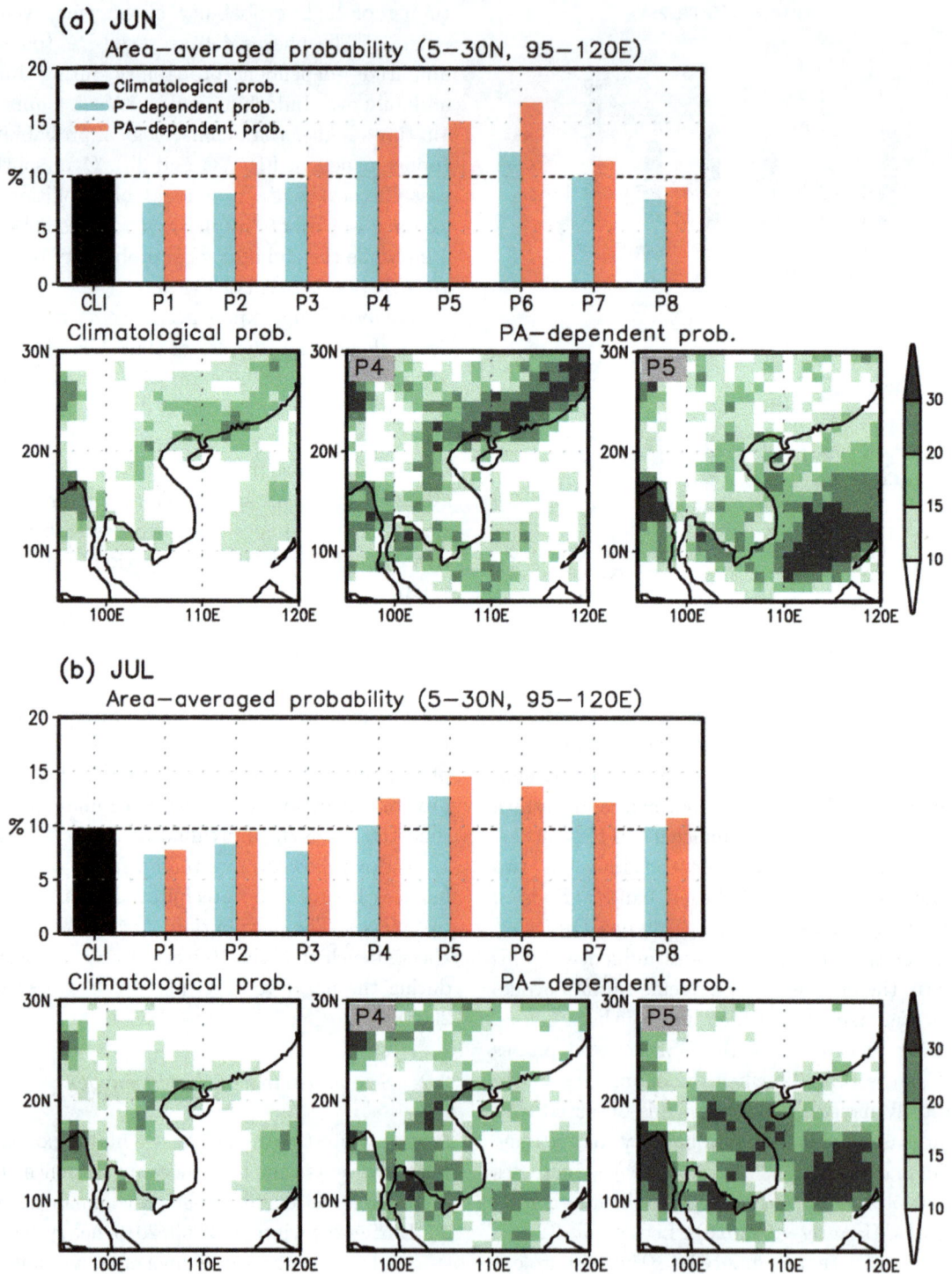

Fig. 3.   Area-averaged (5–30N, 95–120E) probability of extreme wet events (bar, from 90th percentile of monthly gamma distribution) and spatial distributions of climatological and PA (phase-amplitude)-dependent probabilities of extreme wet events. (a) June and (b) July. From Lee *et al.* (2017).

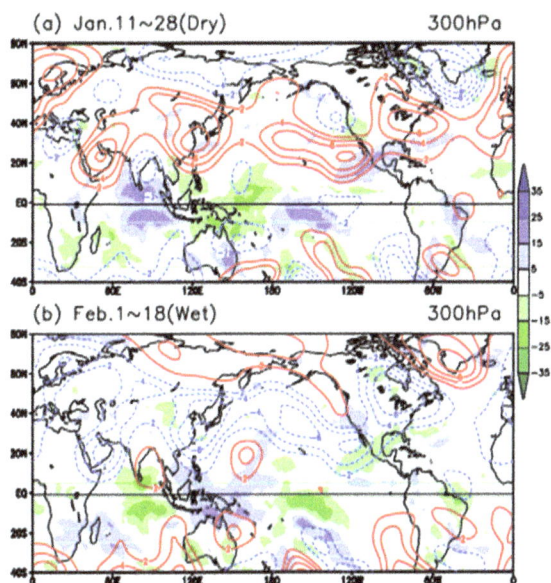

Fig. 4. Intraseasonal anomalies of OLR and 300hPa stream function on the days when convection over the tropical central Pacific in (a) dry and (b) wet phases. The OLR is shaded. The 300 hPa stream function is drawn in contour. From Moon *et al.* (2012).

equatorial central Pacific during an El Niño year (Moon *et al.* 2012).

As the MJO convection reached the equatorial central Pacific, a teleconnection pattern extends from central tropical Pacific to North America, resulting in a westward-tilted deep anomalous trough anchored over the eastern US, producing a low-level pressure dipole anomaly with an anticyclone (cyclone) centered at the US west (east) coast (Fig. 4b). The eastern US was located in a low-level convergence zone between the enhanced cold air from the high-latitude and the warm moist air supplied from the subtropics, resulting in favorable conditions for extremely heavy snowfall.

MJO is found to regulate Atmospheric Reviers (ARs) and associated snow pack over the Sierra Nevada (Guan *et al.* 2012). The ARs are narrow, elongated, synoptic jets of water vapor that plays an important role in the global water cycle and regional weather and hydrology (Guan and Waliser 2015). The ARs stretching from Hawaii to the west coast of North America is the most prominent ARs on Earth, which plays an important role in extreme hydrological events in the west coast (Ralph *et al.* 2006). The AR activity as measured by the high impact ARs and associated snowpack change in Sierra Nevada was shown to be significantly augmented when MJO convection is active over the far western Pacific (Guan *et al.* 2012). This is consistent with the telconnection pattern shown in Fig. 4a: when MJO convection is enhanced over the far western Pacific, the west coast of North America is affected by a low pressure trough and increased precipitation.

Over Southeast Asia, an extreme persistent cold anomaly accompanied by a long-persisting northerly anomaly and a sequence of cold advection occurred in February 2008. The extreme cold anomaly persisted for nearly one month, which not only broke the lowest temperature record for the past 50 years but also resulted in numerous agriculture and fishery losses over Southeast Asia (Hong and Li 2009). The building-up moisture associated with the MJO event is likely to be related to the heavy snowfalls in Middle and South China in January and February of 2008 (Zhang 2013). On 3 February 2008, the MJO have been one of the major factors responsible for the heavy snowfalls in Tokyo metropolitan area (Yamakawa and Suppiah 2009). During the occurrence of the February 2008 cold event, the initiation of the extreme cold anomalies was concurrent with the arrival of an eastward-propagating MJO, which originated from the equatorial western Indian Ocean, at Sumatra. The coincidence between the MJO and the cold air outbreak implies that they might be related (Hsu *et al.* 1990; Chang *et al.* 2005). It was found that the extreme cold anomaly is associated with an enhanced prolonged Siberian high (SH) and a persistent northerly anomaly over Southeast Asia enhanced by MJO. Numerical experiments indicated that the combined effect of the MJO and ENSO maintained the system, which

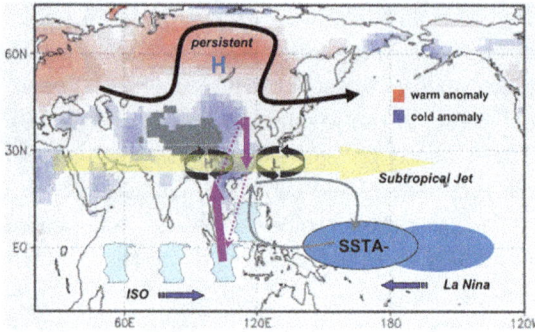

Fig. 5. The schematic diagram to illustrate the physical processes to lead to extreme cold anomaly. The gray arrows represent the vertical overturning circulation. From Hong and Li (2009).

resulted in a persistent northerly anomaly and thus the extreme cold anomaly (Fig. 5).

In East Asia, the surface air temperature anomaly showed distinct spatial variation in association with the MJO-related circulation anomalies through an advective effect at low levels as a primary influence (Jeong *et al.* 2005). In addition, it is suggested that occurrences of extreme cold surges are significantly influenced by MJO such that the synoptic condition related to the MJO provides a suitable condition for amplification of cold surges in MJO phases 2–3.

The MJO is also found to modulate the surface air temperature over Canada and the United States. Over the United States, the strongest and most significant MJO effects on surface air temperature are found during the northern winter seasons. When the MJO convection center is over the Indian Ocean, anomalies in surface air temperature 2°–5°C above normal tend to occur in central/northern Alaska and central/eastern Canada. Temperature fluctuations over Canada are related to warm advection associated with the Rossby wave train emanated from enhanced convection of the MJO (Vecchi and Bond 2004; Lin and Brunet 2009). When the MJO convection center is over the Indian Ocean, below-average surface air temperature tends to occur in New England and the Great Lakes region; as enhanced tropical convection shifts

over the Maritime continent, above average SAT appears in the eastern states of the US from Maine to Florida (Zhou *et al.* 2012).

On a global scale, the frequency of extreme events during extended boreal winter (Nov–Apr) is significantly modulated (i.e., increased by a factor of more than 2 by the MJO with a time lag over some areas in the extratropics as well as in the tropics). In the extratropics, the modulation of the frequency of the extreme events is roughly associated with midlatitude wave responses to tropical forcing and anomalous lower-level circulation due to the MJO (Matsueda and Takaya 2015).

3.3. *Tropical storms*

Favorable large-scale conditions for genesis, intensification, and longevity of tropical cyclones (TCs) can be altered by the MJO. The modulation of TCs by the MJO was first noted in the western North Pacific and Indian Ocean basins by Nakazawa (1988) and Liebmann *et al.* (1994). These studies demonstrated an approximate match between the enhanced convective phase of the MJO and the increased activity of TCs. A similar modulation has also been found in the eastern North Pacific (Molinari *et al.* 1997; Maloney and Hartmann 2000a; Maloney and Hartmann 2001; Aiyyer and Molinari 2008; Barrett and Leslie 2009), the Gulf of Mexico (Maloney and Hartmann 2000b), Atlantic (Barrett and Leslie 2009; Klotzbach 2010; Ventrice *et al.* 2011), the north Indian Ocean (Kikuchi and Wang 2010a; Krishnamohan *et al.* 2012), the South Indian Ocean (Bessafi and Wheeler 2006; Ho *et al.* 2006; Leroy and Wheeler 2008), the northwest Pacific (Liebmann *et al.* 1994; Li and Zhou 2013), the South Pacific (Chand and Walsh 2010; Ramsay *et al.* 2012), and the Australian region (Hall *et al.* 2001). Possible mechanisms for the MJO influences on TCs include reduced vertical wind shear, enhanced low-level convergence, cyclonic relative vorticity, deep convection, increased mid-level moisture,

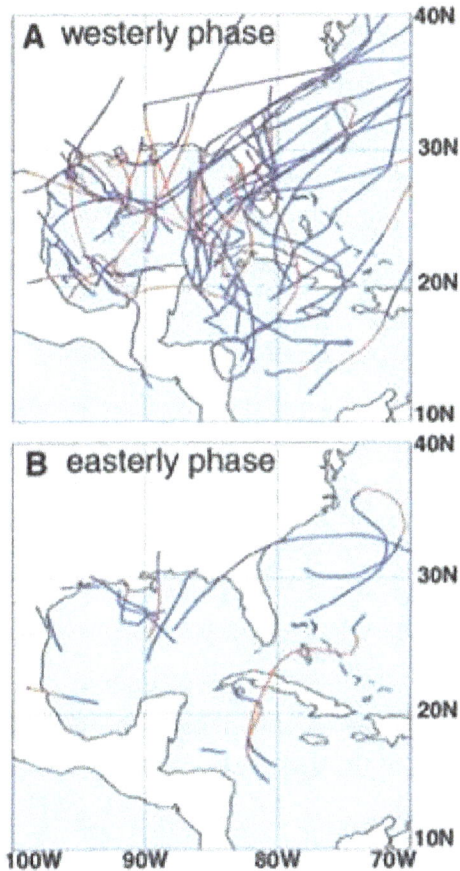

Fig. 6. Tracks of Gulf of Mexico, Caribbean Sea, and western Atlantic tropical cyclones that form to the west of 77.5W when the MJO index has a magnitude greater than 1SD from zero. From Maloney and Hartmann (2000).

small eddies and synoptic disturbances serving as embryos for TCs (Liebmann *et al.* 1994; Mo 2000; Maloney and Hartmann 2001; Maloney and Shaman 2008; Camargo *et al.* 2009).

Figure 6 shows that over the Gulf of Mexico, there are more TCs (hurricanes) in MJO phases 8 to 3 than in phases 4 and 5 (the phase diagram was shown in Fig. 1a). TCs vary coherently with the low-level zonal wind anomalies of the MJO and there are over four times more hurricane-strength storms during westerly phases of the MJO than its easterly phases. Not only cyclone genesis but also all tropical cyclones over the Gulf of Mexico and western Caribbean

Sea during westerly MJO events are enhanced (Mo 2000; Maloney and Hartmann 2000b).

Wang and Zhou (2008) analyzed the number of occurrences of RI (Rapid Intensification) during positive (WNPMI > 1m/s) and negative (WNPMI < −1m/s) phases of ISO, where WNPMI is western North Pacific monsoon index that is used to measure the influence of MJO on the western North Pacific monsoon trough. It was shown that during the positive phases of ISO, the equatorial buffer zone and associated northward cross-equatorial flows as well as the southwest monsoons are all considerably strengthened which is favorable for RI to occur in the Philippine Sea and northern South China Sea (Fig. 7). The number of TCs formed during positive and negative phases are 164 and 67, respectively, yielding a ratio of more than two.

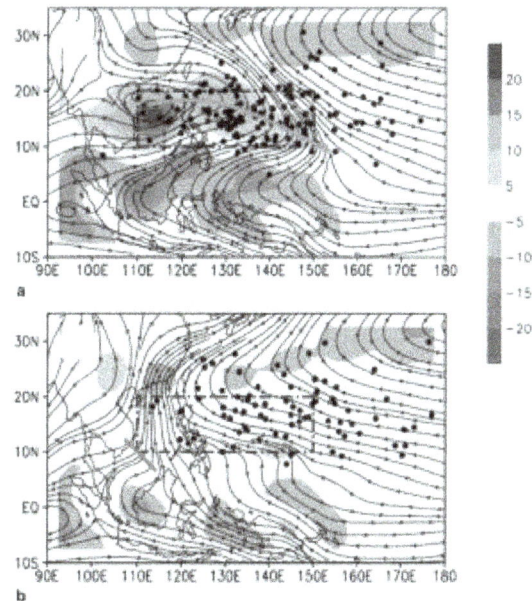

Fig. 7. The composite 850 hPa wind (streamlines) and vorticity (shading) fields for (a) positive (WNPMI > 1m/s, enhanced monsoon trough) and (b) negative (WNPMI < −1m/s, relaxed monsoon trough) phases of ISO during July–August–September–October. Dots represent locations of the first occurrence of RI reported. The dashed box highlights regions in which the largest differences between dry and wet phases of ISO are observed. The thick curves represent the location of the monsoon trough. From Wang and Zhou (2008).

3.4. *Wild fire, aerosols, tornado and hail days*

Possible MJO effects on fire, aerosols, tornado outbreaks, and hail days have been analyzed in recent years. An example is the MJO influence on wild fire over the Maritime Continent, where biomass burning is part of land management practice (Reid *et al.* 2012). When an active convection phase of MJO is located over the Indian Ocean and the Maritime Continent, abundant rainfall minimizes frequency of fire occurrence. The reduced fire count during phases 1–3 when the maximum convection is located over the central Indian Ocean could be explained by precipitation developing ahead of the main MJO convective "envelope" — either from fast propagating Kelvin waves or the orographic precipitation related to prevailing easterlies to the east of MJO major convection. The maximum fire frequency occurs when precipitation is a minimum after the major convection of MJO propagates into the western Pacific.

MJO can modulate large-scale aerosol environment and its associated tropical cyclone activity. Advancement of the MJO from phase 3 to 6 with accompanying cyclogenesis during the cruise period (17 September to 30 September 2011) strengthened the flow patterns in the South China Sea/East Sea that modulated aerosol life cycle (Reid *et al.* 2015).

Thompson and Roundy (2013) documented that the violent tornado outbreak days (VTDs) during March, April, and May are more than twice as frequent during MJO phase 2 as during other phases (Fig. 8). Atmospheric conditions favorable for tornado formation can be provided by combined intraseasonal and seasonal anomalous patterns in upper-tropospheric troughs and upper- and lower-tropospheric winds.

Recently, Barrett and Henley (2014) analyzed the relationships between spring season (April–June) hail days and MJO. They noted that different phases of MJO can modulate synoptic-scale circulation across the United

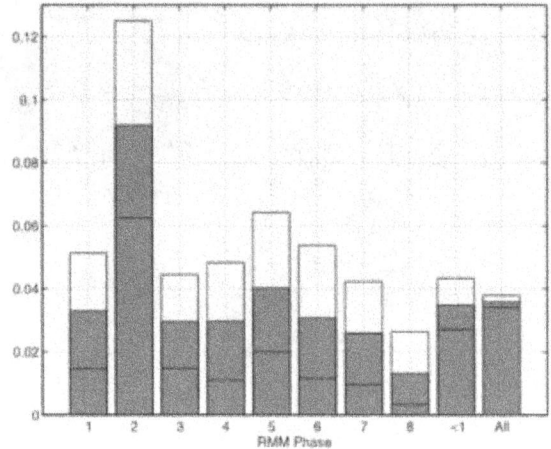

Fig. 8. Fraction of days categorized as VTDs by RMM phase >1.0 amplitude (solid bars). Periods when RMM amplitude was <1 appear in the <1 bar, and the full climatology appears in the "All" bar. Clear bars and lines within the solid bars together represent the 90% confidence interval obtained from a Monte Carlo test. From Thompson and Roundy (2013).

States, and these circulation anomalies lead to synoptic-scale circulation favorable (or unfavorable, for the case of negative hail-day anomalies) for hail producing convective storms.

4. Intraseasonal Forecast of Tropical Cyclone Activity

In recent years, substantial progress made in understanding and numerical simulation of intraseasonal variations of TC activity, has made it possible to forecast TC activity 2–4 weeks ahead of time. The intraseasonal variability includes MJO, BSISO, and convectively coupled equatorial waves. Now, a variety of operational centers and forecast agencies routinely issue intraseasonal forecasts of TC activity for various oceanic basins.

Since October 2011, ECMWF (European Centre for Medium Range Weather Forecast) began to make monthly forecasts using a 51-member ensemble integrated forward for 32 days. The TC subseasonal forecasts are produced by tracking the TC produced by the

32-day subseasonal forecasts described above. The forecasts are produced twice a week (every Monday and Thursday) and cover four calendar weekly periods (Monday to Sunday). The products include grid point strike probability maps (probability of a TC passing within 300 km), and TC statistics (Accumulative Kinetic Energy and number of TCs, hurricanes, depressions) over a whole basin. The forecasts cover all the ocean basins. Verification of the forecasts and more details are discussed in Vitart *et al.* (2010). Elsberry *et al.* (2014) evaluated the performance of ECMWF's 32-day TC forecasts for 2012 in the WNP and found that the forecasts provide skillful subseasonal TC formation and track forecasts during the 2012 season, with high probabilities of detection (>85%) and only a small number of misses through four weeks in advance. The forecast system still features a much larger number of false alarms relative to the number of hits, which reduces its overall reliability.

In 2013, the National Oceanic and Atmospheric Administration (NOAA)'s Climate Prediction Center (CPC) in US initiated an experimental prediction project on tropical storm (TS) activity for week 1 to week 4 periods in support of the CPC's Global Tropics Hazards and Benefits Outlook. Predictability of intraseasonal tropical storm (TS) activity has been assessed using the 1999–2010 CFSv2 hindcast suites. Weekly TS activity in the CFSv2 45-day forecasts was determined using the tropical storm detection and tracking method devised by Camargo and Zebiak (2002). The forecast periods are divided into weekly intervals for week 1 through week 4 and for the 30-day mean. The National Hurricane Center (NHC) participates in the bi-weekly Global Tropical Hazards Assessments by providing guidance on the likelihood of TC genesis in the eastern North Pacific Ocean and the Atlantic basin for week 1 and week 2 forecasts.

These subseasonal forecasts are often based on a combination of statistical, dynamical, and heuristic approaches that leverage the extended

predictability of intraseasonal variability such as the MJO or equatorial wave modes (e.g., convectively coupled Kelvin waves). The more skillful predictions of subseasonal TC activity often feature a combination of statistical–dynamical approaches that use post-processing methods to remove systematic climatological biases or incorporate *a priori* information including the seasonal TC climatology, current TC activity, and subseasonal TC forecast performance provided by reforecast datasets (WMO 2014).

5. Summary and Discussion

In this chapter, the influence of intraseasonal oscillation on many types of weather and climate extreme events in the global tropics and extratropics have been reviewed. ISO not only regulates precipitation and temperature extremes, but also other phenomena such as heavy snowfall, flooding, drought, tropical cyclones, aerosols, wild fire, hail days either directly or indirectly under the influence of other climate variability.

The present review provides a useful reference for when and where the ISO has significant impacts, and can be used as a prediction tool for the ISO to be accurately forecasted. An improved understanding of mechanisms linking the global tropics to the extratropical circulation patterns is critical to developing a better understanding of the forecast potential of the ISO (Zhou *et al.* 2012). There is hardly another phenomenon that is globally connected to both weather and climate as broadly as the ISO (Zhang 2013). The broad connections between ISO and many types of weather and climate events are strong testaments to the realization that the societal need of weather and climate prediction cannot be met without advancing better understanding of the ISO and without improving our ability of predicting the ISO (Waliser *et al.* 2003).

Considerable efforts have been made on the prediction of the MJO, leading to the improvement of MJO forecast skill beyond 20 days (Kang and Kim 2010; Rashid *et al.* 2010; Vitart and Molteni 2010; Wang *et al.* 2014) compared to 7–10 days about 10 years ago (Waliser 2011). Some models evaluated within the recent ISVHE (IntraSeasonal Variability Hindcast Experiment) show predictive skill up to 20–30 days and potential predictability between 31 and 45 days (Mani *et al.* 2014). This progress can be attributed to model improvements, better initial conditions, and the availability of historical reforecasts to calibrate the forecasts (WMO 2015; Vitart 2014; Vitart *et al.* 2014).

However, the main issues remain to be resolved include insufficient ensemble spread, natural predictability barrier over the Maritime Continent, horizontal resolution and improving the stochastic physics in the MJO prediction system (Vitart *et al.* 2012; Vitart and Molteni 2010, Weaver *et al.* 2011; Robertson *et al.* 2015).

Since the ISO contributes to the predictability of extreme events, the realistic representation of the ISO and atmospheric responses to the ISO in forecast models is important for improving early warning information about extreme events (Matsueda and Takaya 2015).

To better predict and understand the ISO modulation of extreme events, we need to establish more of the ISO effects on extreme events together with other factors; explore the extent to which observed ISO-extreme events relationships are reproduced in real-time prediction and climate simulations/projections. This include understanding the way by which the extratropical response to the MJO forcing can be predicted, including the strength and location of the upper level tropical divergence, the vertical structure of tropical diabatic heating, and the wave response to an evolving MJO heating anomalies with an appropriate time delay. Continuous improvement of general circulation models' capability of simulating the variability of regional-scale mean and extreme climate must

be among the first priority tasks for the community (Wang *et al.* 2012).

Acknowledgments

This work is supported by the support from the National Science Foundation of (Climate Dynamics Division Award No. AGS-1540783) and the NOAA/CVP #NA15OAR4310177, as well as the Global Research Laboratory (GRL) Program of the Korean Ministry of Education, Science and Technology (MEST, #2011-0021927). The authors thanks Dr. S-S Lee for data assistance.

References

Aiyyer, A. and J. Molinari, 2008: MJO and tropical cyclogenesis in the Gulf of Mexico and eastern Pacific: Case study and idealized numerical modeling. *J. Atmospheric Sci.*, **65**, 2691–2704.

Aldrian, E., 2008: Dominant factors of Jakarta's three largest floods. *J. Hidrosfir. Indones.*, **3**, 105–112.

Barlow, M., M. Wheeler, B. Lyon, and H. Cullen, 2005: Modulation of daily precipitation over Southwest Asia by the Madden–Julian oscillation. *Mon. Weather Rev.*, **133**, 3579–3594.

Barrett, B. S. and B. N. Henley, 2014: Intraseasonal Variability of Hail in the Contiguous United States: Relationship to the Madden–Julian Oscillation. *Mon. Weather Rev.*, **143**, 1086–1103.

———, and L. M. Leslie, 2009: Links between tropical cyclone activity and Madden-Julian oscillation phase in the North Atlantic and northeast Pacific basins. *Mon. Weather Rev.*, **137**, 727–744.

Bessafi, M. and M. C. Wheeler, 2006: Modulation of South Indian Ocean tropical cyclones by the Madden–Julian oscillation and convectively coupled equatorial waves. *Mon. Weather Rev.*, **134**, 638–656.

Camargo, S. J. and S. E. Zebiak, 2002: Improving the detection and tracking of tropical cyclones in atmospheric general circulation models. *Weather Forecast.*, **17**, 1152–1162.

———, M. C. Wheeler, and A. H. Sobel, 2009: Diagnosis of the MJO modulation of tropical cyclogenesis using an empirical index. *J. Atmospheric Sci.*, **66**, 3061–3074.

Carvalho, L. M. V., C. Jones, and B. Liebmann, 2004: The South Atlantic convergence zone: Intensity, form, persistence, and relationships with intraseasonal to interannual activity and extreme rainfall. *J. Climate*, **17**, 88–108.

Chand, S. S. and K. J. E. Walsh, 2010: The influence of the Madden-Julian oscillation on tropical cyclone activity in the Fiji region. *J. Climate*, **23**, 868–886.

Chang, C.-P., P. A. Harr, and H.-J. Chen, 2005: Synoptic disturbances over the equatorial South China Sea and western Maritime Continent during boreal winter. *Mon. Weather Rev.*, **133**, 489–503.

Elsberry, R. L., H.-C. Tsai, and M. S. Jordan, 2014: Extended-Range Forecasts of Atlantic Tropical Cyclone Events during 2012 Using the ECMWF 32-Day Ensemble Predictions. *Weather Forecast.*, **29**, 271–288.

Emanuel, K. A., 1987: Air-sea interaction model of intraseasonal oscillations in the Tropics. *J. Atmospheric Sci.*, **44**, 2324–2340.

Gottschalck, J., M. Wheeler, K. Weickmann, F. Vitart, N. Savage *et al.*, 2010: A framework for assessing operational Madden-Julian oscillation forecasts: a CLIMVAR MJO working group project. *Bull. Am. Meteorol. Soc.*, **91**, 1247–1258.

Guan, B., D. E. Waliser, N. Molotch, E. Fetzer, and P. Neiman, 2012: Does the Madden-Julian Oscillation Influence Wintertime Atmospheric Rivers and 1 Snowpack in the Sierra Nevada?. *Mon. Weather Rev.*, **140**, 325–342.

Guan, B. and D. E. Waliser, 2015: Detection of atmospheric rivers: Evaluation and application of an algorithm for global studies. *J. Geophys. Res.*, **120**, 12,514–512,535.

Hall, J. D., A. J. Matthews, and D. J. Karoli, 2001: The modulation of tropical cyclone activity in the Australian region by the Madden–Julian oscillation. *Mon. Weather Rev.*, **129**, 2970–2982.

Hendon, H. H. and M. L. Salby, 1994: The life cycle of the Madden–Julian oscillation. *J. Atmospheric Sci.*, **51**, 2225–2237.

Herring, S. C., M. P. Hoerling, T. C. Peterson, and P. A. Stott, Eds., 2014: Explaining Extreme Events of 2013 from a Climate Perspective. *Bull. Am. Meteorol. Soc.*, **95 (9)**, S1–S96.

Ho, C.-H., J.-H. Kim, J.-H. Jeong, H.-S. Kim, and D. Chen, 2006: Variation of tropical cyclone activity in the South Indian Ocean: El Niño–Southern Oscillation and Madden–Julian oscillation effects. *J. Geophys. Res.*, **111**, D22101. doi:10.1029/2006JD007289.

Hong, C.-C. and T. Li, 2009: The extreme cold anomaly over Southeast Asia in February 2008: Roles of ISO and ENSO. *J. Climate*, **22**, 3786–3801.

Hsu, H.-H., B. J. Hoskins, and F.-F. Jin, 1990: The 1985/86 intraseasonal oscillation and the role of extratropics. *J. Atmospheric Sci.*, **47**, 823–839.

Hsu, P.-C., J. Y. Lee, and K.-J. Ha, 2016: Influence of boreal summer intraseasonal oscillation on rainfall extremes in Southeast China. *Int. J. Climatol.*, **36**, 1403–1412.

Jeong, J.-H., C.-H. Ho, B.-M. Kim, and W.-T. Kwon, 2005: Influence of the Madden-Julian Oscillation on wintertime surface air temperature and cold surges in East Asia. *J. Geophys. Res.*, **110**, D11104. doi:10.1029/2004JD005408.

Jones, C., 2000: Occurrence of extreme precipitation events in California and relationships with the Madden–Julian Oscillation. *J. Climate*, **13**, 3576–3587.

———, D. E. Waliser, K. M. Lau, and W. Stern, 2004: Global occurrences of extreme precipitation events and the Madden–Julian oscillation: Observations and predictability. *J. Climate*, **17**, 4575–4589.

——— and L. M. V. Carvalho, 2012: Spatial–Intensity variations in extreme precipitation in the Contiguous United States and the Madden–Julian Oscillation. *J. Climate*, **25**, 4989–4913.

Kang, I.-S. and H.-M. Kim, 2010: Assessment of MJO predictability for boreal winter with various statistical and dynamical models. *J. Climate*, **23**, 2368–2378.

Karl, T. R. and D. R. Easterling, 1999: Climate extremes: Selected review and future research directions. *Clim. Change*, **42**, 309–325.

Kikuchi, K. and B. Wang, 2010a: Formation of tropical cyclones in the northern Indian Ocean associated with two types of tropical intraseasonal oscillation modes. *J. Meteorol. Soc. Jpn.*, **88**, 475–496.

———, and B. Wang, 2010b: Spatio-temporal wavelet transform and the multiscale behavior of the Madden–Julian Oscillation. *J. Climate*, **23**, 3814–3834.

Klotzbach, P. J., 2010: On the Madden-Julian oscillation–Atlantic hurricane relationship. *J. Climate*, **23**, 282–293.

Krishnamohan, K. S., K. Mohanakumar, and P. V. Joseph, 2012: The influence of Madden-Julian oscillation in the genesis of north Indian Ocean tropical cyclones. *Theor. Appl. Climatol.*, **109**, 271–282.

Krishnamurti, T. N., D. K. Oosterhof, A.V. Mehta, 1988: Air-sea interaction on the time scale of 30 to 50 days. *J. Atmospheric Sci.*, **45**, 1304–1322.

Lee, J.Y., B. Wang, M. C. Wheeler, X. Fu, D. E. Waliser, and I. S. Kang, 2013: Real-time multivariate indices for the boreal summer intraseasonal oscillation over the Asian summer monsoon region. *Clim. Dynam.*, **40**, 493–509. doi:10.1007/s00382-012-1544-4.

Lee, S.-S., J.-Y. Moon, B. Wang, and H.-J. Kim, 2017: Probabilistic subseasonal prediction of extreme precipitation events over Asian-Pacific region. *J. Climate*, **30**, 2849–2865.

Leroy, A. and M. C. Wheeler, 2008: Statistical prediction of weekly tropical cyclone activity in the Southern Hemisphere. *Mon. Weather Rev.*, **136**, 3637–3654.

Li, R. C. Y. and W. Zhou, 2013: Modulation of western North Pacific tropical cyclone activity by the ISO. Part I: Genesis and intensity. *J. Climate*, **26**, 2904–2918.

Liebmann, B., H. H. Hendon, and J. D. Glick, 1994: The relationship between tropical cyclones of the western Pacific and Indian Oceans and the Madden–Julian oscillation. *J. Meteorol. Soc. Jpn.*, **72**, 401–412.

Lin, H. and G. Brunet, 2009: The influence of the Madden–Julian oscillation on Canadian wintertime surface air temperature. *Mon. Weather Rev.*, **137**, 2250–2262.

Madden, R. A. and P. R. Julian, 1971: Detection of a 40–50 day oscillation in the zonal wind in the tropical Pacific. *J. Atmospheric Sci.*, **28**, 702–708.

———, and ———, 1972: Description of global-scale circulation cells in the tropics with a 40–50 day period. *J. Atmospheric Sci.*, **29**, 1109–1123.

Maloney, E. D. and D. L. Hartmann, 2000a: Modulation of eastern North Pacific hurricanes by the Madden–Julian oscillation. *J. Climate*, **13**, 1451–1460.

———, and ———, 2000b: Modulation of hurricane activity in the Gulf of Mexico by the Madden–Julian oscillation. *Science*, **287**, 2002–2004.

——— and ———, 2001: The Madden–Julian oscillation, barotropic dynamics, and North Pacific tropical cyclone formation. Part I: Observations. *J. Atmospheric Sci.*, **58**, 2545–2558.

——— and J. Shaman, 2008: Intraseasonal variability of the West African monsoon and Atlantic ITCZ. *J. Climate*, **21**, 2898–2918.

Mani, J. M., J. Y. Lee, D. Waliser, B. Wang, and X. Jiang, 2014: Predictability of the Madden–Julian oscillation in the Intraseasonal Variability Hindcast Experiment (ISVHE). *J. Climate*, **27**, 4531–4543. doi:10.1175/JCLI-D-13-00624.1.

Martin and Schumacher, 2011: Modulation of Caribbean precipitation by the Madden–Julian Oscillation. *J. Climate*, **24**, 813-824.

Matsueda, S. and Y. Takaya, 2015: The global influence of the Madden–Julian Oscillation on extreme temperature events. *J. Climate*, **28**, 4141–4151.

Mo, K. C., 2000: The association between intraseasonal oscillations and tropical storms in the Atlantic basin. *Mon. Weather Rev.*, **128**, 4097–4107.

Molinari, J., D. Knight, M. Dickinson, D. Vollaro, and S. Skubis, 1997: Potential vorticity, easterly waves, and Eastern Pacific tropical cyclogenesis. *Mon. Weather Rev.*, **125**, 2699–2708.

Moon, J.-Y., B. Wang, and K.-J. Ha, 2012: MJO modulation on 2009/10 winter snowstorms in the United States. *J. Climate*, **25**, 978–991.

Nakazawa, T., 1988: Tropical super clusters within intraseasonal variations over the western Pacific. *J. Meteorol. Soc. Jpn.*, **66**, 823–839.

Ralph, F. M., P. J. Neiman, G. A. Wick, S. I. Gutman, M. D. Dettinger, D. R. Cayan, and A. B. White, 2006: Flooding on California's Rissian River: The roles of atmospheric rivers. *Geophys. Res. Lett.*, **33**, L13801. doi:10.1029/2006 GL026689.

Ramsay, H. A., S. J. Camargo, and D. Kim, 2012: Cluster analysis of tropical cyclone tracks in the Southern Hemisphere. *Clim. Dynam.*, **39**, 897–917.

Rashid, H. A., H. H. Hendon, M. C. Wheeler, and O. Alves, 2010: Predictability of the Madden-Julian oscillation in the POAMA dynamical seasonal prediction system. *Clim. Dynam.*, doi:10.1007/s00382-010-0754-x.

Reid, J. S. and Coauthors, 2012: Multi-scale meteorological conceptual analysis of observed active fire hotspot activity and smoke optical depth in the Maritime Continent. *Atmospheric Chem. Phys.*, **12**, 2117–2147. doi:10.5194/acp-12-2117-2012.

———, and Coauthors, 2015: Observations of the temporal variability in aerosol properties and their relationships to meteorology in the summer monsoonal South China Sea/East Sea: the scale-dependent role of monsoonal flows, the Madden–Julian Oscillation, tropical cyclones, squall lines and cold pools. *Atmospheric Chem. Phys.*, **15**, 1745–1768.

Robertson, A. W., A. Kumar, M. Pena, and F. Vitart, 2015: Improving and promoting subseasonal to seasonal prediction. *Bull. Am. Meteorol. Soc.* doi:10.1175/BAMS-D-14-00139.1

Rui, H. and B. Wang, 1990: Development characteristics and dynamic structure of tropical intraseasonal convective anomalies. *J. Atmospheric Sci.*, **47**, 357–379.

Seventh WMO International Workshop on Tropical Cyclones (IWTC-VII), Saint-Gilles-Les-Bains, La Réunion, France, 15–20 November 2010 (WMO TD No. 1561) (WWRP 2011-1).

Thompson, D. B. and P. E. Roundy, 2013: The relationship between the Madden–Julian oscillation and U.S. violent tornado outbreaks in the spring. *Mon. Weather Rev.*, **141**, 2087–2095.

Vecchi, G. A. and N. A. Bond, 2004: The Madden-Julian oscillation (MJO) and northern high latitude wintertime surface air temperatures. *Geophys. Res. Lett.*, **31**, L04104. doi:10.1029/2003GL018645.

Ventrice, M. J., C. D. Thorncroft, and P. E. Roundy, 2011: The Madden-Julian oscillation's influence on African easterly waves and downstream tropical cyclogenesis. *Mon. Weather Rev.*, **139**, 2704–2722.

Vitart, F., A. Leroy, and M. C. Wheeler, 2010: A comparison of dynamical and statistical predictions of weekly tropical cyclone activity in the Southern Hemisphere. *Mon. Weather Rev.*, **138**, 3671–3682.

——, and F. Molteni, 2010: Simulation of the Madden–Julian Oscillation and its teleconnections in the ECMWF forecast system. *Q. J. Roy. Meteor. Soc.*, **136**, 842–855. doi:10.1002/qj.623.

——, A. W. Robertson, and D. L. T. Anderson, 2012: Subseasonal to Seasonal Prediction Project: bridging the gap between weather and climate. *WMO Bulletin*, **61**(2), 23–28.

——, 2014: Evolution of ECMWF subseasonal forecast skill scores. *Q. J. Roy. Meteor. Soc.*, **140**, 1889–1899.

Vitart, F., G. Balsamo, R. Buizza, L. Ferranti, S. Keeley, L. Magnusson, F. Molteni, and A. Weisheimer, 2014: Sub-seasonal predictions. ECMWF technical memorandum no. 738. Reading, Berkshire, UK: European Centre for Medium-Range Weather Forecasts.

Waliser, D. E., K. M. Lau, W. Stern, and C. Jones, 2003: Potential predictability of the Madden–Julian oscillation. *Bull. Am. Meteorol. Soc.*, **84**, 33–50.

——, 2011: Predictability and Forecasting. In Intraseasonal Variability in the Atmosphere-Ocean Climate System, W.K.M. Lau and D.E. Waliser, Eds. (Springer, Heidelberg), 2nd Ed. DOI 10.1007/978-3-642-13914-7.

Wang, B., 1988: Comments on "An air-sea interaction model of intraseasonal oscillation in the tropics". *J. Atmospheric Sci.*, **45**, 3521–3525.

——, and H. Rui, 1990: Dynamics of coupled moist Kelvin-Rossby waves on an equatorial beta-plane. *J. Atmospheric Sci.*, **47**, 397–413.

——, P. Webster, K. kikuchi, T. Yasunari, and Y. Qi, 2006: Boreal summer quasi-monthly oscillation in the global tropics. *Clim. Dynam.*, **27**, 661–675.

——, and X. Zhou, 2008: Climate variation and prediction of rapid intensification in tropical cyclones in the western North Pacific. *Meteorol. Atmos. Phys.*, **99**, 1–16 (2008) doi:10.1007/s00703-006-0238-z.

Wang, H.-J., J.-Q. Sun, H.-P. Chen, Y.-L. Zhu, Y. Zhang, D.-B. Jiang, X.-M. Lang, K. Fan, E.-T. Yu, and S. Yang, 2012: Extreme climate in China: facts, simulation and projection. *Meteorologische Zeitschrift*, **21**(3), 279–304.

Wang, W. and co-authors, 2014: MJO prediction in the NCEP Climate Forecast System version 2. *Clim. Dynam.*, **42**, 2509–2520.

Weaver, S. J., W. Q. Wang, M. Y. Chen, and A. Kumar, 2011: Representation of MJO variability in the NCEP climate forecast system. *J. Climate*, **24**, 4676–4694.

Wheeler, M. C. and H. H. Hendon, 2004: An all-season real-time multivariate MJO index: development of an index for monitoring and prediction. *Mon. Weather Rev.*, **132**, 1917–1932.

——, H. H. Hendon, S. Cleland, H. Meinke, and A. Donald, 2009: Impacts of the Madden-Julian oscillation on Australian rainfall and circulation. *J. Climate*, **22**, 1482–1498.

WMO, 2014: 8th International workshop on tropical cyclones (IWTC-VIII). Topic 5.3. Intraseasonal timescales.

WMO, 2015: Seamless prediction of the earth system: From minutes to months. WMO-No. 1156.

Xavier, P., R. Rahmat, W. K. Cheong, and E. Wallace, 2014: Influence of Madden-Julian oscillation on Southeast Asia rainfall extremes: Observations and predictability. *Geophys. Res. Lett.*, **41**, 4406–4412. doi:10.1002/2014GL060241.

Yamakawa, S. and R. Suppiah, 2009: Extreme climatic events in recent years and their links to large-scale atmospheric circulation features. *Global Environ. Res.*, **13**, 69–78.

Zhang, C., 2005: Madden-Julian Oscillation. *Rev. Geophys.*, **43**, RG2003. doi:10.1029/2004RG000158.

———, 2013: Madden-Julian Oscillation — Bridging the gap between weather and climate. *Bull. Am. Meteorol. Soc.*, DOI:10.1175/BAMS-D-12-00026.1, 1849–1870.

Zhou, S., M. L'Heureux, S. Weaver, and A. Kumar, 2012: A composite study of the MJO influence on the surface air temperature and precipitation over the continental United States. *Clim. Dynam.*, **38**, 1459–1471. doi:10.1007/s00382-011-1001-9.

Zhu, B. and B. Wang, 1993: The 30–60 day convection seesaw between the tropical Indian and western Pacific Oceans. *J. Atmos. Sci.*, **50**, 184–199.

Zhu, C., T. Nakazawa, J. Li, and L. Chen, 2003: The 30–60 day intraseasonal oscillation over the western North Pacific Ocean and its impacts on summer flooding in China during 1998. *Geophys. Res. Lett.*, **30**, 1952. doi:10.1029/2003GL017817.

Chapter 4

Climate Services: For Informing Decisions and Managing Risk

Neil Plummer[*,§], Agata Imielska[*], Karl Braganza[*], David Jones[*],
Janita Pahalad[*], Scott Power[†], Martin Schweitzer[‡], Andrew Watkins[*],
David Walland[*], Perry Wiles[*]

[*]*Climate Information Services Branch*

[†]*Research and Development Branch*

[‡]*Environmental Information Systems Branch*
Bureau of Meteorology,
Melbourne, Victoria 3001, Australia
[§]*n.plummer@bom.gov.au*
www.bom.gov.au

Climate services are becoming increasingly important due to increases in the frequency, intensity and exposure to weather and climate extremes, with stakeholders seeking to better manage their risks and opportunities. The El Niño-Southern Oscillation (ENSO) is one of the main contributing factors to Australia having one of the most variable climates in the world. High variability and climate change magnify the need for environmental intelligence to support weather and climate sensitive sectors such as agriculture, water management, emergency services and health. The Australian Bureau of Meteorology (Bureau) has been developing its climate services since the 1980s. These include data services, climate monitoring and prediction, climate advice and international activities, and are heavily reliant on strong information systems and research support. Services are delivered through a range of online products and tools as well as through briefings, training and social media platforms. The Bureau has shifted to user-centric delivery, and engages strongly with its user communities. Understanding user needs and decision-making processes informs the development of value-added products and services that can capture the full benefit of climate data, information, advice and research for our communities. In our journey of continuous improvement we are continuing to strive to better meet user needs. Having skilled and motivated staff with strong science and IT capability and a service ethic are essential. Improving science and modelling will also help close the gap between user needs and existing climate services. Fundamentally, success will depend on how well climate services assist users to improve their important decisions and the outcome of those decisions.

1. Introduction

Climate services are becoming increasingly important to society. Their benefits are more apparent each year and, with several weather and climate extremes increasing in frequency and intensity, we can expect demand for these services to increase. The past five years alone have seen Australia heavily affected by global warming and the extremes of the El Niño-Southern Oscillation (ENSO) phenomenon.

The hottest year and wettest two-year periods on record were preceded by one of the most severe periods of drought on record. After Australia's wettest two-year period on record through 2010–12, many parts of the country returned to drought.

The purpose of this paper is to describe the development of climate services within the Australian Bureau of Meteorology (Bureau). This paper will show how these services have been shaped by the impacts of climate variability and change on Australian society, rapid improvements in climate science and technologies, and the drive to meet the needs of stakeholders for their climate-sensitive decisions.

1.1. *What are climate services?*

The World Meteorological Organization (WMO) describes climate services as:

> The dissemination of climate information to the public or a specific user. Climate services involve strong partnerships among providers, such as National Meteorological and Hydrological Services (NMHSs), and stakeholders, including government agencies, private interests, and academia, for the purpose of interpreting and applying climate information for decision-making, sustainable development, and improving climate information products, predictions, and outlooks (WMO 2015).

Vaughan *et al.* (2014) regard the aim of climate services as being able to provide people and organisations with timely, tailored climate-related knowledge and information that they can use to reduce climate-related losses and enhance benefits, including the protection of lives, livelihoods, and property. The authors distinguish between climate services and climate research. The former focuses on serving user needs while the latter aims to further our understanding of the climate system. Aside from definitional issues, there are strong linkages between the two, and most benefits flow when they are tightly coupled.

1.2. *Australia's variable climate*

Australia has one of the most variable climates in the world. It is a large island continent containing several distinct climate zones, stretching from cool temperate regions in the south to humid tropics in the north. The tropics are impacted by the northern Australian monsoon, tropical cyclone season and include rainforest, savanna and marine ecosystems such as the Great Barrier Reef. The continental interior (with its long hot summers) includes grassland and deserts, with temperate regions in Australia's south characterised by cooler temperatures and extra-tropical weather systems. The subtropical regions are at the intersections of the Australian tropics and the temperate climate zones. Further information on Australia's climate is available on the Bureau's website at http://www.bom.gov.au/jsp/ncc/climate_avera ges/climate-classifications/index.jsp.

Vast areas of the continent are arid or semi-arid, while large interannual and decadal rainfall variability — characterised by frequent prolonged droughts interspersed by flooding rainfall — affects most of the mainland. In this way, water availability and natural climate variability are major limiting factors for ecosystems and human activity across large parts of Australia.

Australia is heavily dependent, historically, on its agriculture and natural resources sectors for a significant proportion of its national wealth and well-being. Hence a significant portion of Australia's gross domestic product variability can be attributed to the variability in weather and climate. Climate sensitivity to coal mining and offshore oil and gas in Australia is estimated at approximately A$3,000 million each, while agriculture is estimated at A$2,593 (CIE 2014). This has led to a long-standing partnership between the agricultural sector and the Bureau in relation to climate services.

1.3. *A history of Bureau's climate services*

Prior to the mid-1980s the focus of climate in the Bureau was very much on the management and provision of climate data with little "value adding" for external stakeholders. This changed significantly in 1985 with the establishment of the Bureau's National Climate Centre (NCC), including its regional climate centres. The key factors in the evolution of Australian climate services were:

- Improved understanding from the early 1980s of the role of ENSO on Australian climate.
- The development of the World Climate Programme in the 1980s, to improve our understanding of the climate system and its impact on society.
- Concerns about the influence of human activity on our climate and the creation of the Intergovernmental Panel on Climate Change (IPCC) and its influence on strengthening NCC's climate monitoring capability.
- The development of Australian operational seasonal prediction services from the late 1980s.
- Concerns over the impacts of climate change and variability and increased investment in tools and technologies from governments and the agricultural sector, including, from the early 1990s, investors such as the Climate Variability in Agriculture and the Managing Climate Variability programs.
- Development of the Australian Data Archive for Meteorology (ADAM) in the mid-1990s.
- The automation of climate analyses from the mid-1990s and, in particular the ability for the Bureau to readily produce thousands of maps — a significant improvement in efficiency and service delivery.
- Improved outreach and international collaboration through the development of the Internet from the mid-1990s.
- The Millennium drought from 1997 to 2010 and widespread concerns over water security in Australia.

1.4. *International context*

International collaboration has been a priority for climate information service providers and researchers within the Bureau, and it continues to heavily influence Australian climate services and research. The NCC has evolved into the

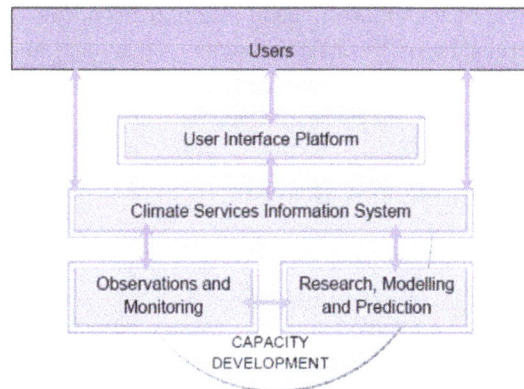

Fig. 1. Schematic illustration of the five pillars of the Framework and their links to various user communities. (GFCS 2014).

broader Climate Information Services Branch (CISB) and a recent restructure increased attention on user needs. The third World Climate Conference, held in 2009, endorsed the concept of a Global Framework for Climate Services (GFCS 2014) to strengthen production, availability, delivery, and application of science-based climate prediction and other services, particularly in developing countries. The Framework is being built upon five components, or pillars (Fig. 1).

Four priority areas were chosen for early attention under the Framework: Agriculture and Food Security (including fisheries and aquaculture), Disaster Risk Reduction, Health, and Water. Energy has been a recent addition. The User Interface Platform is widely considered as the least developed pillar of the GFCS around the world and yet it is critical to achieving the framework's goals.

The Bureau's organisational structure for climate services anticipated the five pillars of the GFCS. Close internal collaboration is essential between CISB and areas handling observations, IT, Research and Development, communication and business development. A significant challenge for both the GFCS and the Bureau is to ensure that information is accessible and enables users' decision-making. In Australia, the Bureau created a Climate Liaison function primarily to strengthen external stakeholder engagement and help ensure that service outputs have uptake and impact for decision-makers.

1.5. *Aims and structure*

The sections that follow describe the components of the Australian climate service — how they have evolved and how we can expect them to do so in the future. Section 2 starts with our transition from data to information services while Sec. 3 focuses on climate monitoring. Section 4 describes our climate prediction function and its rapid improvements to respond to user needs. Our increased emphasis on climate liaison and on engaging governments, industry and communities are described in Sec. 5. The next three sections (6 to 8) cover the underpinning and overarching areas of international collaboration, research and information systems. Finally, Sec. 9 explores the Bureau's aspirations for climate services, including considerations for best practice heading into the future.

2. Transitioning from Data to Information Services

The history of instrumental weather observations in Australia stretches back to the very first European settlement, evolving to systematic observation practices undertaken by the Bureau following its founding in 1908. Over the next century, amateur and official meteorologists continued taking observations in settlements dotted around the continent, providing documentary evidence of climate variability in Australia.

The Bureau continues to develop and advance its network of observing sites. In 2014, the Bureau had 775 temperature recording sites and 6,844 rain gauges operating across Australia.

Most climate service organisations have a fundamental responsibility in collecting and curating meteorological observations into a climate record. However, the utility of the climate record can only be realised if it is readily available to users. Hence, a priority for the Bureau has been to provide open access to its climate record. To the extent that resources allow, this is now through a free online tool that enables users to download data.

The Bureau's tool, called Climate Data Online (http://www.bom.gov.au/climate/data/), has served this purpose over the past decade and caters for approximately three million online requests per year. This now far exceeds the approximately 8,000 additional requests for data that still need to be handled manually by staff.

Climate factors influence decision-making within a wide range of sectors and businesses, many of whom have a strong need for timely climate information tailored to their business and decision-making processes. This type of user is often time poor, and needs complex climate information synthesised into a service bespoke to their business. There is significant benefit in providing value-added information, beyond that of a basic product suite. Examples of such users include insurance companies, agriculture, energy providers, hydrological engineers, and oil and gas exploration companies.

Accordingly, the Bureau offers a range of value-added analysis products to various sectors, many provided as a fee-for-service. Analysis of the climate record to quantify risk and the probability of the occurrence of an event is a commonly requested product. Examples include design rainfall that informs the engineering safety of hydrological structures, and design wave information used for the safe design and operation of offshore exploration infrastructure. Evidence of storm activity and risk of storm activity are products sought after by insurance companies who are assessing claims for damage as well as determining actuarial risk around insurance policies. Energy companies seek information about the occurrence of heatwaves during which demand for energy increases multifold whilst at the same time transmission losses increase significantly.

One particular service that the Bureau has recently developed at the request of the (now) Australian Department of Agriculture and Water Resources is an analysis of recent rainfall to identify the occurrence of periods of extremely low rainfall, based on gridded data (Jones *et al.* 2009). The service, including a tool termed the Rainfall Deficiency Analyser, was set up to support the Department's policy of providing concessional loans to farmers suffering the effects of severe drought. The Rainfall Deficiency Analyser is used as the first objective criterion as part of determining an applicant's eligibility for a loan. The product visually presents regions that experienced droughts of order 1 in 100 years through to 1 in 10 years. A sample is presented at Fig. 2. Additional case studies of sector specific climate and weather services can be viewed on the Bureau's website: http://www.bom.gov.au/business-solutions/.

A further value-add to existing data services will extend the online data portal to go beyond solely providing data for download, to include visualisation capability and basic forms of analysis. In addition, the Bureau intends to progress towards making even more data accessible including station and gridded analyses, model reanalyses and remotely sensed datasets. This will address the evolving needs of our clients who are becoming increasingly more climate literate and are seeking climate services and products to improve their ventures.

3. Climate Monitoring in a Warming World

The increased frequency and intensity of weather and climate extremes have raised the visibility and importance of climate monitoring. Understanding the impacts of climate change by contextualising extremes, both in terms of historical records and projections, has been sought out by governments, industry, the general public and the research community. It has also contributed significantly to increasing the confidence of climate change projections and that of decision-makers. The Bureau's practices in collecting and preparing temporally homogenised temperature data (Australian Climate Observations Reference Network — Surface Air Temperature, ACORN-SAT) have been assessed as amongst the best in the world by an international panel of expert scientists (Bureau of Meteorology 2011).

In addition to a range of near-real time station-based weather observations, the Bureau provides analyses to place recent conditions into historical context. The terrestrial monitoring program is split into a real-time Climate Monitor — that analyses and provides climate data each day for the preceding 24 hours — and a Climate Change tracker that makes use of curated and homogeneous datasets for analysing climate trends. Many analyses are updated daily, published on the web and made freely available, while more tailored services are available to clients upon request.

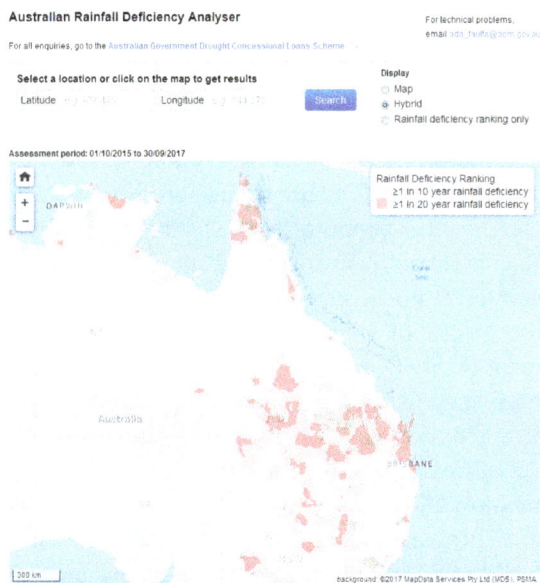

Fig. 2. The Bureau's Rainfall Deficiency Analyser depicting areas experiencing extreme rainfall deficiencies.

The automation of monitoring services has provided a suite of products for users, with the Bureau publishing hundreds of thousands of maps each year. The service was underpinned by key research developing gridded analyses for Australia (Jones *et al.* 2009) and integrated into an automated and real-time monitoring service from which a number of new services have been, and will continue to be, developed (e.g., the Rainfall Deficiency Analyser, Fig. 2).

The real-time Climate Monitor includes daily gridded fields for rainfall, temperature, humidity, solar exposure, vegetation greenness, and soil moisture. The Climate Change tracker includes rainfall, temperature, pan evaporation, cloud and atmospheric pressure.

The monitoring service provides a range of regular reports such as summaries for nationally-significant weather events (e.g., heat-waves and floods), a national drought monitoring service, and summaries of regional and national climate at monthly to annual timescales. It also provides perspective for significant events as they unfold, discussing the contributing synoptic factors and climate influences, tracking records which may be exceeded, and giving historical context to those which are broken (Bureau of Meteorology 2013).

Detailed climate analysis and research can identify underlying trends and improve understanding of the drivers behind these. Such intelligence is communicated to our stakeholders through a range of mediums including a *State of the Climate* publication which reports on key climate indicators (Bureau of Meteorology and CSIRO 2010; 2012; 2014; 2016). An example is the current rainfall decline in southern wet season (April–November) rainfall in southeast Australia and southwest Western Australia (Fig. 3).

The Bureau also provides a comprehensive ENSO monitoring service that complements its climate outlook service (see Sec. 4). Australia is an island continent influenced by the oceans around it and not simply through ENSO. It is

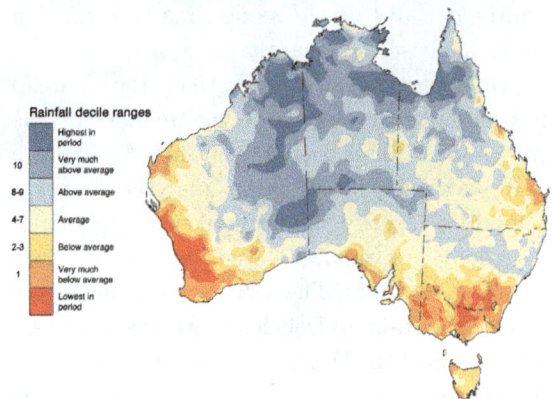

Fig. 3.   Southern wet season (April–November) rainfall deciles from 1996 to 2013. The decile map shows the extent that rainfall is above average, average or below average for the specified period, in comparison with the rainfall record from 1900 to 2013. The southern wet season is defined as April to November by the Bureau of Meteorology.

therefore important to understand the long-term behaviour and response of the oceans given they are a driver of weather and climate. In working with other national and international agencies, the Bureau supports global ocean observing programs that are essential for climate monitoring.

The Bureau maintains some of the world's longest running tide gauges, critically important given the paucity of observations in the Southern Hemisphere, and participates in the Argo float program which has aided efforts in seasonal predictions and climate change projections. The Bureau also supports sea level monitoring for Pacific Island countries. The monitoring of sea surface temperatures and marine water quality for health of reefs provides valuable environmental intelligence which also supports industries such as tourism and marine conservation. Our monitoring extends to participation in the volunteer Ship-of-Opportunity program to help monitor the state of the oceans.

Automation and improved technologies will enable the Bureau to draw on an increasing volume of data and information. In the future, the Bureau aspires to provide an even more comprehensive climate monitoring program that makes use of new observing technologies and scientific methods, including remote sensing. In particular, the Bureau aims to provide improved datasets that make use of observations, as well as satellite and model data.

4. Climate Prediction for Decision-Support

Australia's large year-to-year swings in climate can lead to significant and widespread flooding, drought and bushfires, while tropical cyclones can cause substantial damage to coastal communities. This makes accurate day through to seasonal forecasting integral to the Bureau's product suite.

The first Commonwealth Meteorologist, Henry Hunt, recognised the potential for seasonal forecasting in Australia as far back as 1929 (Hunt 1929) but it was not until June 1989 that the Bureau first issued a seasonal outlook. The outlook was based on the link between the Southern Oscillation Index and rainfall over the continent (McBride and Nicholls 1983). By 1997, the Bureau had upgraded to a statistical model based upon sea surface temperature patterns in the Pacific and Indian Oceans. However, over the following decade, warming in these basins led to a "phase locking", where the statistical model was increasingly producing very similar outlooks despite obvious changes in tropical ocean temperatures.

Conversely, the Australian dynamical model, Predictive Ocean and Atmosphere Model for Australia (POAMA) (Hudson *et al.* 2013), was increasing in accuracy with every iteration. In May 2013 the Bureau of Meteorology switched to using POAMA as the basis for its climate outlook service.

Having a good seasonal forecasting model does not necessarily equate to a good climate service and so, through 2013–14, the Bureau embarked on a major rethink of how the climate outlook service would interface with the public. A five-stage market research process resulted in a user-centred design tailored to meet as many user needs as possible. The new website (Fig. 4) includes zoom and clickable maps. These include probability maps of seasonal and monthly above median temperature and rainfall, probability of exceedance maps for several key rainfall levels, comprehensive information on the key climate drivers, and a four-minute video where a climatologist explains the outlook (e.g., https://www.youtube.com/watch?v=gGDdTV-yowxI). Users are also increasingly mobile, and hence the new service focused on being accessible via tablets and smartphones.

The Bureau receives regular positive feedback from its users, which is further reinforced by the increasing number of website hits and video views. The climate website has an approximate two million page views per month with information on ENSO and outlooks specifically ranging from 150,000–400,000 per month. Users are increasingly accessing information via twitter, Facebook and YouTube which have contributed to some of the increase in uptake of climate information.

The service has proved successful, with increased hit rates and better access to the seasonal outlooks. The videos are now presented by the national TV broadcaster in a nation-wide rural affairs program every month, potentially exposing the outlooks to an audience of at least half a million people. Likewise, other social media, such as Facebook, YouTube and Twitter, have been used to publicise the outlooks when they are issued, with media interest subsequently increasing.

In recent years, the Bureau has introduced several new products to assist decision-making. A tropical cyclone outlook is produced every October and a "Northern Rainfall Onset" service assists users wanting to know about early northern wet season (October–April). An international model summary has been developed in consultation with international agencies in order to give an eight-model ENSO outlook and the Bureau's flagship "ENSO Wrap-Up" which is

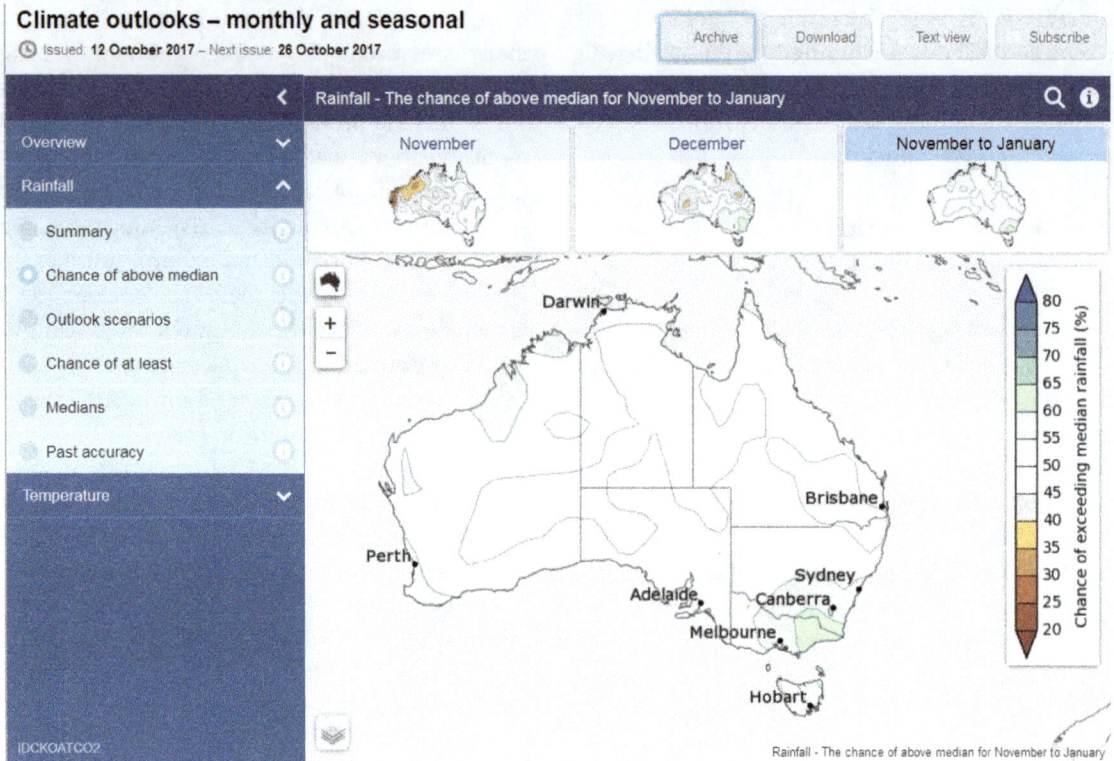

Fig. 4.   Climate outlook interface showing the likelihood of exceeding maximum temperature for November 2017 to January 2018 http://www.bom.gov.au/climate/outlooks/. Verification information on 30-year model hindcasts, as well as archives of all previous seasonal outlooks, are provided to ensure users can develop an understanding of the model accuracy and potential value to their requirements.

produced fortnightly to provide a comprehensive understanding of the Bureau's thoughts on El Niño and La Niña as well as the Indian Ocean. The ENSO Wrap-Up is often quoted directly by international media.

While all these products and services are providing a far stronger view of the climate looking forward than ever before, there are still areas for improvement. One of the areas for improvement is the gap between deterministic weather forecasts (generally up to seven days), and longer term probabilistic climate outlooks. In the deep tropics, multi-week (or "intraseasonal") forecasting has been practiced for some decades through monitoring of tropical atmospheric waves, and particularly the

Madden-Julian Oscillation (MJO, Wheeler and Hendon 2004). These outlooks have qualitative skill for multi-week forecasting of rainfall, tropical cyclone and monsoon development. A Weekly Tropical Climate Note (e.g., see http://www.bom.gov.au/climate/tropical-note/), together with an MJO monitoring website, provides users with probabilistic language to help their planning.

A country-wide, service addressing this gap is being developed through a recent Australian government funded project. This initiative will not only introduce fortnightly outlooks to bridge this gap, but also move to a higher resolution model with potentially even more skill in forecasting Australian conditions in the weeks and

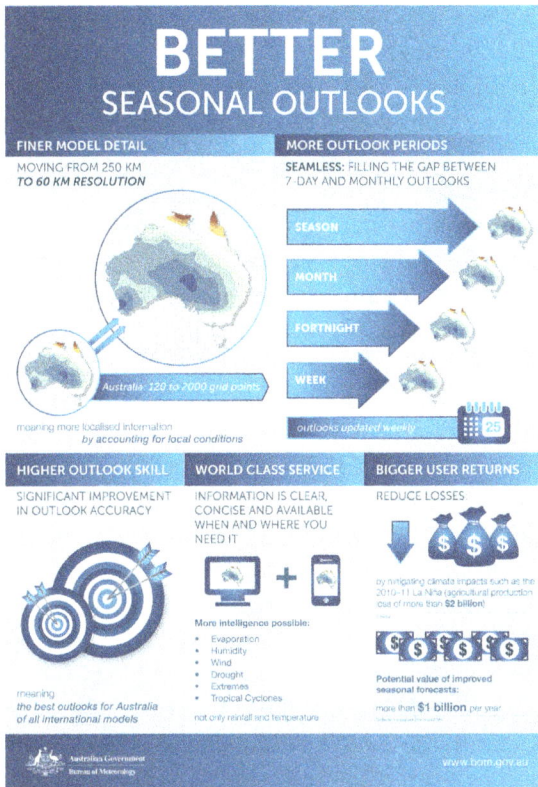

Fig. 5. The range of improvements in climate outlooks planned by the Bureau and potential value to users.

months ahead (Fig. 5). The new outlooks will also be better suited for input into decision support tools.

5. Engaging Governments, Industry and Communities

As has been described earlier, climate services have evolved rapidly and are becoming increasingly complex. Awareness of products, services and their limitations is crucial in capturing the full benefit of using climate information to inform decisions. Understanding user needs and engaging with stakeholders from the very beginning is essential to capturing the full value of climate services. Stakeholder engagement needs to be viewed as "business as usual" and be part

Fig. 6. Senior climatologist, Agata Imielska, presenting the climate influences on Sydney and the state of New South Wales at a Bureau of Meteorology Water Information Seminar, 13 November 2014.

of the on-going activities and functions of a service provider.

The stakeholders of the Bureau's climate, products and services include the agriculture, mining and energy, tourism and infrastructure sectors, as well as government agencies, with water managers and severe-weather-related emergency managers. The sectors interested in climate services continue to grow as impacts increase and capacity to engage effectively is enhanced (Fig. 6).

Historically, the agricultural sector has been the main user of climate services in Australia. A farmer can choose different crops or livestock that will be better suited to the season or the changing climate of the region.

Increasingly, federal, state and local governments are using climate information as a basis for policy decisions. For example, the use of rainfall analyses in determining the initial eligibility of farmers for various financial assistance schemes (Sec. 2) as previously highlighted. The high exposure of farmers to the extremes of climate stimulated the engagement between the Bureau and the agriculture sector. The Bureau engages regularly with the agricultural sector as well as water managers, from the highest levels down to local regions. This led to the development of sector-focused products, training and regular engagement which set the foundations for the Bureau's shift to user-centric design and broader engagement. The Managing Climate Variability program, and the Department of Agriculture and Water Resources, have been strong supporters of this shift in focus.

Tailored advice and information on climate matters are provided to a range of stakeholders and user groups through face-to-face or online climate briefings. Engagement with the Australian national government includes a monthly National Climate and Water Briefing (NCWB), open to policy and decision-makers from a range of stakeholders, delivering information on recent and current climate, water conditions (e.g., storage volumes and streamflows) and outlooks. Engagement is regular and based on stakeholder needs, ranging from briefings or meetings around key decision windows or to stimulate engagement with new sectors through "Executive breakfasts" or engagement forums. Structured interactions, such as the NCWB, are accompanied by stakeholder surveys which underpin future development and improvement of engagement services.

Liaison has deepened as organisations have increasingly understood and captured the benefits of climate services to their operations and planning. In turn, our approach has broadened to involve stakeholders in product development and evaluation, with increasing interacting through new communication channels such as social media.

Climate services are valuable only if they benefit decision-making. This value can only be realised if the stakeholder has access to this information. Thus, making products and services as accessible as possible is a necessary pre-requisite to capturing the full benefit of the services the Bureau provides. Focusing on multi-channel communication and engagement, that matches user preferences, is key in this regard. For many, this means providing information on the web that is intuitive to use and available for use at any time of day. As described in Sec. 1.4, the establishment of a Climate Liaison Section within the Bureau was an important step in addressing the needs of users and enhancing engagement.

Capturing the true value of climate services also relies upon the uptake and correct interpretation of the information used. Thus, keeping our stakeholders in mind, and involving them in the lifecycle of product development, education, delivery and application is crucial. Both service providers and users need to understand that there is no one-size-fits-all. Stakeholders need to be part of the process from the beginning with no expiry date placed on their feedback. Evaluation and improvement of services needs to be ongoing. This not only ensures the best chance of developing suitable products, but also engages stakeholders to increase uptake and application of climate services, and in turn realise the value latent in the service. Addressing and closing the gap between what providers can easily produce and what users actually need has to be incorporated from the start.

6. International Collaboration and Support

International cooperation is critical to the operational success of the Bureau (Bureau of Meteorology 2014). Climate monitoring and prediction requires participation in international programs for the collection and exchange of global data and information. International collaboration also ensures that the Bureau can benefit from scientific, technological and operational developments and expertise from other countries. Much of this collaboration takes place within the framework of the WMO and, in particular the technical teams within the Commission for Climatology (CCl) and Commission for Basic Systems. There are significant benefits from international engagement, especially through access to meteorological data from WMO members and meteorological satellite operators.

Key engagements also include those through the Global Climate Observing System and the growing GFCS. The Bureau also engages widely with other NMHSs, international climate centres, and bodies with broader regional interests such as the South Pacific Regional Environment Programme, Secretariat of the Pacific Community (SPC), the University of the South Pacific, the APEC Climate Center, and the International Research Institute for Climate and Society.

Observations and analyses produced by the Bureau are made available to international meteorological agencies and form a core input to global climate monitoring datasets maintained by agencies around the world. The Bureau also provides climate data, information, reviews and support for a number of WMO products and contributes to, and benefits greatly from, the observational and research efforts of other nations. In recent years, a key part of the Bureau's international engagement has been working with Pacific Island countries, building capacity and developing services to manage risk.

Since early 2000, the Department of Foreign affairs and Trade (DFAT — and formerly AusAID), has funded several projects focusing on climate variability and climate change for Pacific Island Countries. This began with the Pacific Climate Change Science Program and led to another major Pacific Australia Climate Change Science and Adaptation Planning Program (PACCSAP) from July 2011 to June 2014. The science component of the PACCSAP was managed by the Department of Climate Change and Energy Efficiency (and now Department of Environment), and delivered collaboratively between the Commonwealth Scientific and Industrial Research Organisation (CSIRO) and the Bureau.

The Climate and Oceans Support Program in the Pacific is a current DFAT funded climate project for Pacific Island Countries. Demonstrable early impacts have been evident through collaboration between the Bureau and 14 NMHSs, including:

- A comprehensive seasonal forecasting system has been extended with drought monitoring capability and in time for alerting stakeholders to the impacts of the 2015–16 El Niño.
- Seasonal forecasts are also "hard wired" to decision-support systems for malaria risk, renewable energy and drought management.
- A comprehensive climate monitoring bulletin is disseminated every month.
- An Ocean Portal is informing Pacific stakeholders in fisheries, shipping, environmental management and tourism.
- Maintenance and expansion of sea level monitoring in partnership with Geosciences Australia and SPC.

Future international weather and climate support could be directed further into Multi-Hazard Early Warning Systems (across all timescales), climate adaptation, and transitioning to delivering a sustainable climate service.

7. Research Supporting Services

Climate products and services provided by the Bureau generally stem from research and development. Engagement between providers of climate services, researchers and decision-makers, is one of the keys to best practice service delivery.

While the Bureau's Research and Development branch conducts strategic and applied research, it also synthesises research and developments conducted elsewhere in Australia and internationally. This is made possible through the extensive links that Bureau scientists have within Australia and overseas through, for example, the IPCC, the World Climate Research Programme including Climate and Ocean — Variability, Predictability and Change and the WMO CCl.

The Bureau's advances in seasonal forecasting, climate monitoring and climate change have been heavily influenced by strategic and applied research into the drivers of Australian climate variability and change. These have been undertaken through collaborations that have attracted additional Federal and State government funding, such as the Australian Climate Change Science Program (2015), the Victorian Climate Initiative (VicCI; Murphy *et al.* 2014), the South Eastern Australian Climate Initiative (SEACI 2012), the Indian Ocean Climate Initiative (IOCI 2011) and the Eastern Seaboard Climate Change Initiative (ESCCI; AdaptNSW 2015).

Three specific areas of research and development have had a significant influence on the Bureau's climate services — seasonal forecasting, the development of datasets for long-term climate monitoring, and climate change projections.

From an operational perspective, a major milestone was the transition from statistical seasonal forecasting systems to a dynamic model — POAMA. This model has demonstrated better skill than the earlier statistical systems (Charles *et al.* 2015) and advances in POAMA (Hudson *et al.* 2013) have unlocked increasingly more detailed information on seasonal timescales. As described in Sec. 4, the next generation model holds great promise for improved services.

Research and development has also improved capability in climate monitoring, especially through the development of "high-quality" long-term datasets and their analyses. These have been developed for use in operational products for temperature (e.g., Torok and Nicholls 1996; Della-Marta *et al.* 2004; Trewin 2001; 2013), rainfall (Lavery *et al.* 1992; 1997), evaporation (Jovanovic *et al.* 2008) and cloud cover (Jovanovic *et al.* 2011). Research datasets have been developed for dewpoint/humidity and associated variables (Lucas 2010), and an upper air temperature dataset is in the final stages of development (Jovanovic *et al.* 2015).

A long-term dataset of temperature and (for some stations) precipitation has also been developed for Antarctic and remote island stations (Jovanovic *et al.* 2012). These datasets support many of the Bureau's operational climate monitoring products, as well as its contribution to the series of State of the Climate reports, and have also supported a range of other analyses (e.g., Fawcett *et al.* 2012; Trewin and Vermont 2010; Haylock and Nicholls 2000) and Australian contributions to global and regional analyses (e.g., Donat *et al.* 2013; Morice *et al.* 2012).

In terms of climate change projections, modelling has been actively pursued in the Bureau, CSIRO and universities since the late-1980s. A recent joint CSIRO and Bureau project developed the most detailed, comprehensive and scientifically rigorous regional climate projections for Australia ever produced. These projections were based on world standard science of evaluating, combining and projecting models, and included extensive outreach. These projections provide Australians with critical information for the planning, adaptation and mitigation of climate change (CSIRO and Bureau of Meteorology 2015a; 2015b).

Improving climate services will rely on maintaining active research and development within Australia and in collaboration with international partners. Priority research areas to support the Bureau's climate service are:

- Improved understanding of key processes driving climate change and variability in the Australian region, including high-impact weather/climate events.
- Improving climate models, e.g. through:

  ○ Leveraging from more powerful supercomputers
  ○ Better representation of physical processes
  ○ Coupling earth-system components
  ○ Improving observations and advanced data assimilation
  ○ Quantifying uncertainty and confidence

○ Improving ensemble methods
○ Better understanding the processes responsible for significant model biases.

- Advancing seamless prediction: from days to decades and with a focus on extremes.
- Better interfacing forecasts to applications, e.g., for agriculture, water management, energy.
- Event attribution and improving our understanding of past variability and change.
- Improving multi-decadal climate change projections.
- Investigating multi-year to decadal prediction
- Improving datasets for long-term climate monitoring.
- Better understanding communication and decision-making in the broader community.

Progress in these research areas will support current and future climate services.

8. Information Systems Underpinning Climate Services

GFCS (2014) defines the Climate Services Information System as the mechanism through which information about climate (past, present and future) will be routinely collected, stored and processed to generate products and services that inform often complex decision-making across a wide range of climate-sensitive activities and enterprises. Climate information systems can be considered the "engine room" of climate services and are key to efficiency and delivery. While the GFCS definition encompasses a multi-institutional framework spanning global, regional and national domains, the focus here will be on national systems.

ADAM which was implemented in the mid-1990s, has become the main repository for Bureau station data extending back to the mid-1800s. As well as supporting the national climate record, it is also the starting point for many of the products generated by the Bureau. Quality assurance and quality control are essential for climate data management and quality management extends to other areas of climate services. Data are served through the Climate Data Online system (Sec. 2) and, downstream from the station data, gridded data are generated (Sec. 3).

Some Bureau climate information systems have been built for specific users. The Rainfall Deficiency Analyser (Sec. 2) is one example and the "Storm Confirmer" is another; the latter allowing the public and insurance companies to automatically assess if an area was affected by a storm. Our success in climate outlooks (Sec. 4) stems partly from the extensive engagement and analysis that took place with stakeholders coupled with excellent collaboration between climate scientists and IT staff. The increased focus on weather and climate extremes has resulted in the development of an Australian Climate Monitor tool to allow ready reporting and visualisation of those events.

As discussed in Sec. 6, the Bureau has collaborated with Pacific Island countries in developing climate services. At the forefront of these is the Climate Data for the Environment which is a small, self-contained Climate Data Management System. The Seasonal Climate Outlooks in Pacific Island Countries tool generates seasonal outlooks and also supports skill tests, hindcasts, statistical analysis and drought monitoring.

Further development of climate information systems will assist the Bureau in improving both its outputs and efficiencies and transitioning to enhanced enterprise architecture and platforms will support this endeavour. Strong partnerships between IT, research and climate services is key in developing a service that is valuable to stakeholders and remains accessible and sustainable into the future. The Bureau is moving towards a new IT architecture where all the gridded climate data will be available via OPeNDAP (Open-source Project for a Network Data Access Protocol, 2015), allowing users to more readily access all the latest climatologies, model output and observation grids.

9. Towards Best Practice Climate Services

The purpose of this paper is to describe the development of climate services within the Australian Bureau of Meteorology. However, the paper would not be complete without some aspirational perspectives on what will constitute a best practice climate service into the future.

Answering this question inevitably risks ignoring some important elements, though the following should capture most. Key elements of a best practice climate service going forward are:

- A deeper understanding of customer needs and how to inform decisions and risk management.
- Understanding existing and emerging policy needs of government and the ability to respond to those needs.
- High scientific/technical capability to produce and adapt relevant products to serve user needs.
- Strong service ethic and trust with users and partners.
- Timely information, being responsive to high impact weather and climate events.
- An ability to extend outreach and build capability through partnerships.

Climate services will continue to increase in importance as a result of the increasing frequency and intensity of weather and climate extremes and users wanting to better manage their risks and capture opportunities. Having skilled and motivated staff with strong science and IT capability and service ethic is essential. Improving science and modelling will help close the gap between user needs and existing climate services. However, and as is evident from the key elements listed above, success will ultimately depend on how well climate services assist users to improve their key decisions and outcomes.

Acknowledgments

The authors would like to thank Dr Doerte Jakob who provided a review of an earlier version of this manuscript as well as Felicity Gamble and Ian Muirhead for proofs of the document. We are also grateful to Dr Andrew Tupper for some useful additions.

References

AdaptNSW, 2015: Eastern Seaboard Climate Change Initiative. Accessed 1 November 2015. [Available online at http://www.climatechange. environment.nsw.gov.au/Impacts-of-climate-chan ge/East-Coast-Lows/Eastern-Seaboard-Climate-Change-Initiative]

Bureau of Meteorology — CSIRO, 2010: State of the Climate Report. [Available online at http://www. bom.gov.au/state-of-the-climate/documents/CS IRO-State-of-Climate-2010.pdf]

Bureau of Meteorology, 2011: The Australian Climate Observations Reference Network — Surface Air Temperature (ACORN-SAT) Data-set Report of the Independent Peer Review Panel. [Available online at: http://www.bom.gov.au/ climate/change/acorn-sat/documents/ACORN-SAT_IPR_ Panel_Report_WEB.pdf]

Bureau of Meteorology — CSIRO, 2012: State of the Climate Report. [Available online at http:// www.csiro.au/state-of-the-Climate-2012]

Bureau of Meteorology, 2013: Special Climate Statement 46 — Australia's warmest September on record. [Available online at: http://www.bom. gov.au/climate/current/statements/scs46.pdf]

Bureau of Meteorology, 2014: Bureau of Meteorology Annual Report 2014–15. Accessed 1 November 2015. [Available online at http://www. bom.gov.au/inside/eiab/reports/ar14-15/index. shtml]

Bureau of Meteorology — CSIRO, 2014: State of the Climate Report. [Available online at http://www.bom.gov.au/state-of-the-climate/do cuments/state-of-the-climate-2014_low-res.pdf? ref=button]

Centre for International Economics, 2014: Report: Analysis of the benefits of improved seasonal climate forecasting for agriculture. [Available online at: http://www.managingclimate.gov.au/wp-con tent/uploads/2014/06/MCV-CIE-report-Value-of-improved-forecasts-agriculture-2014.pdf]

Charles, A., R. Duell, W. Wang, and A. Watkins, 2015: Seasonal forecasting for Australia using a dynamical model: Improvements in forecast skill over the operational statistical model. *Australian Meteorological and Oceanographic Journal* (in press).

CSIRO and Bureau of Meteorology 2015a: Climate Change in Australia Information for Australia's Natural Resource Management Regions: Technical Report, CSIRO and Bureau of Meteorology, Australia. [Available online at http://www.clima techangeinaustralia.gov.au/en/publications-libra ry/technical-report/]

CSIRO and Bureau of Meteorology 2015b: Climate Change in Australia. Accessed 1 November 2015. [Available online at http://www.climate changeinaustralia.gov.au/en/]

Della-Marta, P., D. Collins, and K. Braganza, 2004: Updating Australia's high-quality annual temperature dataset. *Australian Meteorological Magazine*, **53**, 75–93.

Donat, M. G., L. V. Alexander, H. Yang, I. Durre, R. Vose, and J. Caesar, 2013: Global land-based datasets for monitoring climatic extremes. *Bull. Am. Meteorol. Soc.*, **94**, 997–1006. doi: http://dx.doi.org/10.1175/ BAMS-D-12-00109.1

Fawcett, R. J. B., B. C. Trewin, K. Braganza, R. J. Smalley, B. Jovanovic, and D. A. Jones, 2012: On the sensitivity of Australian temperature trends and variability to analysis methods and observation networks. Bureau of Meteorology, CAWCR Technical Report No. 050. [Available online at http://www.cawcr.gov.au/technical-reports/CTR_050.pdf]

Global Framework for Climate Services, 2014: Implementation Plan of the Global Framework for Climate Services. WMO, 2014. Accessed 1 November 2015. [Available online at http://www. gfcs-climate.org/sites/default/files/implementa-tion-plan//GFCS-IMPLEMENTATION-PLAN-FINAL-14211_en.pdf]

Haylock, M. and N. Nicholls, 2000: Trends in extreme rainfall indices for an updated high quality data set for Australia, 1910–1998. *Int. J. Climatol.*, **20**, 1533–1541.

Hudson, D., A. Marshall, Y. Yin, O. Alves, and H. Hendon, 2013: Improving intraseasonal prediction with a new ensemble generation strategy. *Mon. Weather Rev.*, **141**, 4429–4449. doi:10.1175/MWR-D-13-00059.1

Hunt, H. A., 1929: A basis for seasonal forecasting in Australia. *Q. J. Roy. Meteor. Soc.*, **55**, 323–334. doi:10.1002/qj.49705523201

Indian Ocean Climate Initiative, 2011: Synthesis reports: Stage 1, Stage 2 and Stage 3. [Available online at http://www.ioci.org.au/publications. html]

Jones, D. A., W. Wang, and R. Fawcett, 2009: High-quality spatial climate data sets for Australia. *Australian Meteorological and Oceanographic Journal*, **58**, 233–248.

Jovanovic, B., D. A. Jones, and D. Collins, 2008: A high-quality monthly pan evaporation dataset for Australia. *Clim. Change*, **87**, 517–535.

Jovanovic, B., K. Braganza, D. Collins, and D. Jones, 2012: Climate variations and change evident in high-quality climate data for Australia's Antarctic and remote island weather stations. *Australian Meteorological and Oceanographic Journal*, **62**(4), 247–261.

Jovanovic, B., D. Collins, K. Braganza, D. Jakob, and D. A. Jones, 2011: A high-quality monthly total cloud amount dataset for Australia. *Clim. Change*, **108**(3), 485–517. doi:10.1007/s10584-010-9992-5.

Jovanovic, B., R. J. Smalley, B. Timbal, and S. Siems, 2015: Homogenised monthly upper-air temperature dataset for Australia. *Int. J. Climatol.* (submitted).

Lavery, B., A. Kariko, and N. Nicholls, 1992: A historical rainfall data set for Australia. *Australian Meteorological Magazine*, **40**, 33–39.

Lavery, B., G. Joung, and N. Nicholls, 1997. An extended high-quality historical rainfall dataset for Australia. *Australian Meteorological Magazine*, **46**, 27–38.

Lucas, C., 2010: On developing a historical fire weather data-set for Australia. *Australian Meteorological and Oceanographic Journal*, **60**, 1–14.

McBride, J. L. and N. Nicholls, 1983: Seasonal relationships between Australian rainfall and the Southern Oscillation. *Mon. Weather Rev.*, **111**(10), 1998–2004.

Morice, C. P., J. J. Kennedy, N. A. Rayner, and P. D. Jones, 2012: Quantifying uncertainties in global and regional temperature change using an ensemble of observational estimates: The Had-CRUT4 data set. *J. Geophys. Res.*, **117**, D08101. doi:10.1029/2011JD017187.

Murphy, B., B. Timbal, H. H. Hendon, and M. Ekstrom, 2014: Available online at http://www. cawcr.gov.au/projects/vicci/publications/.

OPeNDAP, 2015: Accessed 1 November 2015. [Available online at http://www.opendap.org/]

The South Eastern Australian Climate Initiative, 2012: Synthesis report: Phase 1 and

Phase 2. [Available online at http://www.seaci. org/publications/synthesis.html]

Torok, S. J. and N. Nicholls, 1996: A historical annual temperature dataset for Australia. *Australian Meteorological Magazine*, **45**, 251–260.

Trewin, B. C., 2001: Extreme temperature events in Australia, PhD thesis, School of Earth Sciences, University of Melbourne.

Trewin, B. and H. Vermont, 2010: Changes in the frequency of record temperatures in Australia, 1957–2009. *Australian Meteorological and Oceanographic Journal*, **60**(2), 113–120.

Trewin, B., 2013: A daily homogenized temperature data set for Australia. *Int. J. Climatol.*, **33**: 1510–1529. doi:10.1002/joc.3530.

Vaughan, C. and S. Dessai, 2014: Climate services for society: origins, institutional arrangements, and design elements for an evaluation framework. Wiley Interdisciplinary Reviews: *Clim. Change*, **5**(5), 587–603.

Wheeler, M. C. and H. H. Hendon, 2004: An all-season real-time multivariate MJO index: Development of an index for monitoring and prediction. *Mon. Weather Rev.* **132**, 1917–1932.

World Meteorological Organization (WMO), 2015: Climate Services Introduction. Accessed: 1 November 2015. [Available online at http://www.wmo.int/pages/themes/climate/climate_services.php#top]

Chapter 5

Early Warning, Resilient Infrastructure and Risk Transfer

David P. Rogers*, Haleh Kootval and Vladimir V. Tsirkunov

Global Facility for Disaster Risk Reduction and Recovery,
World Bank, Washington DC, USA
*drogers@bluewin.ch
www.gfdrr.org

Strengthening the capacity of countries and communities to cope with the adverse effects of natural hazards and at the same time support development objectives requires investment in early warning; resilient infrastructure and financial risk transfer mechanisms. By shifting emphasis from hazard warnings to impact-based warnings, hazard, vulnerability and exposure information become central to all three components and can be used to improve the utility of each. A safe society depends on timely warning of the impact of a hazard, the ability to protect and transform lives and livelihoods, and the financial tools to recover quickly if losses are incurred.

1. Introduction

Agriculture, water resources, trade, tourism and many other aspects of national economies are highly sensitive to natural hazards. Disruption to lives and livelihoods are manifest in loss of life, loss of income, food and water shortages, disease and destruction of infrastructure. While avoiding hazards is impossible, reducing the impact of disasters is attainable. Emphasis has shifted from responding to disasters to trying to avoid them altogether. While this will probably never be fully achieved, significant progress is feasible if people and their livelihoods are secure and mechanisms exist to rapidly recover from any adverse impact of a hazard. Catastrophe risk insurance, for example, can provide financial resources to speed up recovery, while helping to avoid subsequent disasters, such as social, water and food insecurity, which collectively result in compromised health and well-being. This works best if lives and livelihoods are protected from the initial hazard, which requires access to resilience infrastructure — homes, schools and

other public buildings, roads and bridges, communication, water and electricity, and food. In turn, ensuring the safety of people requires that they be well prepared to take appropriate action to secure their livelihoods (e.g., livestock) and their own safety, which requires effective early warning.

These three components, which contribute to the overall resilience of a society, are often treated independently; however, under resourcing any one of these "pillars" can compromise the effectiveness of the whole (Fig. 1). For example, inadequate warnings or lack of understanding of the appropriate response to warnings often results in people not seeking shelter, even if it exists, until it is too late as evinced in the Philippines during Tropical Cyclone (TC) Haiyan (WMO 2014) (Box 1). The absence of "resilient" infrastructure limits the options of those at risk, even if warnings are provided in time, which was the case in Uttarakhand, India during the devastating floods of 2013 (Box 1). Conversely, the presence of a financial instrument, for example,

Fig. 1. Schematic of relationship between early warning, risk transfer and infrastructure.

the Pacific Catastrophe Risk Assessment and Finance Initiative (PCRAFI) response to TC Ian in the Kingdom of Tonga, has demonstrated that it is possible to reduce or limit the impact of a natural hazard and accelerate recovery (Box 2).

These three components are not simply pillars of a resilient society, but are also intimately linked through a focus on quantifying and conveying information of the impact of hazards on people and assets. It is recommended that adopting an impact-based warning system and risk transfer strategy would better inform planning and investment in resilient infrastructure, which in turn would be used to best effect to protect and transform lives and livelihoods.

2. Hydrological, Meteorological and Geophysical Hazards

Meteorological, hydrological and geophysical hazards include heavy rain events, high winds, storm surges, flash floods, river floods, tsunamis, earthquakes, volcanic eruptions and volcanic ash. To a greater or lesser extent they affect all parts of the world with differing impacts depending on the coping and adaptive capacity

of a country or region. Disasters, which ensue from hazards, are deaths and damage that result from human acts of omission and commission (World Bank 2010). In Asia and the Pacific, for example, meteorological hazards are dominated by monsoons and tropical cyclones, which result in heavy rainfall, high winds, and periods of extended drought when the monsoons fail. While providing most of the precipitation that supports rain fed agriculture, failure of the monsoons in onset and intensity can have devastating consequences in terms of food security. The majority of the more than 500 million rural poor in the region are subsistence farmers depending on rain-fed agriculture controlled by the monsoons, highly sensitive to climate variations. The monsoons largely determine the annual distribution of rainfall. In Vietnam, for example, the annual rainfall is between 1,800 and 2,500 millimetres with about 70 percent occurring in the summer monsoon season between May and September/October (Rogers and Tsirkunov 2013a).

In the Philippines, the average annual rainfall ranges from 5,000 to less than 1,000 millimetres in some sheltered valleys with the maximum also occurring during the summer monsoon. The Pacific Island Countries (PICs) are especially vulnerable to tropical cyclones and their resulting storm surges and flash floods because of their relatively low adaptive capacity. PICs rely on subsistence agriculture and fishing as a primary means of livelihood, vulnerable to external shocks, including severe weather events that are not necessarily associated with tropical cyclones. PICs are also at risk from extended periods of drought.

Two of the extreme meteorological events of 2013 are described in Box 1. These incidents are indicative of the need to redouble efforts to improve early warning and response. As pointed out in the report on Typhoon Haiyan (WMO 2014), the use of current methods may be inappropriate for building resilient societies in the future because historical, statistical return periods no longer adequately represent the occurrence of severe events. Rather they need to be complemented by projecting future changes based on seasonal and climate models.

This begs the question — what is resilient infrastructure if we are unsure of the future severity of hazards?

3. Resilient Infrastructure Systems and Services

3.1. *What is resilience?*

The Royal Society (2014) report on resilience to extreme weather and the Rockefeller Foundation (2009) define resilience as "the capacity of individuals, communities and systems to survive, adapt, and grow in the face of stress and shocks, and even transform when conditions require it". This recognizes that a resilient society must not merely cope with hazards, but must also be able to adapt and progress in the face of extremes. Thus resilient infrastructure must contribute to positive adaptation and transformation, rather than simply surviving. In simple terms, a building designed as shelter from a storm surge would help people survive, but one that also provided additional services, such as economic or community development opportunities could enhance quality of life and would be considered transformative.

There is disproportionately high societal disruption caused by infrastructure system failures compared with physical damage in disasters. Thus building resilience is not only an engineering issue, but also one that requires the collaboration of social scientists to design resilient infrastructure systems (Chang 2009). It is not sufficient to simply design infrastructure to withstand extreme forces, but to build infrastructure systems that are resilient to disasters; that is, take into consideration all aspects of human behaviour. Resilience entails three interrelated dimensions: lower probabilities of failure; less severe negative consequences when

failures do occur; and faster recovery from failures (Bruneau *et al.* 2003), which as Chang (2009) points out involves both technical solutions and social dimensions.

Figure 2, based on Moteff (2012), illustrates conceptually the resilience of two hypothetical systems: one less prepared than the other. In the example, in the absence of an event, the system operates at a maximum performance level of 100 over a nominal time period. In the system with higher resilience disruption occurs, but this is offset by warning and preparedness, which helps minimize mortality and morbidity; resilient infrastructure, which prevents further losses; and access to financial support, which accelerates recovery efforts. Normal performance of 100 is regained by time 7. The area above the line represents loss of capacity during that time. The system with lower resilience affected by a similar event at time 1 fails abruptly with performance dropping to 0, largely due to the impact on an ill-prepared and exposed population that suffers significant losses. Economic assistance aids recovering some of its performance, but in the absence of existing resilient infrastructure, does not return to the original performance level in the time recorded. In this scenario, there is a significant and sustained loss of capacity. This low resilience situation is more commonly observed in developing countries, although disasters, such as Hurricane Sandy and Hurricane Katrina in the US, highlight the vulnerability of urban infrastructure systems and slow recovery even in developed economies. This interpretation of a resilient society serves to highlight the three key elements — warning and preparedness, resilient infrastructure and risk transfer. Significant post-disaster investment, well beyond the financial assistance provided by risk transfer mechanisms, would be required to bring the lower resilient society to full capacity. "Build back better" is no panacea; early warning and infrastructure investment prior to experiencing a major meteorological or hydrological hazard

Box 1 — Some of the Meteorological Hazards in 2013

Uttarakhand Floods

In the Indian State of Uttarakhand, the monsoon in June 2013 arrived almost two weeks earlier than expected. From June 15 to 17, 2013, heavy rainfall, which was 375 percent above the amount the state would have expected to receive, hit several parts of the higher reaches of the Himalayas. This resulted in a rapid increase in water levels that gave rise to flash floods in the Mandakini, Alakananda, Bhagirathi and other river basins, causing extensive landslides. Continuous rains caused Chorabari Lake to rise and the lake's weak moraine barrier gave way causing a huge volume of water along with large boulders to come down the channel to the east, devastating the towns of Kedarnath, Rambara, Gaurikund and others in its wake. According to official sources, over 900,000 people have been affected by the event in Uttarakhand with official estimates of fatalities in excess of 5,700, the majority of them tourists on pilgrimage to the State's Sikh and Hindu holy sites.

Typhoon Haiyan in Philippines

Typhoon Haiyan (also called "Yolanda" in the Philippines), one of the strongest typhoons ever recorded, crossed the Philippines on 8 November 2013, caused significant casualties and disruption to socio-economic activities in the Philippines although the accurate and timely forecasts and warnings on its movement and intensity had been provided days in advance (WMO 2014). Typhoon Haiyan generated large storm surges in many islands of the Philippines, which was the cause of most of the damage and loss of life. The City of Tacloban on Leyte Island, located in the inner part of Leyte Gulf received a surge of 4–5 m and up to 7 m in places causing extensive damage and fatalities in excess of 6,000.

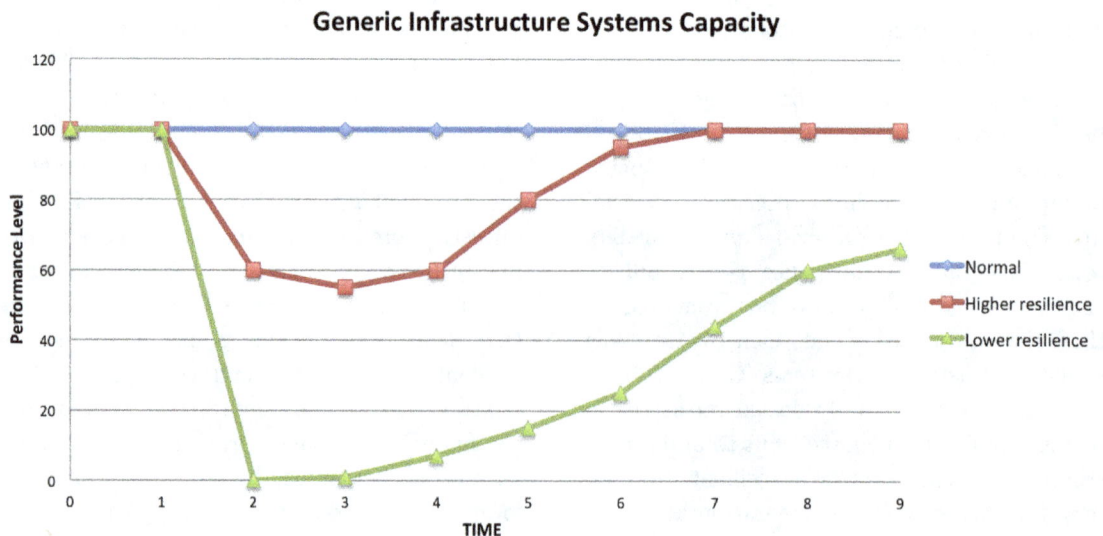

Fig. 2.   Idealization of the impact of a hazard on higher and lower resilient societies (after Moteff 2012).

is needed to avoid catastrophic long-term social disruption.

3.2. *Social implications of infrastructure failures*

The social implications of infrastructure failures are often the key factor in slow recovery and have begun to play a more important role in resilience. Chang (2009) asks how will the failure of one bridge, for example, affect businesses throughout an urban area that relies on the transport system? How will the failure of one infrastructure system disrupt other infrastructure systems? How can repairs following a disaster be planned so they minimize social and economic losses? An important concept is that of *infrastructure services*, which link physical damage to societal impacts. How loss of power, for example, leads to disruptions in water, transportation and healthcare systems.

Failure to share data between different infrastructure organizations, such as highways agencies, and meteorological organizations, highlights the difficulties in identifying and reducing vulnerabilities in urban systems. The gridlock on the roads that occurred in Shanghai, China following the impact of Tropical Cyclone Fitow in October 2013 is one example where data sharing could have helped manage the transportation system more effectively by linking vulnerable road intersections with likelihood of flooding.

Infrastructure is also vulnerable to multiple hazards, e.g., floods, earthquakes and deterioration of the infrastructure itself, which increases disaster risk. For example, during Typhoon Haiyan some people evacuated to unsafe shelters that were subsequently destroyed. This happened because of a lack of appreciation of the exceptional severity of the storm surge with evacuation orders based on past vulnerability maps that indicated maximum storm surge heights of up to 4 meters,

severely underestimating the weaknesses in the infrastructure. In other cases, the limited number of shelters and lack of adequate resources and services (power, sanitation and fresh water) to keep people comfortable and safe resulted in people refusing to evacuate.

Human behaviour is an important element in developing resilience infrastructure systems and services, and behavioural change is a necessarily and often neglected aspect of infrastructure development. This requires that all stakeholders (National Meteorological and Hydrological Services (NMHSs), disaster managers, planners, communities, economic sectors, etc.) work together on early warning and preparedness and resilience infrastructure activities. If people better understand the impact of a hazard, they are more likely to require the infrastructure systems and services to be resilient. This would encourage communities to take effective actions to improve the resilience of their communities by helping to build and sustain local critical infrastructure (e.g., protecting communications, building to code, avoiding construction in high risk zones, maintaining shelters, and protecting supply chains).

"Self-rescue" is a term sometimes associated with the abrogation of the responsibility of government to protect the public; however, given the often overwhelming scale of natural hazards, it is an essential component of any community's capacity to reduce its vulnerability. This requires, among other things, investment in local public education and training, reliable two-way communication for warnings and community feedback, and the means to shelter in place or access to safe evacuation routes.

To be effective, building resilience infrastructure systems and services must consider both physical and social dimensions. Unfortunately, this is not often the case, with the enthusiasm to build hazard-proof structures outpacing the human capacity to adapt to their use.

3.3. *Application of resilient infrastructure to risk*

Assessing the resilience of infrastructure to current and future hazards is a critical step in adaption. The UK Government has developed a comprehensive approach to the risk climate change presents to resilient infrastructure interdependencies, adaption investment and potential economic opportunities (UK Gov. 2011). The UK Government defines national infrastructure as nine sectors comprising systems, sites and networks necessary for the delivery of essential services upon which daily life depends. These sectors include communications (ICT), finance, health, emergency services, food, transport, energy, water, and government. Additional elements include social (schools, hospitals, prisons) and domestic infrastructure. It highlights how current failures in single systems have had widespread repercussions (e.g., the flooding of a single water treatment plant in 2007 cut off water supply to 350,000 people for 17 days) and the likelihood that current vulnerability of infrastructure to extreme weather will worsen in the future. It considers disruptions to supply chains caused by severe weather in other countries and the economic losses and service failures from poorly adapted assets.

3.4. *Infrastructure interdependencies*

Understanding infrastructure interdependencies is critical to developing resilience infrastructure systems (Table 1). These interdependencies span all natural hazard risks as well as the impact of climate change on resilience infrastructure systems and services. Coordination between infrastructure operators and others is essential to reduce vulnerabilities. This is probably the weakest link in achieving climate resilient infrastructure and by inference resilience to current hazards also. Resilience in one sector is highly dependent on resilience in another and therefore attention is needed to ensure that vulnerabilities in one sector do not compromise another. Developing a framework for resilience infrastructure investment with appropriate metrics for measuring resilience is one way to achieve this. As noted above, sharing of information between sectors is critical.

Balancing hazard risks with efficiency and value for money considerations is essential. A first step is to evaluate existing infrastructure systems for their fitness for purpose. This should be based on historical hazard data, and especially present and future scenarios, bearing in mind the need to be particularly cautious about relying on historical data as an indicator of the future climate.

Another important consideration is the increased risk that insurers will bear (Insurance and reinsurance) and the role of government as the risk bearer of last resort.

3.5. *A framework for resilient infrastructure*

Determining where infrastructure investments should be made requires a full understanding of sectorial interdependencies described earlier. It also requires the ability to measure each individual project against the risks of high impact, low probability hazards. The example of Fukushima Daiichi nuclear power plant disaster following the 2011 Tohoku earthquake and tsunami highlights the danger of not considering extreme events, which may have a very low probability but are within the realm of the possible.

For hydro-meteorological hazards, the Royal Society (2014) advocates the testing and adoption of the "triple stress test" for sectors by placing a *value* on resilience, which is equally applicable to homeowners, farmers or corporations. This comprises the following (Royal Society 2014):

- 1 in 100 (1%) risk per year (a stress test for an organization's solvency in an extreme event scenario)

Table 1. Examples of sectorial interdependencies for 7 sectors (after UK Gov. 2011).

Sector	Dependencies on infrastructure	Impacts on other sectors
Food	Water for irrigation Transport infrastructure for agricultural activities and food supply Energy for storage and agricultural activities	Domestic is dependent on food supply
Energy	Water for cooling in power stations, fuel refining and energy production Transport for fuel supply and workforce ICT for control and management systems of electricity	Transport is dependent on energy Food production is dependent on energy Water is dependent on energy for pumping, treatment and supply Domestic is dependent on energy for heating and cooling and many other functions
Social and domestic	Food, Water ICT, Transport, Energy for all aspects of life and livelihoods Emergency services providing continuity to operate while recovering from an event	All sectors dependent on workers and efficient domestic consumption of sectorial resources Health depends on general well-being of population to avoid overwhelming sector Water depends on well-managed sanitation systems to avoid contamination of water supply Emergency services infrastructure depends on people for effective response
ICT	Energy for all services Transport for maintenance workers	All sectors dependent on ICT
Transport	Domestic infrastructure for travel to and from work, school, etc. Energy infrastructure for fuel and electricity Drainage infrastructure to prevent flooding Internal dependencies with and across modes (road, rail, sea, and air)	All sectors depend on transport
Water	Energy for treating, pumping and processing ICT for control systems Transport for workers and supplies for processing	All workplaces and domestic homes require water for people and sanitation Cooling water for some energy infrastructure Energy infrastructure may depend on water for generation Food production requires water
Emergency services	Transport (all modes) for safe and rapid evacuation, and emergency supplies Energy to manage emergency pumps to relieve flooding and operate flood controls Health infrastructure to respond to emergency situations Water infrastructure to extinguish fires ICT to respond effectively to emergency situations Domestic infrastructure to provide security for population	All sectors depend on emergency services for safety and security during emergency situations Health infrastructure for emergency response

- 1 in 20 (5%) risk per year (a stress test for an organization's annual earnings or profits)
- Annual Average Loss (AAL) (a standardized metric for an organization's exposure to extreme events)

The 1% stress test implies a 10% chance of an organization being affected once a decade. In the case of geophysical hazards, such as earthquakes and tsunamis, this stress test would need to consider even lower probability events.

Bruneau *et al.* (2003) define the properties (robustness, redundancy, resourcefulness and rapidity) and dimensions (technical, organizational, social and economic) of resilient systems, which can be used to determine underlying weaknesses and therefore where vulnerabilities to hazards may exist. This can be used to identify policies to address weaknesses in the system, which can be prioritized using cost-benefit analysis.

Measuring resilience, however, is not a straightforward process. Consistent metrics are needed to determine the effective of different resilience options. Keating *et al.* (2014) propose the development of a comprehensive set of metrics to help guide the exploration of potential sources of resilience and test their effect on outcomes in order to drive an evidence-based understanding of flood resilience. They define the five capitals framework of potential resilient indicators, namely: Physical capital — the number of access roads and bridges (source) and the number of households with uninterrupted access to utility services post flood (outcome); Social capital — the number (or percentage) of stakeholder groups represented on a planning board discussing ways to reduce losses from future disasters and the amount of times they meet (source) and the number of community members engaged in aiding others in recovery (outcome); Human capital — diversity of skills/training in the community (source) and the number of days children are displaced from schooling (outcome); Financial capital — the average household savings in the community (source) and the amount of days of lost income (outcome); and Natural capital — the degree of soil absorption (or ability for natural runoff) (source) and the percentage of protective barriers eroded (outcome). Such an approach may be adapted to other hazards and is an active area of ongoing research and development.

4. Risk Transfer Mechanisms

4.1. *Catastrophe risk financing*

The financing of disasters in developing countries has, for decades, relied on a reactive approach, consisting of the diversion of funds from domestic budgets and extensive financing from international donors (Arnold 2008). For a number of years, efforts have been undertaken to develop a more proactive approach. The World Bank has developed a disaster risk-financing and insurance (DRFI) framework for understanding and improving the financial resilience of states against disasters (World Bank 2012). The DRFI framework promotes a dual approach to increasing overall financial resilience based on financial disaster risk assessment and modelling (Mahul and Boudreau, 2013). This approach includes: (1) sovereign disaster risk financing, which entails identification and assessment of the government's contingent liabilities associated with natural hazards and financial strategies to increase their financial response capacity in the aftermath of a disaster while protecting their long-term fiscal balance, and (2) catastrophe risk market development, which increases the transfer of public and private risks to the insurance sector. Examples include the Caribbean Catastrophe Risk Insurance Facility (CCRIF) (Ghesquiere and Mahul 2010), which was the first regional institution, which allows countries to pool risks and save on individual premiums, and the Turkish Catastrophe Insurance Pool (TCIP), which is a mandatory earthquake insurance scheme for homeowners.

The CCRIF is also the model for the PCRAFI, which was piloted in 2012 and 2013 in 15 Pacific Island Countries to test the viability of sovereign parametric insurance in the region (World Bank 2013). The scheme paid out in Tonga following the impact of TC Ian, which caused extensive destruction in Ha'apai islands in January 2014. The World Bank's Pacific Resilience Program (PREP) is continuing and expanding this scheme based on the positive feedback from participating countries.

These initiatives provide much needed, immediate liquidity after a disaster for more effective government response, and some relief of the fiscal burden placed on governments due to disaster impacts. They constitute critical steps in promoting a more proactive risk management strategy that includes preparing for disaster impacts and planning for the response. What remains unclear is how well do these mechanisms reach the poorest communities, which are consistently the most affected by disasters.

4.2. *Micro-insurance*

Micro-insurance complements these formal mechanisms by providing low-income households, farmers and businesses with rapid access to post-disaster liquidity, thus protecting their lives and providing for reconstruction. As insured households and farms are credit worthier, insurance can also promote investments in productive assets and higher-risk/higher-yield crops. In addition, insurance has the potential to encourage investment in disaster prevention if insurers offer lower premiums to reward risk-reducing behaviour.

The situation in parts of Bangladesh highlights the problem (BCCRF 2014). Seasonal flooding was a way of life, but nowadays these floods are more frequent and more persistent inundating freshwater wells and damaging sanitation systems. Crops are lost and the risk of waterborne disease increases with each day of the flood. With little cash, money must be

borrowed to buy food. In the past, the villagers of Kurigram district were able to recover but the frequency of the floods makes this increasingly difficult. Families become permanently indebted. The need to earn money forces many to seek employment in Dhaka. This creates additional stress on families and prevents local investment in activities that would increase resilience of the community. Insurance would remove the dependence on borrowing and risk of exploitation and allow reinvestment in communities. Cultivating flood tolerant rice would, for example, address some of the food security issues and may encourage the male family members to remain in the community.

5. Early Warning

5.1. *Hazard-based warning systems*

Despite the availability of high quality weather, climate and hydrological forecasts and warnings from National Meteorological and Hydrological Services (NMHSs), hydro-meteorological hazards continue to cause significant losses and damage. One problem is that people don't know how to translate hazard-based meteorological and hydrological forecasts and warnings into actionable information, and meteorologists and hydrologists, who may not work closely with disaster managers, are often unaware of the vulnerability of their countries social and economic sectors to hydro-meteorological hazards. Consequently, they continue to focus on hazard-based warnings.

If this gap is to be closed, then an all-encompassing approach to observing, modelling and predicting severe hydro-meteorological events, and the consequent cascade of hazards through to impacts, is needed. Tackling this problem requires a multi-disciplinary approach to access the best possible science, and the optimum services, to manage multi-hazard events today, and to provide the best possible evidence base on which to make the costly decisions on infrastructure investments to protect the population in the future. This is a key component to achieving the goals of the Sendai framework for disaster risk reduction (United Nations 2015). This requires much closer working relations among meteorologists, hydrologists, disaster managers, first responders, and the sectors at risk.

All countries should provide their citizens and economic sectors actionable information that wherever possible identifies the timing and anticipated impacts of specific hazards. An informed population that fully understands what a hazard will do is more likely to take the necessary actions that protect their lives and livelihoods.

Historically, all NHMSs have featured forecasting of the weather as central to their mission, and most also issue weather warnings in the case where hazardous weather is expected. In the case of both weather forecasts and warnings, the focus is on what the weather will *be*. It is now advocated that this weather-based paradigm evolve to one, which is focused primarily on forecasting impacts. In other words, the focus should evolve to what the weather will *do* (Fleming *et al.* 2015).

Many NMHSs have in place warning services to address significant hazards that are expected to threaten life or property. This level of information includes messages that are provided on a non-routine and as-needed basis. In general, these products feature provision of specific headline messages, a colour-coded or numbering system, and/or activation of specialized public messaging systems that are only used during extreme events (Fleming *et al.* 2015). Hydrometeorological events for which such warnings are provided may include flooding, winter storms, severe convective weather, extreme temperatures and poor air quality.

While messaging within warnings often describes expected impacts to the public and disaster managers, as pointed out by Fleming *et al.* (2015), the driver for issuing these early warnings is normally weather-based factors only (e.g., winds speeds of at least "X" km/hour, rainfall totalling at least "Y" mm.) and can often be expressed as a probability of a fixed threshold being reached or exceeded (e.g., there is a 60% probability of winds speeds of at least "X" km/hour being reached).

Public surveys following extreme events highlight the lack of understanding of these kinds of messages. For example, following TC Ian, the Tonga National Meteorological Service surveyed the worst affected by the event. They found that people evacuated only when their homes were destroyed putting themselves at great risk of injury, which highlighted their lack of understanding of the severity of the weather warning ('Ofa Fa'anunu, Permanent Representative of Tonga with WMO, personal communication).

The floods in Uttarakhand in June 2013 were well forecast by the Indian Meteorological Department and timely warnings of extremely heavy rainfall were issued. However, lack of understanding of how to interpret the information led to an inadequate response and significant loss of life. The published reponse from the Vice Chair of the National Disaster Management Authority typifies the problem "We get a copy of the IMD bulletin but action has to be taken by state government only. They put out bulletin (this time) and said "very heavy rain". What does "heavy rain" mean? "Very heavy rain" means very heavy rain. But it doesn't mean that in such a short time so much rain".[a]

Clearly, unless warnings are issued with adequate knowledge of the impact of the hazard, the desired response from the public or disaster

---

[a]Interview by Rediff.com with M. Shashidhar Reddy, Vice Chair of NDMA, immediately following the Uttarakhand flood (see complete transcript here — http://www.rediff.com/news/slide-show/slide-show-1-uttarakhand-more-than-4000-deaths-are-expected/20130705.htm#1)

managers will not happen. This means that more effort is needed to use impact forecasts and analyses to build scenarios, and to use these to train and educate the public, communities and sectors on what actions must be taken to avoid disasters.

5.2. *Impact-based warning systems*

The WMO Guidelines on Multi-Hazard Impact-Based Forecast and Warning Services (Fleming *et al.* 2015) defines probability and magnitude of harm attendant on human beings, their livelihoods and assets because of their exposure and vulnerability to a hazard. The magnitude of harm may change due to response actions to either reduce exposure during the course of the event or reduce vulnerability to relevant hazard types in general.

The set of risks may be mathematically expressed as:

$$|Risk\ of\ Impact(x,t)|$$
$$\equiv |Hazard(x,t)|$$
$$\times |Vulneability(x,t)|$$
$$\times |Exposure(x,t)|$$

A hazard is defined as a weather-, climate- or hydrological-based, geophysical or human-induced element that poses a level of threat to life, property or the environment (see, for example, Rogers and Tsirkunov 2013b).

Vulnerability refers to the susceptibility of exposed elements, such as human beings, their livelihoods and property to suffer adverse effects when impacted by a hazard. Vulnerability is related to the lack of resilience of the infrastructure systems, that is the predisposition, sensitivities, fragilities, weaknesses, deficiencies, or lack of capacities that favour adverse effects on the exposed elements. Vulnerability is situation specific, interacting with the hazard to generate risk. Therefore, vulnerability may also be time and space dependent.

Exposure refers to who and what may be impacted in an area in which hazardous events may occur. If the population and economic resources were not located in (exposed to) potentially dangerous settings, no problem of disaster risk would exist.

Exposure is a necessary, but not sufficient, determinant of risk. It is possible to be exposed, but not vulnerable; for example, by living on a floodplain, but having sufficient means to modify building structure and behaviour to mitigate potential loss. However, to be vulnerable to a hazard, it is also necessary to be exposed. Exposure is time (t) and space (x) dependent.

Risks can be connected to each other and their effects can be compounded. Several or many risks can occur simultaneously within the same area. This requires an ability to compare them and to make trade-offs, assessing the relative importance of one risk in comparison to another, which may not necessarily be hydro-meteorological in character. They are not always easy to identify, quantify and categorize, and sometimes identification occurs long after serious adverse consequences have been felt; and they are evaluated differently in social terms. Thus, a risk considered serious in one place may be considered less so in another, or there is flexibility in accepting the risk.

This has led to a more systematic approach to gathering vulnerability and exposure data, albeit, often as a consequence of a post-disaster needs assessment, focused on rebuilding a community. Nevertheless, these data can be used to provide actionable warnings to avoid disasters.

6. Datasets

Based on the above, three data sets are required to develop impact-based forecast and warnings. Hazard forecasts (weather, hydrological, climate and geophysical); vulnerability (roads, bridges, major crops, residential, commercial and industrial buildings, airports, rail and maritime ports); and exposure (people (individuals

Box 2. Pacific Risk Information System (PacRIS)

PCRAFI established the Pacific Risk Information System (PacRIS), one of the largest collections of geospatial information for the PICs (World Bank 2013). PacRIS contains detailed, country-specific information on assets, population, hazards, and risks. The exposure database leverages remote sensing analyses, field visits, and country specific datasets to characterize buildings (residential, commercial, and industrial), major infrastructure (such as roads, bridges, airports, ports, and utility assets), major crops, and population. More than 500,000 buildings were digitized from very-high-resolution satellite images, representing 15 percent (or 36 percent without Papua New Guinea) of the estimated total number of buildings in the PICs. About 80,000 buildings and major infrastructure were physically inspected to calibrate satellite-based data.

In addition, about 3 million buildings and other assets, mostly in rural areas, were inferred from satellite imagery. PacRIS includes the most comprehensive regional historical hazard catalogue (115,000 earthquake and 2,500 tropical cyclone events) and historical loss database for major disasters, as well as state-of-the art country-specific hazard maps for earthquakes (ground shaking) and tropical cyclones (wind). PacRIS contains risk maps showing the geographic distribution of potential losses for each PIC as well as other visualization products of the risk assessments, which can be accessed, with appropriate authorization, through an open-source web-based platform.

Country risk profiles were developed for each of the 15 PICs from the data contained in PacRIS. They can be used to draw attention to not only the risk that is faced by each country but also to give an indication of the frequency of these hazardous events and their associated economic and fiscal losses. Under this analysis, it was established that the average annual loss caused by natural hazards across all 15 PICs is estimated at USD 284 million, or 1.7% of the regional GDP. Vanuatu, Niue and Tonga experience the largest Average Annual Losses (AAL) from natural disasters in the region equivalent to 6.6%, 5.8% and 4.4% of their national GDP, respectively. This places them among those countries that experience the highest levels of AAL globally. There is a 2% chance that the Pacific region will experience disaster losses in excess of USD 1.3 billion from tropical cyclones and earthquakes in a given year.

and demographic data) and moveable assets (animals, vehicles, etc.)). In general, hazard information is available for past and present hazards. This is of high quality and high spatial resolution for weather hazards. Historical data has provided a baseline for what might be expected in the present, but as stated earlier, may no longer be a good indication of what to expect as more and more extreme weather records are broken.

Access to vulnerability and exposure data is more problematic; however, recent efforts to acquire vulnerability data in support of pilot insurance programs, provides an opportunity. For example, these data, based on satellite and *in situ* observations, have been used to develop the PCRAFI (World Bank 2013). PCRAFI has established the Pacific Risk Information System (PacRIS), one of the largest collections of geospatial information for Pacific Island Countries (PICs). Combined with historical hazard data, risk profiles of potential losses have been developed and used to develop insurance schemes (Box 2). Another example of using high-resolution satellite data for the identification of urban risk can be found in Deichmann *et al.* (2011).

On short time scales, exposure data are mostly related to people and their activities, which place them in harm's way. For forecasting, these data need to be acquired in near real-time and vary with time of day (e.g., location of people on roads, at home, at work at school). In the absence of exposure data, impact forecasts are commonly referred to as impacted-based forecasts; i.e., they consider only the hazard and vulnerability, and do not consider specific individual circumstances. As communication systems evolve, particularly smartphone applications, it is likely that there will be greater emphasis on personalizing warning information. For example, the elderly or infirm may require significantly more time to evacuate than others ahead of a storm surge; drivers of high-sided vehicles may be at risk of overturning due to high winds; and warnings may be directed at people in specific high-risk locations. Much of this is possible today, but more effort is needed to address existing scientific limitations of forecasts to ensure the accuracy of the warnings.

7. Application of Warnings, Resilience and Risk Transfer

In Fig. 2, we showed schematically how low and high resilient infrastructure systems respond to a hazard. The response of each system is interpreted in terms of the three pillars. Operationally, warning/preparedness assures the initial safety of people, which may delay or offset the impact of the hazard by reducing mortality and morbidity. If people are safe, focus can be put on restoring critical infrastructure. If the infrastructure systems and services are resilient, then damage and loss of services will be mitigated. If they are not resilient, further social and economic decline will occur. In the absence of financial intervention, the pace of recovery would be very slow, if at all.

Used effectively, financial risk transfer would have accurately anticipated the unavoidable loss of infrastructure systems and services; the subsequent monetary transfers would enable a full recovery. Knowing what infrastructure investments are needed and the appropriate scale of the risk transfer are essential if the society is to be resilient. Clearly, weaknesses in any element would increase the loss of operations and shift the whole system closer to the low resilience case.

As described above, risk profiles are calculated using historical hazard and vulnerability assessment data. However, if the past is no longer a good representation of future hazards, more emphasis needs to be placed on climate change scenarios and simulations (Royal Society 2014).

Similarly, forecasts of impacts will improve as more vulnerability data is used in the preparation of forecasts and warnings, and then verified against actual impacts. Using the basic methodologies for weather forecasting verification and validation, weaknesses and strengths in infrastructure can be identified based on actual impact forecasts.

In time, the identified weaknesses in infrastructure would be addressed, which in turn could be expected to reduce the cost of catastrophe risk transfer mechanisms.

Maintenance of up to date vulnerability and exposure data is essential for accurate impact forecasts. Therefore, proposed investments in early warning/preparedness need to consider how to contribute to the ongoing data acquisition. By identifying, areas likely to be affected by specific hazards, gaps in vulnerability data can be addressed more efficiently and cost-effectively.

8. Conclusion

Early warning, resilient infrastructure and financial risk transfer mechanisms are three inter-related pillars of a resilient society. A resilient society has the capacity to warn its population and economic sectors to take action to protect them from immediate harm; has the infrastructure systems and services to cope with and adapt to the impact of multiple hazards; and has the financial means to recover quickly at the national level and at the local level.

By taking a holistic approach, investments can be optimized. Risks can be identified based on observed infrastructure, known and expected hazards, and investments in resilient infrastructure systems developed to mitigate these risks, and the excess transferred to the insurance sector based on the vulnerability assessments.

Key to improving the response to warnings is the focus on understanding the impact of hazards, which depends on knowledge of the vulnerability of the infrastructure systems and services. Similarly, these data are required to determine the risk transfer requirements.

Until recently, warning systems have been hazard-based. Vulnerability and exposure information has been collected to increase infrastructure resilience and provide the basis for financial risk transfer. By broadening the application of vulnerability and exposure information, warning services can be developed that focus on the impact of hazards, increasing the likelihood that people would take appropriate action to mitigate their risks.

In practice, in preparing investments, all three elements should be included. We cannot avoid the need to provide timely and actionable information to minimize the impact of meteorological and hydrological hazards on people. Equally, we need infrastructure to protect and shelter people and resist adverse impacts. At present, we often treat early warning systems as separate, independent activities, decoupled from the infrastructure we build and the actions we take to avoid harm. In reality, warning systems are an integral part of the optimal management and use of a country's infrastructure. NMHSs and their partners, who provide information on the impact of hazards, have an integral role to play in adapting this information to inform the design of infrastructure — ensuring sound and effective financing of climate adaptation investments. At the same time, this expanded role for NMHSs needs to be nurtured by WMO to ensure that all of its Members have the capacity to not only forecast and warning, but also support national development objectives.

Acknowledgments

The Global Facility For Disaster Reduction and Recovery and the World Meteorological Organization supported this work.

References

Arnold, M., 2008: The Role of Risk Transfer and Insurance in Disaster Reduction and Climate Change Adaptation. CCCD (Commission on Climate Change and Development), Sweden.

BCCRF (Bangladesh Climate Change Resilience Fund), 2014: Climate Smart Development. Climate Action Newsletter, 4, June 2014. http://site resources.worldbank.org/INTSOUTHASIA/Res ources/223497-1378327471830/bangladesh.html

Bruneau, M., S. E. Chang, R. T. Eguchi, G. C. Lee, T. D. O'Rouke, A. M. Reinhorn, M. Shinozuka, K. Tierney, W. A. Wallace, and D. von Winterfeldt, 2003: A framework to quantitatively assess and enhance the seismic reliance of communities. *Earthquake Spectra*, **19**(4), 733–752.

Chang, S., 2009: Infrastructure Resilience to Disasters. *The Bridge Frontiers to Engineering*, Vol. 39, No. 4, National Academy of Engineering of the National Academies. Washington, DC.

CIF (Climate Investment Funds), 2012: Strategic Program for Climate Resilience for the Pacific Program — Regional Track. https://www.climate investmentfunds.org/cif/sites/climateinvestment funds.org/files/PPCR_7_Strategic_Program_for_ Climate_Resilience_Pacific_Regional_Track.pdf

Deichmann, U., D. Ehrlich, C. Small, and G. Zeug, 2011: Using high resolution satellite data for the identification of urban risk. GFDRR, Washington, DC.

Fleming, G., D. Rogers, P. Davies, E. Jacks, J. Milton, C. Honoré, L. S. Lee, J. Bally, Wang Zhihua, V. Tutis, and P. Goolaup, 2015: WMO Guidelines on Multi-Hazard Impact-Based

Forecast and Warning Services. (WMO-1150), World Meteorological Organization, Geneva.

GFDRR (Global Facility for Disaster Reduction and Recovery), 2014: Strengthening Financial Resilience in the Pacific. http://www.worldbank.org/content/dam/Worldbank/document/drm/gfdrr-stories-of-impact-pcrafi.pdf

Ghesquiere, F. and O. Mahul, 2010: Financial Protection of the State against Natural Disasters. Policy Research Working Paper 5429. World Bank, Washington, DC.

Keating, A., K. Campbell, R. Mechler, E. Michel-Kerjan, J. Mochizuki, H. Kunreuther, J. Bayer, S. Hanger, I. McCallum, L. See, K. Williges, A. Atreya, W. Botzen, B. Collier, J. Czajkowski, S. Hochrainer, and C. Egan, 2014: Operationalizing Resilience Against Natural Disaster Risk: Opportunities, Barriers and A Way Forward, Zurich Flood Resilience Alliance. http://www.iiasa.ac.at/web/home/research/researchPrograms/RiskPolicyandVulnerability/whitepaper.pdf

Mahul, O. and L. Boudreau, 2013: Financial Protection: Risk Financing and Transfer. In *Strong, Safe, and Resilient: A Strategic Policy Guide for Disaster Risk Management in East Asia and the Pacific.* Abhas K. Jha and Zuzana Stanton-Geddes, Editors. Directions in Development. Washington, DC: World Bank.

Met Office, 2014: The recent storms and Floods in the UK. Met Office, Exeter. http://www.metoffice.gov.uk/media/pdf/1/2/Recent_Storms_Briefing_Final_SLR_20140211.pdf

Moteff, J. D., 2012: Critical Infrastructure Resilience: The Evolution of Policy and Programs and Issues for Congress. Congressional Research Service, 7-5700 Washington. D.C. http://fas.org/sgp/crs/homesec/R42683.pdf

Rockefeller Foundation, 2009: Building Climate Change Resilience. http://www.rockefellerfoundation.org/uploads/files/c9725eb2-b76e-42eb-82db-c5672a43a097-climate.pdf

Rogers, D. P. and V. V. Tsirkunov, 2013a: Emergency Preparedness: Weather, Climate, and Hydromet Services. In *Strong, Safe, and Resilient: A Strategic Policy Guide for Disaster Risk Management in East Asia and the Pacific.* Abhas K. Jha and Zuzana Stanton-Geddes, Editors. Directions in Development. Washington, DC: World Bank.

Rogers, D. P. and V. V. Tsirkunov, 2013b: Weather and Climate Resilience: Effective Preparedness through National Meteorological and Hydrological Services. Directions in Development. Washington, DC: World Bank.

Royal Society, 2014: Resilience to Extreme Weather. Royal Society Policy Centre Report. http://royalsociety.org/resilience

UK Government, 2011: Climate Resilient Infrastructure: Preparing for a Changing Climate. Secretary of State for Environment, Food and Rural Affairs, HM Government, Presentation to Parliament.

United Nations, 2015: Sendai Framework for Disaster Risk Reduction 2015–2013. United Nations, New York.

World Bank, 2010: Natural Hazards, Unnatural Disasters: Effective Prevention through an Economic Lens. Washington, DC: World Bank.

World Bank, 2012: World Bank Disaster Risk Financing and Insurance (DRFI) Program. http://www.gfdrr.org/gfdrr/DRFI.

World Bank, 2013: Pacific Catastrophe Risk Assessment and Financing Initiative. Risk Assessment — Summary Report. Washington, DC: World Bank.

WMO (World Meteorological Organization), 2014: Report of the WMO/UN-ESCAP/Typhoon Committee Expert Mission to the Philippines, 7–12 April 2014.

Chapter 6

Climate Services for Sustainable Development

Mannava V. K. Sivakumar[*,‡] and Filipe Lucio[†]

[*]*Acting Secretary, Intergovernmental Panel on Climate Change (IPCC)*

[†]*Director, Global Framework for Climate Services (GFCS),*
World Meteorological Organization, Geneva, Switzerland
[‡]*msivakumar@wmo.int*

The past few decades, have seen an increase in the intensity and frequency of weather and climate extremes around the world with severe impacts on several socio-economic sectors, especially in the developing world. Climate change impacts have increased due to enhanced greenhouse gas emissions and they undermine the ability of all countries to achieve sustainable development. In September 2015, the Member States of the United Nations adopted the new Sustainable Development Agenda, with 17 Sustainable Development Goals (SDGs). Climate services contribute directly to 11 SDGs through collection of climate data; generation and provision of a wide range of information on past, present and future climate; development of products to improve the understanding of climate and its impacts on natural and human systems; and the application of these data, information and products for decision-making in all socio-economic sectors affected by climate, at the global, regional and local scales. The Global Framework for Climate Services (GFCS) established by the World Climate Conference-3 in 2009, is enabling society to better manage the risks and opportunities arising from climate variability and change, through the development and incorporation of science-based climate information and prediction into planning, policy and practice. Climate Services in six socio-economic sectors i.e., Agriculture and Food Security, Disaster Risk Reduction, Health, Water, Energy and Tourism to promote sustainable development have been presented with suitable examples.

1. Introduction

The idea of sustainability was a key theme of the United Nations Conference on the Human Environment in Stockholm in 1972. There was a tangible need for a developmental concept that would allow reconciling economic development with environmental protection. In December 1983, the Secretary General of the United Nations appointed the Brundtland Commission, or more formally, the World Commission on Environment and Development (WCED). The Brundtland Commission promoted the idea that while the "environment" was previously perceived as a sphere separate from human emotion or action, and while "development" was a term habitually used to describe political goals or economic progress, it is more comprehensive to understand the two terms in relation to each other.

Brundtland mentioned that "the 'environment' is where we live; and 'development' is what we all do in attempting to improve our lot within that abode. The two are inseparable." The term sustainable development was coined in the paper *Our Common Future* (Brundtland 1987). Sustainable development was defined as "development that meets the needs of the present generation without compromising the ability of future generations to meet their own needs."

Cash *et al.* (2003) emphasized that science and technology must play a more central role in sustainable development and provide the necessary environmental information and products to promote sustainable development. Environment by definition is "the complex physical, chemical, and biotic factors (as climate, soil, and living things) that act upon an organism or an ecological community and ultimately determine its form and survival". In this context, climate services are critical as they provide the necessary information and products for the user communities in different socio-economic sectors. This paper describes these aspects in detail.

2. Challenges to Sustainable Development in the 21st Century

According to the report entitled "Back to our Common Future" (United Nations 2012), over the last three decades, human development has seen progress on a global level. Some countries have developed rapidly. Progress has been registered in several areas such as access to education, health, and access to basic services such as water and sanitation. Considerable progress was also achieved in the provision of increased access of people to information and increased participation in decision-making, human rights, indigenous peoples, and gender equality. According to the United Nations (2012), numerous gaps remain on the development agenda. Poverty has not been eradicated. With the exception of China, the absolute number of poor people has remained more or less stable since 1990, with marked regional differences. Basic food insecurity concerns as many people, about 1 billion, as it did in 1970. Income inequality is growing, both across and within countries. At the global level, human activities are showing considerable impacts on the environment. In the 20th century, a 4-fold increase in human population was accompanied by a 40-fold increase in economic output and a 16-fold increase in fossil fuel use, along with a 35-fold increase in fisheries catches and a 9-fold increase in water use. Climate change impacts have increased due to greenhouse gas emissions as carbon dioxide emissions increased 17 times, sulfur emissions by

13 and other pollutants by comparable amounts. Climate change impacts undermine the ability of all countries to achieve sustainable development. Increases in global temperature, sea level rise, ocean acidification and other climate change impacts are seriously affecting coastal areas and low-lying coastal countries, including many least developed countries and Small Island Developing States. According to Munich Re (2013), the average percentage of direct losses per year with respect to GDP is highest in emerging economies at 2.9%, compared with developing economies (1.3%) and industrialized countries (0.8%). Global primary energy use, carbon emissions, biodiversity loss, nutrient loadings, deforestation, global fossil water extraction are all still increasing. The biodiversity losses are enormous with the current rate of species extinction being estimated by some to be 100 to 1,000 times higher than in pre-industrial times and two orders of magnitude above a safe level for humans in the long term. Current rates of ocean acidification in the Atlantic and Pacific oceans exceed those experienced in the last glacial termination by a factor of 100.

According to the United Nations (2012), by 2050, the world population is projected to increase to 9.2 billion, with most of the increase in South Asia, the Middle East and Africa. Poverty and hunger are projected to persist and some of the most vulnerable and poorest economies will remain marginalized. In 2050, a whopping 3.9 billion people (more than 40% of world population) will live in river basins under severe water stress, and 6.9 billion people will experience some water stress. Agricultural land area is expected to increase until 2030, intensifying competition for land and most primary forests might be destroyed by 2050. Global primary energy use will increase by 80% and urban air quality will continue to deteriorate globally. Greenhouse gas emissions are expected to increase at an accelerated rate at least until 2030 and most of these emissions increase will be due to large emerging economies.

3. Development and Adoption of Sustainable Development Goals (SDGs)

The United Nations Conference on Environment and Development (Earth Summit) held in 1992, led to the adoption of a broad sustainable development agenda in the form of Agenda 21. Twenty years later, the United Nations Conference on Sustainable Development (Rio+20 Conference) held in June 2012, recognized that while some progress had been made in the implementation of sustainable development since the Earth Summit in 1992, implementation was still a challenge for many countries. Reasons for this lack of implementation lie, *inter alia*, in insufficient progress and setbacks in the integration of the various dimensions of sustainable development (economic, social and environmental). The Rio+20 outcome document entitled "The future we want" agreed on the need to devise Sustainable Development Goals (SDGs). In September 2014, the Open Working Group (OWG) on Sustainable Development Goals — a United Nations intergovernmental group — proposed 17 Sustainable Development Goals (SDGs) and 169 associated targets to be achieved by the year 2030 (OWG-SDG 2014). These were adopted by the 193 Member States of the United Nations at the start of a three-day Summit on Sustainable Development in September 2015. Sustainable development goals could assist in focusing the broad international sustainable development agenda at a practical level.

4. Relevance of Climate Services for Sustainable Development

In recent decades, changes in climate have caused impacts on natural and human systems on all continents and across the oceans (IPCC 2014). The number of hydrometeorological hazards in particular (such as droughts, floods, tropical storms and wild fires) which were measured on an average of 195 per year in 1987–1998 increased to 365 per year in 2000–2008 (WMO 2014b). Impacts from recent climate-related extremes, such as heat waves, droughts, floods, cyclones, and wildfires, reveal significant vulnerability and exposure of some ecosystems and many human systems to current climate variability. Impacts of such climate-related extremes include alteration of ecosystems, disruption of food production and water supply, damage to infrastructure and settlements, morbidity and mortality, and consequences for mental health and human well-being. Climate-related hazards exacerbate other stressors, often with negative outcomes for livelihoods, especially for people living in poverty. Climate-related hazards affect poor people's lives directly through impacts on livelihoods, reductions in crop yields, or destruction of homes and indirectly through, for example, increased food prices and food insecurity.

All countries are having difficulties in coping adequately with the increasing effects of hydrometeorological disasters, whether through a growth in the number of severe events, through increased exposure, heightened vulnerability, or all three. Efforts have to be directed towards strengthening capacities at national and local levels, with international support where necessary. Climate services develop and provide science-based and user-specific information relating to the past, present and potential future climate and address all sectors affected by climate, at global, regional and local scale. They connect natural and socio-economic research with practice.

Climate services consist of the collection of climate data; generation and provision of a wide range of information on the past, present and future climate; development of products that help improve the understanding of climate and its impacts on natural and human systems; and the application of these data, information and products for decision-making in all walks of life and at all levels in the society. Depending on the user's needs, these data and

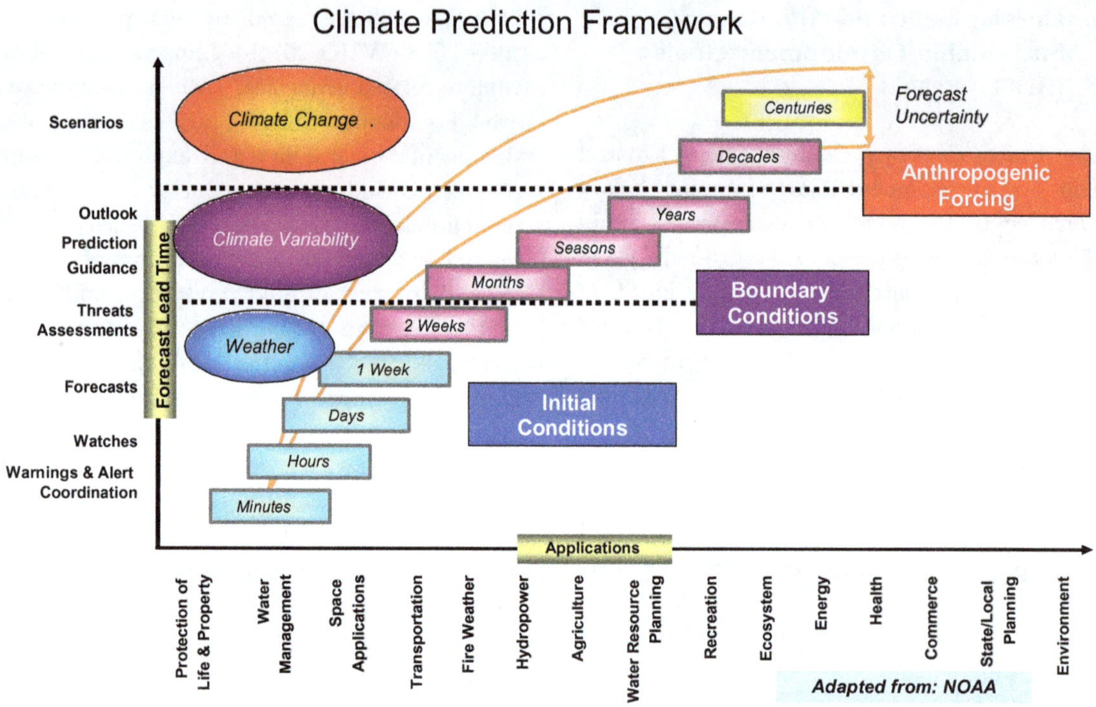

Climate Prediction Framework

Fig. 1.   Seamless hydrometeorological and climate services.

information products may be combined with non-meteorological data, such as agricultural production, health trends, population distributions in high-risk areas, road and infrastructure maps for the delivery of goods, and other socio-economic variables. International cooperation in seamless research on hydrometeorological and climate services has led to advances in predictive accuracy and increased lead time facilitating applications in a wide range of user sectors (Fig. 1).

Climate services help society cope with climate variability and change through the transformation of climate-related data — together with other relevant information — into customised products such as projections, trends, economic analysis and services to the user communities in different sectors. For example, the provision of more and better climate services will allow farmers to fine-tune their planting and marketing strategies based on seasonal climate

forecasts; empower disaster risk managers to prepare more effectively for droughts and heavy precipitation; assist public health services to target vaccine and other prevention campaigns to limit climate-related disease outbreaks such as malaria and meningitis; and help improve the management of water resources. These activities all contribute to appropriate adaptation planning to a changing climate.

5. Contribution of Climate Services to Sustainable Development Goals (SDGs)

Of the 17 SDGs, climate services can contribute to 11 goals:

Goal 1. *End poverty in all its forms everywhere*: Seasonal and long-term climate outlooks and predictions would empower the poor to reduce their exposure and vulnerability to climate-related extremes, thus addressing one of the root causes of poverty (instantaneous poverty caused by hazards).

Goal 2. *End hunger, achieve food security and improved nutrition and promote sustainable agriculture*: Information on the start, end and quality of the rainy season will help improve food production, nutrition and food security. Climate services could strengthen the capacity of farmers for adaptation to climate change, extreme weather, drought, flooding and other disasters.

Goal 3. *Ensure healthy lives and promote well-being for all at all ages*: Vector-borne diseases such as malaria transmitted by mosquitoes which thrive under certain climatic conditions can be anticipated through the provision of climate information with several months lead time. This can allow well-targeted response strategies that provide advanced distribution of medication, insecticides and bed nets to vulnerable communities.

Goal 6. *Ensure availability and sustainable management of water and sanitation for all*: Seasonal and multi-year predictions help dam and reservoir managers to plan ahead of time so as to ensure the availability and sustainable management of water resources. Climate services could help protect and restore water-related ecosystems, including mountains, forests, wetlands, rivers, aquifers and lakes.

Goal 7. *Ensure access to affordable, reliable, sustainable and modern energy for all*: Climate services can inform the development and use of renewable energy sources such as wind, solar and hydropower, thus contributing to a low carbon development path.

Goal 9. *Resilient infrastructure*: Selection of appropriate sites for infrastructure and the standards for infrastructure development based on knowledge of the past characteristics of hazards, current and future trends will ensure that infrastructure is not placed at sites at risks and is developed taking into account the frequency and intensity of hazards.

Goal 11. *Make cities and human settlements inclusive, safe, resilient and sustainable*: Design of cities and human settlements taking into account weather and climate patterns will ensure natural flow of air and minimize the use of air conditioning for example. Climate services could help in the implementation of integrated policies and plans towards mitigation and adaptation to climate change, resilience to disasters, and holistic disaster risk management at all levels.

Goal 12. *Ensure sustainable consumption and production patterns*: Climate services could promote the sustainable management and efficient use of natural resources. Early warnings and information for adaptation such as seasonal to multi-year predictions will strengthen resilience to climate vulnerability and change.

Goal 13. *Take urgent action to combat climate change and its impacts*: Climate services could strengthen resilience and adaptive capacity to climate related hazards and natural disasters in all countries. The Global Framework for Climate Services (GFCS) assists with information and knowledge that supports use of renewable sources of energy such as solar, wind, hydropower, thus reducing emissions of GHGs.

Goal 14. *Conserve and sustainably use the oceans, seas and marine resources for sustainable development*: Climate services could assist with the sustainable management and protection of marine and coastal ecosystems to avoid significant adverse impacts, including by strengthening their resilience.

Goal 15. *Protect, restore and promote sustainable use of terrestrial ecosystems, sustainably manage forests, combat desertification, and halt and reverse land degradation and halt biodiversity loss*: Climate services can ensure the conservation, restoration and sustainable use of terrestrial and inland freshwater ecosystems and their services, in particular forests, wetlands, mountains and drylands and combat desertification.

Goal 17. *Strengthen the means of implementation and revitalize the Global Partnership for Sustainable Development*: GFCS is enhancing North-South, South-South and triangular regional and international cooperation on and access to climate services and the capacity building support to developing countries, including for LDCs and SIDS, to increase significantly the availability of high-quality, timely and reliable climate data and products.

6. Global Framework for Climate Services

The World Climate Conference-3 (WCC-3) held in Geneva in 2009 unanimously decided to establish a Global Framework for Climate Services (GFCS), a UN-led initiative spearheaded by WMO to guide the development and application of science-based climate information and services in support of decision-making in climate sensitive sectors. Thirteen heads of state or government, 81 ministers and 2,500 scientists unanimously agreed to develop the GFCS. Following the decision of WCC-3, a High Level Taskforce (HLT) was appointed through an intergovernmental process to prepare a report that was to include recommendations on the proposed elements of the GFCS and the next steps for its implementation. The HLT produced the report "Climate Knowledge for Action: A Global Framework for Climate Services" as the basis for GFCS (WMO 2011).

The vision of the GFCS is to enable society to better manage the risks and opportunities arising from climate variability and change, especially for those who are most vulnerable to such risks. This will be done through development and incorporation of science-based climate information and prediction into planning, policy and practice. The vision of GFCS supports major agendas such as the Sendai Framework for Disaster Risk Reduction (SFDRR), which was adopted at the Third World United Nations Conference on Disaster Risk Reduction (WCDRR-III) held in Sendai, Japan, in March 2015. The vision supports the SFDRR's overarching goal to "Prevent new and reduce existing disaster risk through the implementation of integrated and inclusive economic, structural, legal, social, health, cultural, educational, environmental, technological, political and institutional measures that prevent and reduce hazard exposure and vulnerability to disaster, increase preparedness for response and recovery, and thus strengthen resilience." The vision of GFCS also supports the Sustainable Development Goals (SDGs) adopted by world leaders at the UN General Assembly in New York, in September 2015.

To ensure that the GFCS provides the greatest benefit to those who are most in need of climate services, the HLT recommended that the following eight principles be adhered to in its implementation:

Principle 1: All countries will benefit, but priority shall go to building the capacity

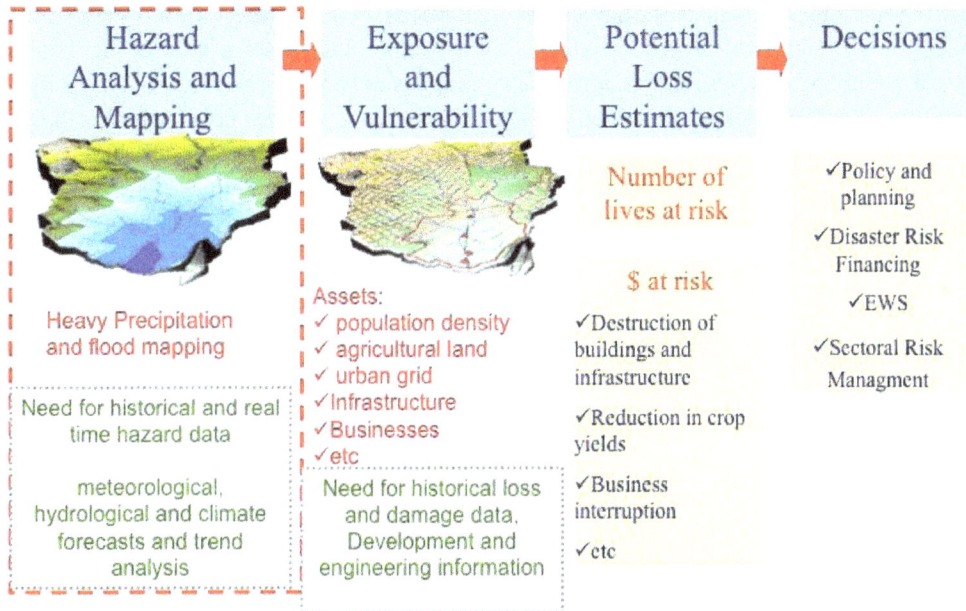

Fig. 2. GFCS facilitates hazard analysis and risk assessment to promote sectoral risk management and societal resilience.

of climate vulnerable developing countries

Principle 2: The primary goal of the Framework will be to ensure greater availability of, access to, and use of climate services for all countries

Principle 3: Framework activities will address three geographic domains; global, regional and national

Principle 4: Operational climate services will be the core element of the Framework

Principle 5: Climate information is primarily an international public good provided by governments, which will have a central role in its management through the Framework

Principle 6: The Framework will promote the free and open exchange of climate-relevant observational data while respecting national and international data policies

Principle 7: The role of the Framework will be to facilitate and strengthen, not to duplicate

Principle 8: The Framework will be built through user-provider partnerships that include all stakeholders

As shown in Fig. 2, GFCS places emphasis on hazard analysis through the use of historical and real time hazard data and meteorological, hydrological and climatological forecasts and trend analysis. This analysis in combination with the analysis of exposure and vulnerability helps in effective risk assessment which facilitates effective decision making to promote societal resilience.

7. Climate Services in Different Socio-Economic Sectors to Promote Sustainable Development

The scope and thrusts of the GFCS include five initial priority sectors, namely agriculture and food security, disaster risk reduction, energy, health, and water. GFCS therefore focuses considerable early attention on these areas. In this paper, an additional sector on tourism is added.

The following sections seek to outline some specific guidelines on how these sectors might be supplied with climate services at the national level including how providers, intermediaries and end-users might interact.

7.1. *Agriculture and food security*

It is very well known that agriculture is inherently sensitive to climate conditions and is among the sectors most vulnerable to weather and climate risks. The relationships between weather, climate and production risk are well recognized (George *et al.* 2005). More frequent and intense extreme weather events and increasing uncertainty in rainy season patterns are already having significant impacts on food production, food distribution infrastructure, livelihood assets, human health and food emergencies, in both rural and urban areas (FAO 2008). Battisti and Naylor (2009) concluded that projected changes in the frequency and severity of weather and climate events, including higher growing season temperature, have significant consequences for food production, farm income, and food security. Despite the impressive advances in agricultural technology over the last half a century, climate variability has a large influence on agriculture, which is heavily dependent on rainfall, sunshine and temperature. Of the total annual crop losses in world agriculture, many are due to direct weather and climatic effects such as droughts, flash floods, untimely rains, frost, hail, and severe storms (Hay 2007). Adverse weather and climate conditions directly affect agricultural productivity, livelihoods, water security, land use, agricultural marketing systems, market instability, food prices, trade and economic policies; and small-holder farmers, fishers, livestock herders and forest dependent communities are often highly vulnerable to these impacts. Climate change will exacerbate existing threats to food security and livelihoods from a combination of increasing frequency of climate hazards, diminishing agricultural production in vulnerable regions, expanding health risks, increasing water scarcity, and intensifying conflicts over scarce resources, which will likely lead to new humanitarian crises, as well as increasing displacement. Climate change is expected to affect all of the components that influence food security: availability, access, stability and utilization.

Climate services in agriculture extend to where it can help develop sustainable and economically viable agricultural systems, improve production and quality, reduce losses and risks, decrease costs, increase efficiency in the use of water, labour and energy, conserve natural resources and decrease pollution by agricultural chemicals or other agents that contribute to the degradation of the environment (WMO 2014a). Climate services are critical for the food availability and stability which are two of the four aspects of food security. The agriculture and food security community relies on appropriate and timely phenological, environmental, and climate information at relevant spatial and time-scale data points to make informed decisions. Available, accessible, comprehensive and useful weather and climate data can help agriculture and food security decision-makers improve their understanding of climate's impact on agricultural development and food systems, and their estimates of populations at risk (risk mapping). Weather and climate data can be particularly helpful to anticipate, prepare for and respond to agriculture or food security risks, on both short timescales to address problems triggered by climate extremes (i.e., droughts, thermal extremes) as well as longer term risks associated with climate change (e.g., increased frequency of cyclones, desertification, etc.).

Intra- and inter-seasonal variability has a major impact on agriculture and food security. Seasonal climate outlooks can influence decisions on which varieties to plant and when, or the best timing for spraying where plant disease outbreaks are likely to occur, or perhaps estimate of the quantity of water needed for irrigation or whether to reduce livestock numbers if a drought is forecast. Farmers may be unprepared for expected weather conditions and make decisions based on an understanding of general climate patterns in their regions. Better climate predictions three to six months in advance can help shape appropriate decisions, reduce impact and take advantage of forecasted favourable conditions. Seasonal forecasts provide probability distribution for monthly to seasonal means of climate parameters (in terms of their departures from long-term averages), such as rainfall and temperature, several months in advance that can be used for crop yield estimates. Yet, information about growing season weather beyond the seasonal average is also needed, such as growing degree days, chill days, and changes in the growing season (WMO 2014a).

7.2. *Disaster risk reduction*

Natural hazards involving weather, climate and water are a major source of death, injury and physical destruction. Natural hazards become natural disasters when people's lives and livelihoods are destroyed. Human and material losses caused by natural hazards are a major obstacle to sustainable development as every year, natural hazards cause significant loss of life and erode gains in economic development. During the past five decades, disasters of hydrometeorological origin such as droughts, floods, storms and tropical cyclones and wildland fires have caused major loss of human lives and livelihoods, the destruction of economic and social infrastructure, as well as environmental damages. According to the Centre for Research on the Epidemiology of Disasters (CRED 2015), 11,938 disasters occurred during 1970 to 2014, leading to a total loss of 3.48 million lives and economic losses amounting to US$2.69 trillion.

Figure 3 shows that the frequency of disasters increased steadily from 1970 with the

peak frequencies in the year 2000. Such changes in weather and climate extremes, and their related impacts, pose challenges for national and local disaster risk reduction systems. Better climate services can help meet these challenges, in both the short- and the long-term, by giving decision-makers enhanced tools and systems to analyse and manage risk, under current hydrometeorological conditions and in the face of climatic variability and change (WMO 2014b). The value of climate services in reducing disaster risk is broadly recognized, given the preponderance of hydrometeorological hazards in shaping disaster risk, and the fundamental role that climate information plays in disaster risk reduction efforts (WMO 2014b). Disaster risk reduction decisions are taken by a broad group that includes disaster risk managers, as well as government sectors, humanitarian and development agencies and banks, the private sector, non-governmental organizations, communities and individuals. Multiple consultations, meetings and publications have found that these actors need climate information that is tailored to their specific decision-making needs, and provided in appropriate language and formats that facilitate action (Helmuth *et al.* 2011).

Under climate services for disaster risk reduction, there are six priority categories of activities (Fig. 4), that would catalyze provision of related products and services by the National Meteorological and Hydrological Services (NMHSs), and promote widespread implementation of programmes and initiatives that incorporate climate information and services (WMO 2014b). These categories are aligned with existing disaster risk reduction structures, and compatible with other relevant international initiatives, including the Sendai Framework for Disaster Risk Reduction (SFDRR).

GFCS addresses a number of issues identified under the different priorities of SFDRR so that climate services can help the countries in their pursuit towards achieving SFDRR goals and targets. For example, under Priority 1 of SFDRR

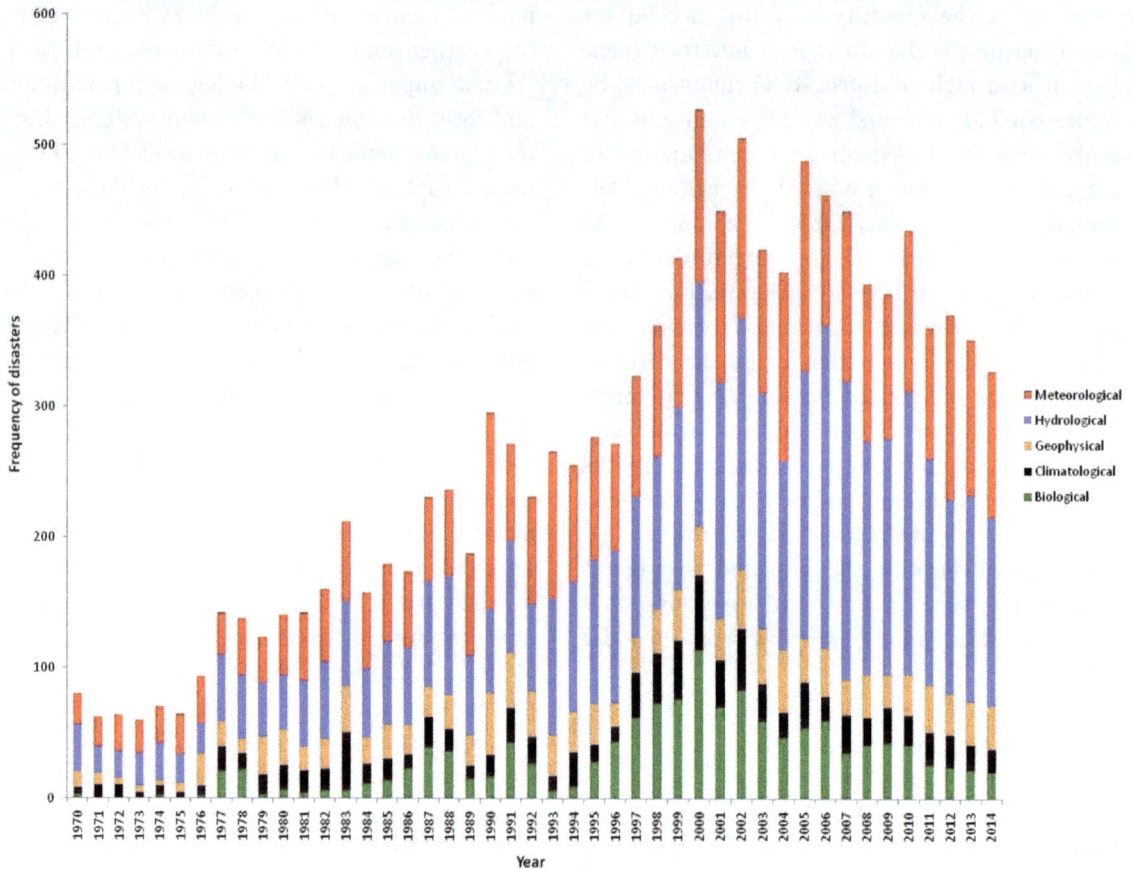

Fig. 3.   Frequency of natural disasters from 1970 to 2014 (*Source of data*: CRED 2015).

on "Understanding disaster risk", emphasis is placed on promoting the collection, analysis, management and use of relevant data and practical information and encouraging the use of and strengthening of baselines and periodically assessing disaster risks, vulnerability, capacity, exposure, hazard characteristics and their possible sequential effects.

Past climate data are essential for quantifying hazard characteristics of a region, in particular the frequency, severity and location of climatic extremes by retrieving and computerizing all available data at the highest temporal and spatial resolution possible in order to capture the characteristic features of particular hazards.

Priority 4 of SFDRR entitled "Enhancing disaster preparedness for effective response and to "Build Back Better" in recovery, rehabilitation and reconstruction" places emphasis on developing, maintaining and strengthening people-centred multi-hazard, multisectoral forecasting and early warning systems. GFCS addresses this important aspect since an essential starting point for reducing risks from disasters is a quantitative assessment that combines information about the hazards with exposures and vulnerabilities of the population or assets, e.g., agricultural production, infrastructure and homes. Climate information is critical for the analysis of hazard patterns and trends. Information on historical and

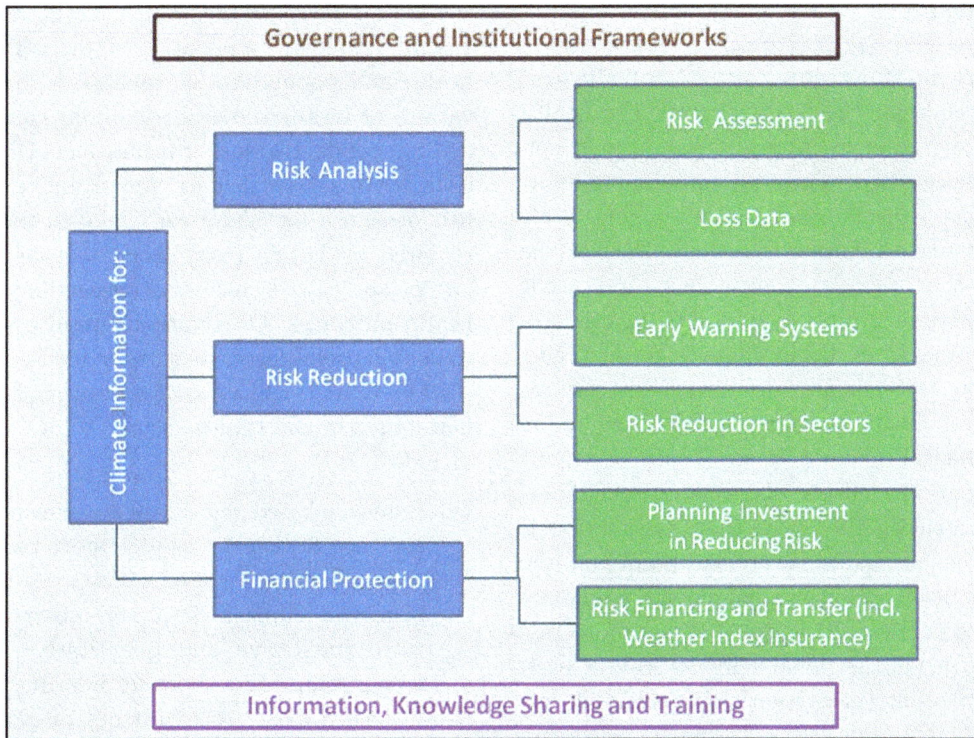

Fig. 4. Priority categories of activity under climate services for disaster risk reduction (in green).

ongoing extreme climate events can help to identify and build processes for integrating this information into loss and damage accounting systems.

In the past five decades, mortality rates from disasters have decreased in some regions as a consequence of the development of multi-hazard early warning systems. NMHSs are working together with the public and private sectors to implement multi-hazard early warning systems, which aim to further significantly reduce the number of fatalities caused by weather, water- and climate-related natural disasters. These systems enable decisions to protect lives and livelihoods in short and longer-term time-frames by extending the lead time for contingency planning and preparation.

Climate services for risk financing and transfer can be provided to inform risk assessments and catastrophe risk analysis which are ideally based on at least 30 years of hydrometeorological data and other asset and vulnerability information. NMHSs can provide climate information for the innovative risk transfer tools, such as weather derivatives or index-insurance in order to determine payout structures, as payouts are not based on actual losses, but triggered by meteorological parameters such as wind, rainfall and temperature. For example, the weather index-based crop insurance provides financial coverage to protect smallholder farmers against the potential impacts of deficit/erratic rainfall, extreme temperatures and other environmental variables. An index insurance contract pays out on the value of an index; in this case the index is based on measured hydrometeorological variables such as rainfall, temperature or river levels. Forecasts of these types of parameters have been used for both portfolio risk management and diversification purposes.

7.3. *Health*

Evidence based decision-making is a fundamental principle for the health sector. The health community relies on appropriate and timely epidemiological, environmental, and climate information at relevant spatial and temporal scale data to make informed decisions. Despite the widely acknowledged connections, and needs for climate information from the health sector (Rogers *et al.* 2010) there is an emerging consensus that climate information and services to inform health decisions is not used to its full potential. Connor *et al.* (2008) point out that given the effects of variations on different timescales, it is clear that information is needed on all levels. They also point out the need for down-scaled regional models, as climate change scenarios and seasonal climate forecasts are modeled at the global–regional scale at best.

Climate-informed health systems and services can not only save lives but help increase the efficient use of limited resources by identifying and targeting the populations most at risk in vulnerable areas and developing the capacity of health and other sectors to manage the risks to health. Available, accessible, and useful weather and climate information can help health decision makers improve, *inter alia*, understanding of the mechanisms of climates impact on disease transmission and occurrence, and estimate populations at risk (e.g., risk mapping). It can help estimate seasonality of disease occurrence and necessary timing of interventions and investments. It can help monitor and predict year-to-year variations in disease incidence (e.g., early warning systems for epidemics), as well as longer term trends of potential impacts (e.g., climate change assessments). Climate information can also improve impact assessments, by removing climate as a confounder of health intervention performance.

Climate services to health sector include those directly related to user needs, such as local measurements of precipitation, soil moisture and surface air temperature, which are needed, for example, to identify malaria risk by correlating health and population information with observations of local ecological conditions conducive or non-conducive for transmission. Observational data are particularly important to establish baselines for historical climate conditions required by health researchers to make correlations and causal linkages between climate and health outcomes. Observations are also needed to enable useful forecasts to be made.

The operational needs for weather and climate data in the health sector (WMO 2014c), include:

(a) Evidence based health risk assessments that are core health decision support processes that require integrated approaches to link historical climate data and observations, with qualitative and quantitative health vulnerability and exposure information. A real need exists to take stock of gaps and needs in: (i) the availability of historical and future hazard data, metadata, tools and methodologies in hazard mapping and human expertise of the technical agencies (on the provider side); but also (ii) availability of health sensitivity, impact, vulnerability and weather related hazard exposure information, and user-capacity to incorporate climate information in routine health decisions.

(b) Health surveillance that is a core function of the health sector and the backbone of decision-making. It is analogous to observations for the climate community. Integrating social indicators from health surveillance with climate and environmental observations will be a core task for most collaborative action. Development of guidance, standards, and tools can assist this process.

Decadal climate projection maps can be provided for human vulnerability assessment and adaptation planning. Such tools and methodologies can help the officials in the health sector to incorporate the information generated in routine health decisions. Such information can be particularly helpful to anticipate, prepare for and respond to health risks on both short timescales to address health problems triggered by climate variability (such as an outbreak, or thermal extremes), as well as longer time frame risk changes associated with climate change (i.e., droughts, sea-level rise and health infrastructure protection). In this context, early warning systems of extreme weather events, particularly for heat and cold, that pose health risks can provide very useful information to the health authorities to take preventive action.

Short-term climate information has a broad range of applications, including adaptation of World Health Organization (WHO)/national response plans based on seasonal information (such as El Niño/La Niña) and development of national/community/health facility response plans for climate-related hazards, including wildfire, flood, storms, landslides, infectious diseases, water shortages, cold weather, heat stress, chemical and radiological hazards and other potential sources of risk, including food security, mass gatherings, population displacement and infrastructure failure. Long-range climate information such as global climate models and climate scenarios anticipate how the conditions of the climate will be decades into the future and is critical for climate adaptation in the health sector. These climate products can provide key information for research, long-range policy, planning, and investment decisions. Collaborative work between the climate service providers and the health sector will be essential to develop the prediction products (monthly, seasonal, decadal) that can be easily used by the health community for their decision needs.

Growing concerns over climate change have brought to the fore three important aspects: adaptation, disaster-risk reduction and the need for climate information and services to support these. Heat-Health Warning Systems bring together these three facets and exemplify an effective demonstration of climate-risk management in practice. These provide meteorological and/or climate-prediction-based information on the likelihood of forthcoming hot weather that may have an effect on health. This information is used to alert decision-makers, health services and the general public to trigger timely action to reduce the effects of hot-weather extremes on health.

A number of countries around the world have successfully developed these early warning systems, which necessitates close coordination between meteorological and health services. The WMO-WHO publication *Heatwaves and Health: Guidance on Warning-System Development* promotes more widespread development and implementation of these warning systems (WMO and WHO 2015). The Guidance considers who is at risk from heat, outlines approaches to assessing heat stress, surveys heat-intervention strategies which are a necessary part of any truly integrated Heat Health Warning System and presents the underpinning science and methodologies.

7.4. *Water*

Water management (both surface water and groundwater) is intrinsically linked to climate variability and change. The major research gap in the climate change as related to water appears to be lack of studies of climate change impact on existing and future water development infrastructures. How the impact of climate would modify the existing water uses such as hydropower production and irrigation, the operation of reservoirs and navigation locks is not yet explored (UNECA 2011). For example, the implication of inconsistent future rainfall scenarios is conflicting effects of climate change on water resources of Africa (Elshamy *et al.* 2008; Conway and Hulme 1996).

Climate data are critical for the assessment of fluctuations and trends and the risks arising from exposure and vulnerability to natural hazards (floods and droughts) and for the sustainable management of the water resources. There is a wide cross-section of users from the water sector, including for example, hydrological characterization, water supply, flood management and control, irrigation and drainage, power generation, fisheries and conservation, navigation and recreation. These users have a need for a range of climate services (WMO 2014d) to support decisions relating for a range of uses related to Integrated Water Resource Management (IWRM) planning.

A wide range of data in different formats, e.g., point or distributed data, instantaneous or averaged over different lengths of time, serve a number of purposes for water management. Many meteorological and hydrological models are now designed to produce probabilistic output for risk analysis, so the interfacing of climate data feeds with predictive water models is a complex matter. There are frequently gaps and mismatches between the nature and distribution of climate observing systems and those networks devised for water monitoring (WMO 2014d). An improved climate-water interface will enhance the structure and development of compatible observation networks, by extending them to meet user needs, and ensure quality assurance of data.

Water security in a variable and possibly changing climate continues to be a key concern at national, regional and global scales. In addressing this concern, the critical importance of ongoing climate data for the assessment of fluctuations, and trends in risks arising from exposure and vulnerability to climate variability and related natural hazards is well recognized, in order to assist countries and communities in optimal adaptation efforts. Past weather and water observations have left an enormous legacy of data that now provide the basis of knowledge on climate variability and change. Water management design depends heavily on historical data, whereas use of operational data may depend on rapid data delivery and assimilation into models.

The improved and targeted delivery of climate information products and open communications can enhance the quality of information available to the water community for the assessment of fluctuations, and trends in risks arising from exposure and vulnerability to climate variability and related natural hazards and can assist countries and communities in optimal adaptation efforts. Improved access to accurate and reliable climate information results in appropriate and robust design and construction of water-related structures such as culverts, bridges and dams and coastal zone infrastructure. Improved climate prediction services on timescales from seasons to decades and spatial scales from local to regional support improved water resources management and prioritized allocation of resources to the wide variety of water demand sectors, including urban water supply, irrigation systems, flood storage capacity, etc.

The Regional Climate Outlook Fora (RCOFs) an innovative concept conceived, developed and supported as part of WMO's Climate Information and Prediction Services (CLIPS) project in partnership with the National Meteorological and Hydrological Services (NMHSs), regional climate institutions and other agencies in 1997, has completed successful operation in different parts of the world. RCOFs promoted the recognition in many parts of the world that short-range climate predictions could be of substantial benefit in adapting to and mitigating climate variations. One important aspect of the Forums is the facility to bring together experts in various fields, local meteorologists and end users of forecasts in an environment that encourages interaction and learning. Advances made in the RCOFs and the establishment of regional climate centres are of interest to the water community, taking specific care of the communication aspects of the scientific content of specific products.

The range of climate services to support decisions by water managers include identification of extreme weather and climate hazards that pose water-related risks; identification of populations vulnerable to weather and climate hazards, including those in the coastal zone; implementation of risk management and emergency preparedness practices and procedures; development and implementation of water and environmental policy; and development and implementation of water and flood management policies and strategies.

7.5. *Energy*

Energy is essential to practically all aspects of human welfare, including access to water, agricultural productivity, health care, education, job creation and environmental sustainability (UNDP 2005). As noted by UN Secretary-General Ban Ki-moon at the launch of the Decade of Sustainable Energy for All (SE4ALL) in September 2011, the world faces two urgent and interconnected challenges related to energy. One is related to energy access. Nearly one person in five on the planet still lacks access to electricity. More than twice that number, almost three billion people, rely on wood, coal, charcoal or animal waste for cooking and heating. This is a major barrier to eradicating poverty and building shared prosperity. As for the second challenge, where modern energy services are plentiful, the problem is different — waste and pollution. Energy sector emissions such as CO_2 account for the largest share of global anthropogenic greenhouse gas (GHG) emissions. In 2010, 35% of direct GHG emissions came from energy production.

Despite the fact that the energy sector is already one of the important users of weather, climate and water information, there are gaps in meeting in the needs and facilitate the communication between the providers and users of climate data (Dubus 2007). One important gap concerns the use of probabilistic forecasts, which are still not exploited in the application models (Dubus 2009). Energy generation and planning of operations are markedly affected by meteorological events and energy systems are increasingly exposed to the vagaries of weather and climate affecting both the availability and energy demand. By taking into account weather and climate information, energy systems can therefore considerably improve their resilience to weather extremes, climate variability and change, as well as their full chain of operations during their entire life-cycle.

Through appropriate partnerships and stakeholder engagement, the application of weather and climate information can provide useful support to energy management decisions and relevant policymaking to achieve optimal balancing of supply and demand as well as to drive behavioral changes in energy saving.

Climate services in the energy sector are needed to support:

- Greater climate resilience and adaptation across the sector, due to its fundamental importance for development;
- The important role of efficiency and reduction of energy consumption with consequent emissions reduction in support to mitigation targets; and
- The growing renewable sub-sector, given the apparent climate sensitivity of renewables on the one hand and the policy priority accorded to them due to their GHG emissions reduction benefits on the other.

There is a need to improve energy sector (and broader) decision making by improving local weather and climate knowledge, regardless of whether large climate changes are expected; by improving access to existing meteorological and hydrological data; and by developing better mechanisms so that local weather and climate data as well as specialized analyses are archived for the public good (Johnston *et al.* 2012). There are critical weather and climate data needs for the different energy sub-sectors i.e., wind, solar, hydro and thermal. In the wind power sub-sector, vertical gradients in mean wind speed and wind direction as well as in turbulence intensity above the surface layer are critical to the construction, planning and operations of wind turbines. Increase in current data needs led to a boom in surface-based remote sensing techniques such as wind lidars (Emeis 2014). For off-shore wind parks, marine boundary-layer weather and climate variables need to be assessed. Accurate measurements of the incoming irradiance are essential to solar power plant project design, implementation and operations.

Hydropower is obviously dependent on river flow which depends on the following weather parameters: precipitation and snow amounts; air temperature, which in particular controls the snow melting process in spring in mountain areas; the altitude of the 0°C isotherm is of particular importance; and evaporation, which plays a strong role in controlling the water level in large areas reservoirs, in particular in tropical and sub-tropical regions. Floods and droughts have a strong impact on hydropower generation. In the area of thermal power, data needs are diverse as the thermal conversion efficiency depends on ambient air temperature. The efficiency of cooling systems depends on several parameters: water temperature (ocean, rivers), river flow (with special emphasis on drought periods, but also in the case of floods), and air temperature and humidity, which control the efficiency of cooling towers. Rising air and water temperature, and lack of water may then lead to reduced power generation or temporary shutdowns.

Hydropower generation management requires essentially river flow forecasts at the different time scales at which power systems are operated: yearly, quarterly, monthly, weekly, daily and intra-daily. The current practice in to use weather forecasts, either deterministic or probabilistic depending on the ability and means of each company, up to one or two weeks. For longer timescales, the more advanced energy companies use intraseasonal to seasonal forecasts but the climatological approach, which uses historical time series of precipitation and/or river flow, is more widely spread. On longer timescales, for planning purposes, the general rule is to use climatological information as well, even if more and more companies have started to use climate change projections.

7.6. *Tourism*

The tourism sector is one of the largest and fastest growing global industries and is a significant contributor to national and local economies around the world. According the United Nations World Tourism Organization (UNWTO), International tourist arrivals were 25 million in 1950. In 2011 this number was up to 980 million and is expected to reach 1.8 billion by 2030. International tourism receipts in 1950 were US$2.1 billion. By the year 2010, the volume of receipts had increased to US$919 billion. While the majority of international tourism currently occurs in developed countries, the sector is a vital contributor to the economy of many developing countries. Between 1995 and 2007, it is estimated that international tourism in emerging and developing markets grew at twice the rate of industrialized countries — by 11% for least-developed countries (LDCs) and 9% for other low- and lower-middle-income economies (UNWTO 2008). With the growth of tourism in developing countries, international tourism is frequently promoted by development organizations and many governments as having an important role in contributing to the SDGs, particularly the alleviation of poverty in LDCs, gender equality, and environmental sustainability.

With its close connections to the environment and climate itself, tourism is considered to be a highly climate-sensitive economic sector similar to agriculture, insurance, energy, and transportation (Wilbanks *et al.* 2007). Tourism operators have adapted to provide tourism services in every climatic zone on the planet and are affected by climate in a number of ways. All tourism destinations are climate sensitive to a degree in that they are influenced by natural seasonality and demand, which are defining characteristics of tourism worldwide (Scott *et al.* 2011). Tourism destinations are affected either positively or negatively by interannual climate variability that brings heat waves, unseasonable cold, drought, storms, and heavy rain, which can affect not only tourist comfort and safety (and thereby satisfaction), but also the products that attract tourists (e.g., snow cover, coral reefs) or deter them (e.g., infectious disease, wildfires, tropical cyclones, heat waves). Climate variability also influences various facets of tourism operations (e.g., water supply and quality, heating-cooling costs, snowmaking requirements). Weather and climate are an intrinsic component of the vacation experience and have been found to be a central motivator for travel (Mintel International Group 1991; Lohmann and Kaim 1999; Kozak 2002; Hamilton and Lau 2005; Gössling *et al.* 2006; Hill 2009; Moreno 2010). There is also evidence that the weather conditions experienced at the destination have important influence on travel and holiday satisfaction (Smith 1993; Scott 2006).

The Second International Conference on Climate Change and Tourism held in Davos, Switzerland, in October 2007 agreed that climate is a key resource for tourism and that the sector is highly sensitive to the impacts of climate change and global warming, many elements of which are already being felt. The projected increase in the frequency and magnitude of weather and climate extremes due to climate change will affect the tourism industry through increased infrastructure damage, additional emergency preparedness requirements, higher operating expenses (e.g., insurance, backup water and power systems, and evacuations), and business interruptions.

For example, coastal and island destinations are highly vulnerable to direct and indirect impacts of climate change (such as storms and extreme climatic events, coastal erosion, physical damage to infrastructure, sea level rise, flooding, water shortages and water contamination), given that most infrastructure is located within short distance of the shoreline. Mountain regions are important destinations for global tourism. Snow cover and pristine mountain landscapes, the principal attractions for tourism in these regions, are the features that are most vulnerable to climate change.

Tourism sector is estimated to contribute some 5% of global CO_2 emissions. Transportation causes around 75% of the CO_2 emissions generated by tourism (UNWTO and UNEP 2008) with aviation representing the bulk part of it (40%). Nature-based tourism relies on a high diversity of tourism resources (landscapes, flagship species, ecosystems, outdoor activities relying on specific resources like water level in rivers for canoeing, etc.). These resources are highly variable in space, and will be affected by climate change in various ways (UNWTO and UNEP 2008).

Despite the growing global economic importance of the tourism sector and the multiple, complex interactions between climate and tourism, there have been very limited evaluations of the use of climate information or assessments of the climate services needs within the sector (Altalo and Hale 2002; De Freitas 2003; McBoyle 2007; Scott 2006).

NMHSs have a critical role in supplying much of the climate information to the mass media and other tourism-specific outlets (e.g., tourist guides, marketing brochures, travel planning websites) and provide the essential historical, current, and forecast information that allows other providers to develop specialized climate products for the tourism sector (Scott *et al.* 2011). Accurate, geographically specific meteorological information is essential for tourism operations. The potential use of climate information within the tourism sector is tremendous given the high number and diversity of end-users. The full temporal scale of weather and climate information, from nowcasting (up to 1 h), to short- and medium-range forecasts (1 and 7–10 days), to multi-decadal climate-change projections, is being utilized in a wide range of decision-making contexts by tourism operators and planners (Scott *et al.* 2011). Early warning systems also reduce the safety risks associated with extreme events, such as storms, cyclones, or avalanches. Seasonal climate predictions can assist the Tourism sector in fuel supply procurement, marketing, setting insurance premiums, inventory management, or cruise line destination planning.

8. Conclusions

Given the current challenges faced by governments around the world, especially in the developing countries, for eradication of poverty, ensuring food security, reducing carbon emissions, arresting biodiversity loss, and controlling deforestation, land degradation etc., there is an urgent need to promote sustainable development at the global, regional and local levels. The timely provision of climate services to the user communities in different sectors and at different levels can support climate change adaptation. Climate data, science, information and knowledge are critical components that can facilitate the implementation of sustainable development goals.

The Global Framework for Climate Services (GFCS), the UN-led initiative spearheaded by WMO to guide the development and application of science-based climate information and services in support of decision-making in climate sensitive sectors will provide the needed tools and information to promote sustainable development through current climate and future climate change scenarios, higher resolution models, and more precise and useful specialized products

for societal benefits, such as decadal climate projection maps for human vulnerability assessment and adaptation planning. Implementation of multi-hazard early warning systems can significantly reduce the number of fatalities caused by weather, water- and climate-related natural disasters, enhance resilience of societies, sustain productivity and economic growth, and reduce damage to property.

GFCS is an end-to-end system that uses observations, technology and scientific understanding as inputs for the development of climate services to meet user requirements. The User Interface Platform of GFCS enables the users to make their voices heard through the Platform and make sure climate services are relevant to their needs.

In the context sustainable development, investment in NMHSs will enable them to better interact with other ministries and sectors to provide integrated climate services which underpin sustainable development at the national level. Appropriate investment in climate services can enhance sustainable development through the provision of necessary information and products in a timely manner.

References

Altalo, M. and M. Hale, 2002: Requirements of the US Recreation and Tourism Industry for Climate, Weather and Ocean Information. Consultants report to NOAA.

Brundtland, H., 1987: Our Common Future. Report for the World Commission on Environment and Development. Oxford University Press, Oxford, UK.

Cash, D. W., W. C. Clark, F. Alcock, N. M. Dickson, N. Eckley, D. H. Guston, J. Jäger, and R. B. Mitchell, 2003: Knowledge systems for sustainable development. *Proc. Natl. Acad. Sci. U.S.A.*, **100**(14), 8086–8091.

Connor, S. J., T. Dinku, T. Wolde-Georgis, E. Bekele, and D. Jima, 2008: A collaborative epidemic early warning & response initiative in Ethiopia. In: *Proceedings of International Symposium on PWS: A Key to Service Delivery,*

Geneva, 3–5 December 2007, World Meteorological Organization, 166–171.

Conway, D. and M. Hulme, 1996: The impacts of climate variability and future climate change in the Nile basin on water resources in Egypt. *Water Resources Development*, **3**, 277–296.

CRED, 2015: EM-DAT: The International Disaster Database 1970–2014. www.emdat.be. Centre on the Epidemiology of Disasters (CRED), Louvain, Belgium: Université Catholique de Louvain.

De Freitas, C. R., 2003: Tourism climatology: evaluating environmental information for decision-making and business planning in the recreation and tourism sector. *Int. J. Biometeorol.*, **48**, 45.

Dubus, L. 2007: Weather, water and climate information in the energy sector. In: *Elements for Life, WMO Publication for the Madrid Conference.* Tudor Rose Ed., World Meteorological Organization, Geneva, Switzerland.

Dubus, L. 2009: Practices, needs and impediments in the use of weather/climate information in the electricity sector. In: *Management of Weather and Climate Risk in the Energy Industry* (Ed. Alberto Trocolli), Springer.

Elshamy, M. E., I. A. Seierstad, and A. Sorteberg, 2008: Impacts of climate change on Blue Nile flows using bias-corrected GCM scenarios. *Hydrology and Earth System Sciences Discussions*, **5**, 1407–1439.

Emeis, S., 2014: Current issues in wind energy meteorology. *Meteorol. Appl.*, **21**, 803–819. DOI: 10.1002/met.1472.

FAO. 2008. Climate change adaptation and mitigation: challenges and opportunities for food security. Information document prepared for the high level conference on World Food Security: the Challenges of Climate Change and Bioenergy, 3–5 June 2008, Food and Agriculture Organization of the United Nations, Rome, Italy.

George, D. A., C. Birch, D. Buckley, I. J. Partridge, and J. F. Clewett, 2005: Surveying and assessing climate risk to reduce uncertainty and improve farm business management. *Extension Farming Systems Journal*, **1**, 71–77.

Gössling, S., M. Bredberg, A. Randow, P. Svensson, and E. Swedlin, 2006: Tourist perceptions of climate change: A study of international tourists in Zanzibar. *Curr. Issues Tourism*, **9**, 419–435.

Hay, J., 2007: Extreme weather and climate events, and farming risks. In: Sivakumar MVK, Motha, R. (Eds.), *Managing Weather and Climate Risks in Agriculture.* Springer, Berlin, Heidelberg.

Hellmuth, M. E., S. J. Mason, C. Vaughan, M. K. van Aalst, and R. Choularton (eds.)., 2011: *A Better Climate for Disaster Risk Management.* International Research Institute for Climate and Society (IRI), Columbia University, New York.

Hill, A., 2009: Holiday deals abroad vanish in rush to flee the rain. *Observer*, London, 9 August. www.guardian.co.uk/business/2009/aug/09/holiday-last-minute-bookings-increase.

Ilbery, B., 1985: Agricultural Geography: A Social and Economic Analysis. London, Clarendon Press.

IPCC, 2014: *Climate Change 2014: Impacts, Adaptation, and Vulnerability.* Part A: Global and Sectoral Aspects. Contribution of Working Group II to the Fifth Assessment Report of the Intergovernmental Panel on Climate Change [Field, C. B., V. R. Barros, D. J. Dokken, K. J. Mach, M. D. Mastrandrea, T. E. Bilir, M. Chatterjee, K. L. Ebi, Y. O. Estrada, R. C. Genova, B. Girma, E. S. Kissel, A. N. Levy, S. MacCracken, P. R. Mastrandrea, and L. L. White (eds.)]. Cambridge University Press, Cambridge, United Kingdom and New York, NY, USA, 1132 pp.

Johnston, P. C., J. F. Gomez, and B. Laplante, 2012: Climate risk and adaptation in the electric power sector. Asian Development Bank publication. Available at: http://www.iadb.org/intal/intalcdi/PE/2012/12152.pdf.

Kozak, M., 2002: Comparative analysis of tourist motivations by nationality and destinations. *Tourism Manage*, **23**, 221–232.

Lohmann, M. and E. Kaim, 1999: Weather and holiday preference — image, attitude and experience. *Rev. Tourisme*, **54**, 54–64.

McBoyle, G., 2007: Approaching tourism through climate, in: Approaching Tourism (G. Wall, eds), Department of Geography Occasional publication 21, University of Waterloo, Waterloo.

Mintel International Group, 1991: Special report — holidays. Leisure Intelligence, Mintel International Group, London, United Kingdom.

Moreno, A., 2010: Mediterranean tourism and climate (change): A survey-based study. *Tourism Hospitality Plan Dev.*, **7**, 253–265.

Munich Re, 2013: Topics Geo. Natural Catastrophes 2012: Analyses, assessments, positions 2013 issue. Munich Reinsurance, Munich, Germany.

OWG-SDG, 2014: Report of the Open Working Group on Sustainable Development Goals established pursuant to General Assembly resolution 66/288. A/68/L.61. http://www.un.org/ga/search/view_doc.asp?symbol=A/68/L.61&Lang=E.

Rogers, D. P., M. A. Shapiro, G. Brunet, J. C. Cohen, S. J. Connor, A. A. Diallo, and W. Elliott, 2010: Health and climate-opportunities. *Procedia Environmental Sciences*, **1**, 37–54.

Scott, D., 2006: Climate change and sustainable tourism in the 21st century. In: Cukier, J. (ed.) *Tourism Research: Policy, Planning, and Prospects*. University of Waterloo, Waterloo, pp. 175–248.

Scott, D. J., C. J. Lemieux, and L. Malone, 2011: Climate services to support sustainable tourism and adaptation to climate change. *Clim. Res.*, **47**, 111–122.

Smith, K., 1993: The influence of weather and climate on recreation and tourism. *Weather*, **48**, 398–404.

United Nations, 2012: Back to Our Common Future, Sustainable Development in the 21st century (SD21) project: Summary for policymakers. United Nations Department of Economic and Social Affairs, Division for Sustainable Development, New York, USA. 44 pp.

UNDP, 2005: Energizing the Millennium Development Goals: A Guide to Energy's Role in Reducing Poverty. United Nations Development Programme, New York, USA. Available at: http://www.undp.org/content/undp/en/home/librarypage/environment-energy/sustainable_energy/energizing_the_mdgsaguidetoenergysroleinreducing poverty.html

UNECA, 2011: Climate Change and Water in Africa: Analysis of Knowledge Gaps and Needs. Working Paper 4, African Climate Policy Center, United Nations Economic Commission for Africa, Addis Ababa, Ethiopia.

UNWTO, 2008: Emerging tourism markets — the coming economic boom. Press release, 24 June. United Nations World Tourism Organization. Madrid, Spain.

UNWTO and UNEP, 2008: Climate Change and Tourism — Responding to Global Challenges. World Tourism Organization and the United Nations Environment Programme. World Tourism Organization, Madrid, Spain.

Wilbanks, T. J. *et al.*, 2007: Industry, Settlement and Society, in: M. L. Parry *et al.* (eds.), *Climate Change 2007: Impacts, Adaptation and Vulnerability*. Contribution of Working Group II to the Fourth Assessment Report of the IPCC, Cambridge University Press, Cambridge and New York, pp. 357–390.

WMO, 2011: Climate knowledge for action: A Global Framework for Climate Services–empowering the most vulnerable. The Report of the High-Level Taskforce for the Global Framework for Climate Services. WMO-No. 1065, World Meteorological Organization, Geneva, Switzerland, 240 pp.

WMO, 2014a: Agriculture and Food Security Exemplar to the User Interface Platform of the Global Framework for Climate Services. World Meteorological Organization, Geneva, Switzerland, 35 pp.

WMO, 2014b: Disaster Risk Reduction Exemplar to the User Interface Platform of the Global Framework for Climate Services. World Meteorological Organization, Geneva, Switzerland, 47 pp.

WMO, 2014c: Health Exemplar to the User Interface Platform of the Global Framework for Climate Services. World Meteorological Organization, Geneva, Switzerland, 35 pp.

WMO, 2014d: Water Exemplar to the User Interface Platform of the Global Framework for Climate Services. World Meteorological Organization, Geneva, Switzerland, 35 pp.

WMO and WHO, 2015: Heatwaves and Health: Guidance on Warming-System Development (G.R. McGregor, P. Bessemoulin, K. Ebi and B. Menne, Eds.). WMO-No. 1142. World Meteorological Organization, Geneva, Switzerland, 96 pp.

Chapter 7

Future Changes of Extreme Weather and Natural Disasters due to Climate Change in Japan and Southeast Asia

Eiichi Nakakita*, Yasuto Tachikawa[†], Tetsuya Takemi*,
Nobuhito Mori* and Kenji Tanaka*

*Disaster Prevention Research Institute,
Kyoto University,
Uji, Kyoto 611-0011, Japan
mori@oceanwave.jp
[†]Department of Civil and Earth Resources Engineering,
Kyoto University,
Nishikyo-ku, Kyoto 615-8540, Japan

Climate change significantly impacts on the occurrence of natural disasters. However, regional impact assessment of climate change on severe weathers, hazards, and water resources is still limited by coarse resolution of global/general circulation models, limited applications of regional downscaling, and few impact assessment models. This chapter summarizes current activity of impact assessment of climate change around Japan and Southeast Asia based on the latest climate projection since the IPCC Fifth Assessment Report and on developed impact assessment models. This chapter mainly covers the future changes of heavy precipitation, river flooding, coastal flooding, and water resources around Japan and Southeast Asia.

1. Introduction

The Intergovernmental Panel on Climate Change (IPCC) WGI Fifth Assessment Report (AR5) (2013) states that climate change exacerbates the vulnerability on regional scales to extreme and impulsive physical processes, such as heavy precipitation and storm surges. However, the impact assessment of natural hazards due to climate change is still difficult on regional scales due to scale difference between global/general circulation models (GCMs) and hazard scales (less than $O(10\text{--}500 \text{ km})$). The impact assessment of climate change on natural hazards was discussed in the IPCC AR5 WGII (2013) but the number of available quantitative results is still limited on individual regional scales. It is highly expected that climate change impact on regional scales for the next step.

Climate change due to global warming is expected to have major impacts on phenomena such as typhoons, monsoons, precipitation, and seasonal winds. Such changes will also impact the occurrence of natural disasters, including the frequency and intensity of river disasters, landslides, storm surges, droughts, and so on.

Preparing for natural disasters (e.g., developing and maintaining infrastructure, evacuation planning) is important for engineering task to ensure a safe society. Consequently, quantitative evaluation of changes in extreme natural phenomena from the viewpoint of natural disasters, water resources, ecosystems, and biodiversity on timescales of ten to one hundred years is crucial to promote the long-term development of social infrastructure, adaptation, and mitigation of climate change. The time length of planning and installation of infrastructure takes upwards of 10–30 years, depending on the size of the project. However, recent severe disaster events, such as Typhoon Haiyan in 2013 (Mori *et al.* 2014; Takayabu *et al.* 2015) and Cyclone Pam in

2015, indicate that the combination of long-term trend and natural variability might be causing severe weather earlier than expected. Therefore it is particularly important to know and use reliable current scientifically established climate projection results and impact assessment models for long-term planning of natural disaster risk reduction. It is also important to know the time scale of natural hazard intensity change if future natural external forces exceed the existing facility planning measures.

The monsoon and tropical cyclone (i.e., typhoons in the Pacific) are major sources of natural hazards in East Asia and Southeast Asia. Thus, it is important to know the changes of these phenomena and the related impacts of climate change on natural hazard intensity. Future changes of monsoon and tropical cyclone characteristics due to climate change is one of the key issues in East Asia and Southeast Asia. Regional impact assessment based on the latest knowledge is highly important to assess the risks of future tropical cyclones (TCs) and related river and coastal hazards, and water resources.

This chapter summarizes the latest findings on climate change and its impacts on natural disasters and water resources in Southeast Asia and Japan. All impact assessment models are driven by the latest Japan Meteorological Agency/Metrological Research Institute climate projections for CMIP5, thus the climate change impact on different sectors are presented by consistent forcing. This chapter starts from rainfall, river discharge, storm surges, ocean waves to water resources. It is convenient to look down both positive and negative impacts of climate change in Japan and Southeast Asia.

2. Meteorological and River Disasters

2.1. *Rainfall during the rainy season*

Extreme rainfall in Japan frequently occurs during the Baiu season, or the rainy season which sometimes spawns severe disasters. In the future climate, it is anticipated that the frequency of extreme rainfall will increase (e.g., Takemi 2012; Takemi *et al.* 2012). In order to advance our understanding of extreme rainfall in the future climate, analysis is needed not only of rainfall characteristics alone but also of the background atmospheric circulation. In this section, we demonstrate how climate simulation data are used for such analyses.

Using precipitation data from 60-km resolution and 20-km resolution Atmospheric Global/General Circulation Models (denoted hereafter by MRI-AGCM3.2H or 60-km AGCMs and MRI-AGCM3.2S or 20-km AGCMs) and a 5-km Regional Climate Model (denoted hereafter as RCM) developed by Japan Meteorological Agency/Metrological Research Institute (Kitoh *et al.* 2009), studies during the Japanese research program of Innovative Program of Climate Change Projection for the 21st Century investigated future changes in precipitation in the Baiu season under SRES A1B scenario. Compared to the present climate, the simulated future climate projection indicates a delayed northward movement of the Baiu front and a significant increase in the daily precipitation later in the Baiu front. However, the analysis of only rainfall in the Baiu season is not sufficient, because future changes in atmospheric circulation that is closely related to the Baiu rainfall are not known. Before discussing the changes in rainfall during the rainy season under future global warming, it is important to investigate the background atmospheric circulation averaged during the Baiu season. The present study investigated the rainfall in the Baiu season from the viewpoint of average atmospheric circulation with the use of the 20-km mesh climate simulation data for the present climate (corresponding to the years of 1979–2003) and the future climate at the end of the 21st century (2075–2099) from 20-km AGCM. Furthermore, the changes in extreme precipitation is also discussed through examination of the relationships with the water vapor field.

The Baiu frontal zone is often recognized by the distribution of the north-south gradients of the equivalent potential temperature. The equivalent potential temperature can also be represented by moist static energy (h). Actually moist static energy is more commonly used to investigate the Baiu frontal activity, and the quantity h divided by isobaric specific heat (C_p) is approximately equal to equivalent potential temperature. Thus, the Baiu frontal zone is diagnosed in terms of the largest latitudinal gradient of moist static energy hereinafter. The climatological analyses of moist static energy field from the Japanese 55-year Reanalysis data (JRA-55; Kobayashi et al. 2015) as well as the AGCM outputs for the present climate indicated that in the present climate the Baiu front is located over the Pacific side of the Kyushu, Shikoku, and Honshu islands. On the other hand, in the future climate it was shown that the Baiu frontal zone exists over the Pacific Ocean just to the south of the Japanese islands. The changes of the Baiu front from the present to the future climate can clearly been demonstrated by the differences of the latitudinal gradient of moist static energy. In Fig. 1, the left panel shows the changes in the north-south gradients of moist static energy in June from the present climate to the future climate. In June the Baiu front typically remains stationary over the Kyushu and Honshu islands and travels north as the rainy season progresses. However, under the future climate the Baiu front in June is projected to remain to the south of Honshu and Kyushu, including Okinawa. The right panel of Fig. 1 shows the spatial distribution of moisture flux, moisture flux convergence, and precipitable water in June in the future climate. A band-shaped moisture flux appears around 30° north latitude, which transports moist air from the southwest. In particular, the region from southern China to southern Kyushu becomes very humid, as demonstrated by the future increase of approximately 8 mm in precipitable water.

Sampe and Xie (2010) investigated the formation processes of the Baiu front by examining the relationship between the Baiu frontal zone and the jet stream. They found that the jet stream predominantly transports the warm air originating from the Tibetan Plateau in the middle and upper troposphere, where precipitating

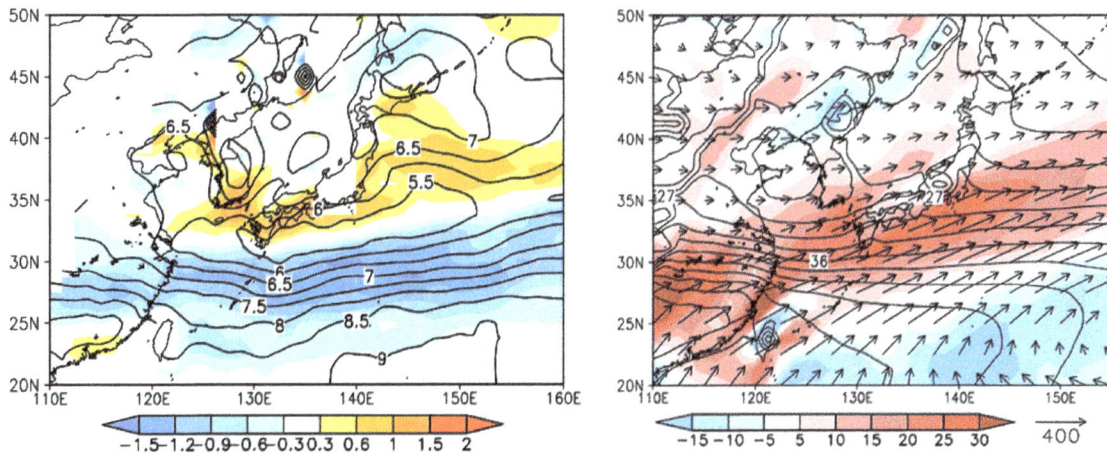

Fig. 1. (*Left*) Future changes in the north-south gradients of moist static energy (h/C_p) (color shading) and the changes in moist static energy (solid line) at the 925 hPa level in June. (*Right*) The spatial distribution of moisture flux (vectors, in kg m^{-1} day^{-1}), convergence/divergence of the moisture flux (color shading, in kg m^{-2} day^{-1}), and precipitable water (solid line, in mm) in June in the future climate.

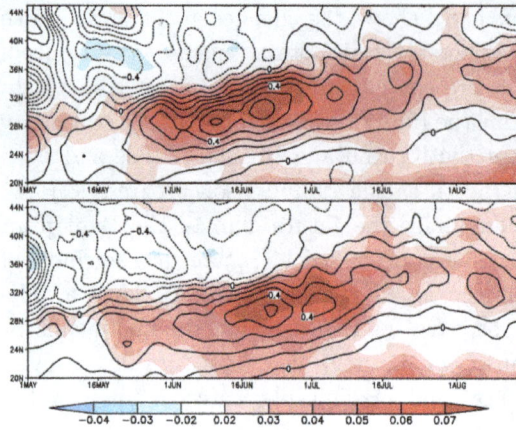

Fig. 2. Time and latitude diagram of horizontal thermal advection (solid line) and vertical velocity (color shading, in Pa s^{-1}) at the 500-hPa level in the present climate (*top*) and the future climate (*bottom*) from 1 May to 15 August. The values are averaged in the longitudinal direction between 125°E and 142°E.

convection develops, and that the development of convection in turn generates and enhances a synoptic-scale disturbance and thus the Baiu front. Figure 2 shows the seasonal march of warm air advection and vertical velocity in the middle troposphere. It is clearly seen that the

warm air advection and ascent corresponding to the Baiu frontal zone move northward with time both in the present and the future climate. However, it is seen that the northward translation of the Baiu frontal zone is delayed overall in Japan in the future climate by approximately five days. Particularly in the future climate, in the southern part of the region shown in Fig. 2 (including the Okinawa region), where the Baiu frontal zone prevails in June (Fig. 1), a northward translation of the Baiu frontal zone is not seen until the middle of July; the northward shift of the Baiu front is also delayed in the southern part of Japan including the Okinawa region. This result is a contrasting feature from the Baiu activity in the southern part of Japan under the current conditions (Okada and Yamazaki 2012).

Figure 3 shows the frequency of days with a daily precipitation of 100/130 mm or more from May to July in the future climate, based on the precipitation data obtained from the outputs of the 20-km AGCM. The frequency is computed by dividing the number of days with the precipitation exceeding the threshold by the total number of days during the period and is represented

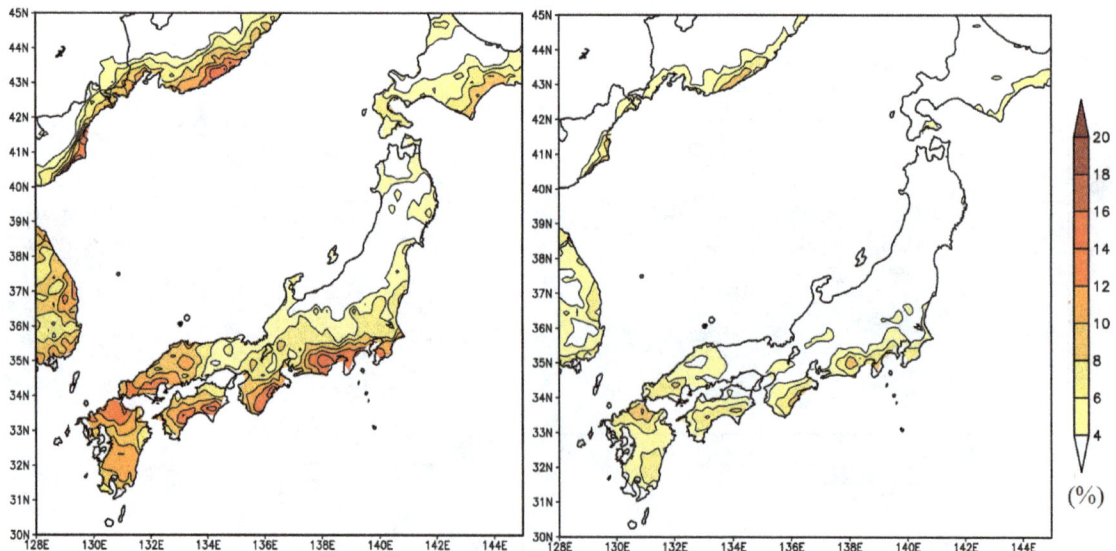

Fig. 3. The spatial distribution of the number of days with a daily precipitation of greater than or equal to 100 mm (*left*)/130 mm (*right*) during May-June-July in the future climate. The number is represented in %.

in %. The frequency is higher in the western part of Japan and the areas along the Pacific coast. In general, a moist tongue and a low-level jet are observed on the south side of the Baiu frontal zone and are accompanied by heavy rain where the moisture flux is significant. As shown in the right panel of Fig. 1, it was found that the amount of precipitable water is greater than 35% in the region from southern China to off the southern coast of Honshu and Kyushu, and that the moisture inflow due to southwesterly winds converges in Kyushu and the areas along the Pacific coast. Thus, the extreme precipitation shown in Fig. 3 can be partly explained in terms of the synoptic-scale moisture flux convergence shown in Fig. 1. The heavy precipitation will be discussed in the next section.

2.2. Heavy rainfall and atmospheric circulation in the Baiu season

The atmospheric circulation in the summer of 2013 is characterized by a moisture flux that is directed along the edge of the strong Pacific High towards the Japanese islands. This atmospheric circulation brought a sufficient amount of moisture that resulted in heavy rainfall in various areas in the Japan Sea side of Chugoku, Tohoku, and other regions in Japan. In this study, the frequency and characteristics of localized heavy rainfall and the relevant atmospheric circulation were analyzed with the use of the 20-km AGCM projection as a following research in the previous section. In addition, an attempt was made to evaluate the difference of the analysis with the single run of 20-km AGCM using the 60-km AGCM ensemble runs.

Both 20-km and 60-km AGCMs used the Yoshimura scheme for a cumulus scheme thus they are consistent model except the spatial resolutions. For the present climate, we use data from two experiments which are conducted through varying the initial conditions, while for the future climate we use five ensembles with various initial conditions and with different

sea surface temperature patterns. The present analysis was conducted on the monthly mean outputs for the present climate (corresponding to the period of 1979–2003) and the climate at the end of the 21st century (the period of 2075–2099), hereafter referred to as the future climate.

The results of both the 20-km and 60-km AGCM outputs were shown around Japan in terms of the anomalies of sea-level pressure and moisture flux in summer both in the present climate and in the future climate from the 25-year means of the present climate. In addition, the years that exhibited the characteristics of the atmospheric circulation in the summer of 2013 were subjectively extracted. In Fig. 4, typical examples of the extracted sea-level pressure and moisture flux are shown.

To quantitatively verify the correspondence with reference heavy rainfall (teaching data), years with localized heavy rainfall were extracted from the 5-km RCM, which is a downscaled model of the 20-km AGCM. The definition of heavy rainfall is (1) 50 mm for 30 mins. within 2 hours, (2) 150 mm for 3 hours, (3) large spatial gradient of equivalent potential temperature on the ground from June to July. If a grid point satisfy (1)–(3), we regard the heavy rainfall occurs. We compared the results between the 5-km RCM and the years subjectively extracted from the AGCMs from the viewpoint of the similarity in the atmospheric circulation with the summer 2013 feature, as shown in Fig. 5.

The bar charts represent the number of years extracted according to the atmospheric circulation similarity with the summer of 2013 from the 20-km and 60-km AGCMs where the blue and red bars represent the present climate and the future climate, respectively. The colors in each map represent a significant increase in the frequency of atmospheric circulation in the future based on the results of the T-test conducted on the difference in the number of years extracted under the present climate and the future climate. The results reveal that the

SLP deviation 199307 20km

(a) Sea level pressure

WVFL deviation 209108 20km

(b) Specific humidity

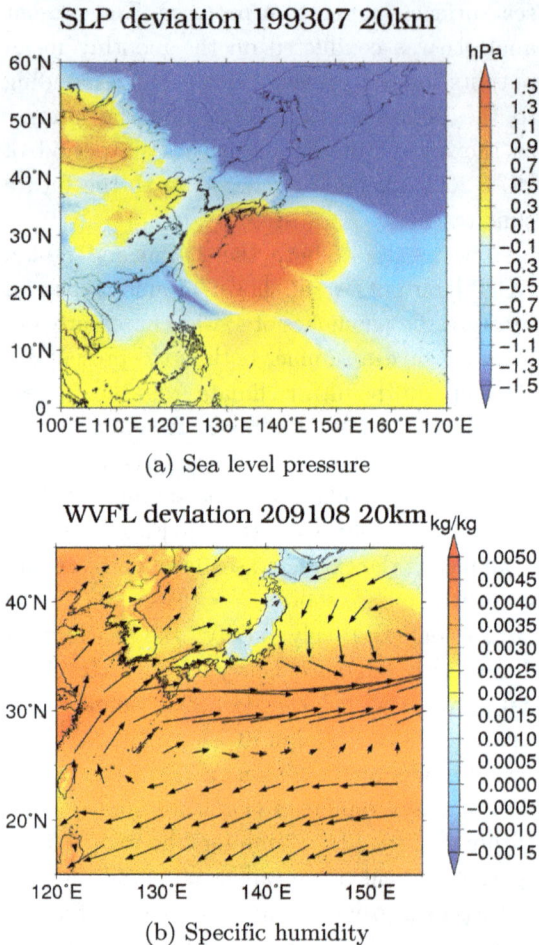

Fig. 4.   Typical examples of the extracted atmospheric circulation. Shades of color represent anomalies of (a) the sea level pressure and (b) the specific humidity. Arrows in (b) indicate the vectors of the moisture flux anomaly.

appearance of the target atmospheric circulation is unanimously more frequent on the Japan Sea side in the future climate. In particular, a significant increase is observed at the 5% significance level on the Japan Sea side of the Tohoku region in July and in all regions on the Japan Sea side in August. In addition, analyzing 5-km RCM, it should be noted that approximately 60% of the years with localized heavy rainfall extracted from the 5-km RCM were extracted based on the characteristics of atmospheric circulation in the AGCMs.

2.3. *River flooding*

A river routing model to estimate the change in river discharge throughout Indochina was developed and applied to detect the change of river discharge for three 25-year periods (present, near future, and future experiments) as an example. The input to the river routing model was the runoff generation data projected by the 20-km AGCM, the 60-km AGCM, and MIROC5 (Duong *et al.* 2014). Figure 6 shows maps of the estimated change ratio of the mean of the annual maximum discharge in the far future climate experiment (2075–2099) to that of the present climate experiment (1979–2003) using the 20-km AGCM and MIROC5. The figure also shows the results of the statistical significance analysis of these changes. The results obtained by using the 20-km AGCM and the 60-km AGCM show similar results; both indicate a statistically significant increase in the annual maximum discharge in the Irrawaddy Rivers. In contrast, the result obtained by using MIROC5 shows a statistically significant increase in discharge in a large region of Indochina.

The annual maximum discharges are going to increase due to warmer and more humid atmospheric environment in the future climate. Although the signal of future change shows consistency in the specific area, there is clear dependence of magnitude of changes on the projections. Therefore the uncertainty of climate projections gives significant impact on quantitative future changes of river discharge.

3. Coastal Hazards

3.1. *Storm surge*

It is important to consider the influence of climate change on coastal area and especially on coastal disasters. To evaluate the impact on coastal disasters, two examples of storm surge effects are provided. One is based on GCM output for regional assessment and the other is

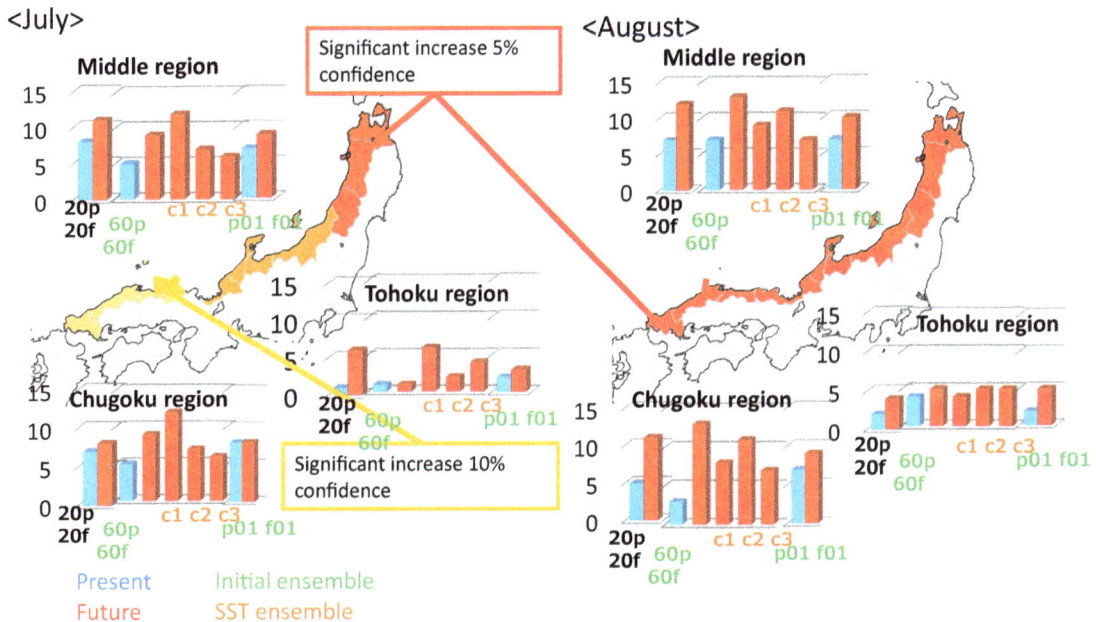

Fig. 5.   Years with the target atmospheric circulation extracted from the 20-km and 60-km AGCM models. Blue and red bars represent the results of the present climate and the future climate, respectively.

based on a pseudo global warming experiment for particular target location. The pseudo global warming experiments is adding future climate condition over particular historical events and gives consistent results to historical extremes. In addition, an example based on the results of a single GCM ensemble experiment using a 20-km AGCM is presented to evaluate the future changes in ocean waves.

The impact of climate change on storm surge was conducted targeting coastal areas of Japan by utilizing the results of climate change projection experiments using 20-km AGCM and downscaling of MRI-NHRCM (5-km RCM) forced by SRES A1B scenario. A storm surge simulation was conducted using a nonlinear shallow water model (Kim *et al.* 2008) to evaluate the intensity of the change in the recurrence probability values forced by wind speed and sea level pressure from 5-km RCM. The future changes of storm surge height were evaluated based on the downscaling of GCM to 5-km RCM

and are discussed on the regional storm surge level.

Figure 7 shows the projected future changes in storm surges forced by 5-km RCM around the western part of Japan (Yasuda *et al.* 2015). The 100 years return value of storm surge height increase in most of the region in the Seto Inland Sea from Shimonoseki to Osaka, Ise Bay including Nagoya and in other areas. The differences of future change of local storm surge may be due to the deviation in the frequency and track of typhoons in the present and future climates, but could be due to the differences in typhoon intensity and track shift (Mori and Takemi 2015). Although many typhoons pass through areas near the Kyushu Island under the present climate, the 20-km AGCM projected that the number of passing typhoons around this area will decrease in the future. In other words, the storm surge deviation decreases in these areas in the future. However, due to the limited number of ensembles and the

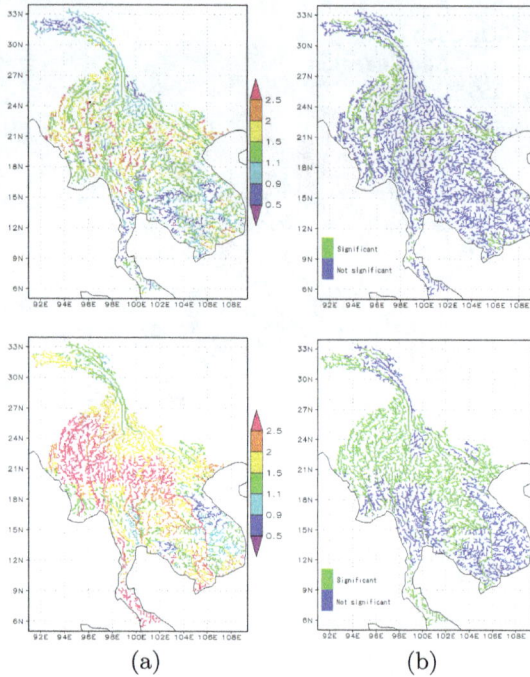

Fig. 6. (a) Change ratio of the mean of the annual maximum discharge of the future climate experiment (2075–2099) to the present (1979–2003) using the 20-km AGCM (MRI-AGCM3.2S) (*top*) and MIROC5 (*bottom*). (b) Statistical significant change at the 5% level (Duong *et al.* 2014).

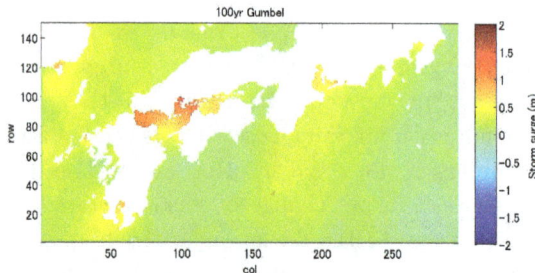

Fig. 7. Changes in the 100 years return values of the storm surge height forced by RCM5 (future climate minus present climate; unit: m) (Yasuda *et al.* 2015).

experimental period, the sample size may be sufficient. Thus, evaluating the reliability and uncertainty of the absolute values is a future challenge.

Figure 8 shows the extreme value analysis of storm surges under the present climate

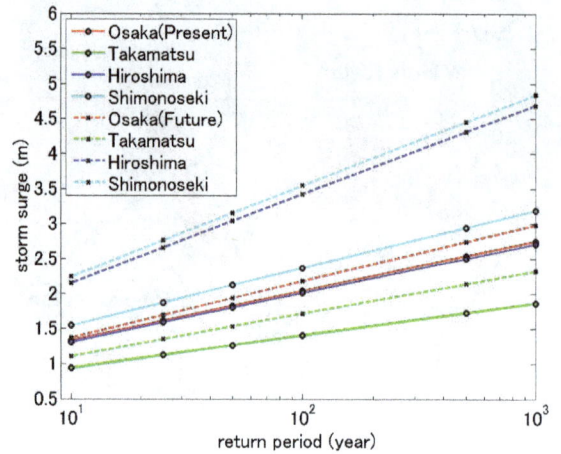

Fig. 8. Deviation of storm surges in the Seto Inland Sea with different recurrence periods. Solid and dashed lines represent the present and future climate, respectively. Each color denotes a different location shown in Fig. 7 (Yasuda *et al.* 2015).

and the future climate in the Seto Inland Sea obtained using 5-km RCM for various return years (10, 25, 50, 100, 500, and 1000 years). Regardless of location, the return period is shorter under the future climate than under the present climate. Thus, the current design level for coastal protection may not provide sufficient strength against storm surges associated with extreme typhoons that may arise under the future climate. Although the qualitative results are reliable in the fact that changes will occur in the future, the sample size for typhoon data in this evaluation is small. For example, the annual mean landed typhoon is only 2–3, therefore a few typhoons can generate significant level of storm surge over 10 years return period. Moreover, a storm surge occurs due to a combination of shape of bay and approaching typhoon track and thus spatial averaging cannot apply to increase sample size. Consequently, the absolute value of the storm surge deviation may change as the sample size increases at target locations.

In addition to studying the changes in storm surges based on the GCM/RCM forcing, the number of storm surge events is limited at regional scale due to low occurrence of extreme storm surges. Another approach is to make impact assessment of storm surge considering the worst-case scenario which analyzes perturbation of typhoon characteristics (e.g. track, intensity) for a specific typhoon effect at a specific location. The worst case scenario was estimated for a historical storm surge caused by Typhoon Vera (1959) using an empirical typhoon model based on a pseudo global warming approach to investigate the changes in inundation area for the urban district of Nagoya in Japan.

The effects of global warming were implemented towards typhoon intensity based on CMIP3 and CMIP5 analysis (Mori and Takemi 2015). The decrease of 10–15 hPa in the central pressure of typhoon is considered in the empirical typhoon model. The perturbation to storm track was conducted by changing the horizontal and rotational motions of the historical typhoon track of Vera. Figure 9 shows the maximum sea storm surge height considering climate change caused by Typhoon Vera like event. Based on the pseudo global warming approach with worst case track, the maximum storm surge height is 4.67 m, which is an increase of 1.14 m and 0.48 m in the storm surge deviation compared with historical record and excluding the worst case track. Therefore, both the global warming and the worst case scenario have significant impact on maximum storm surge height. The ratio of global warming and worst case track can be changed depending on the target region and event. It is important to know the worst-case scenario of storm surge using this type of perturbation approach due to the sensitivity of storm surge to the typhoon track.

3.2. *Ocean waves*

Study of ocean wind waves in stormy conditions is important for coastal, ocean, and environment engineering. The wave hindcasts in the Atlantic Ocean show more significant wave height increase in the region off the Canadian coast and the northwest of Ireland but less significant change in the North Sea and in the region off the Scandinavian coast. Although, the dynamic projection of wave climate change was discussed in IPCC AR5, the number of projection is quite limited and uncertainty of projection is large at present (e.g., Hemer *et al.* 2012).

We have conducted future changes of wave climate since 2007. An example of an ensemble

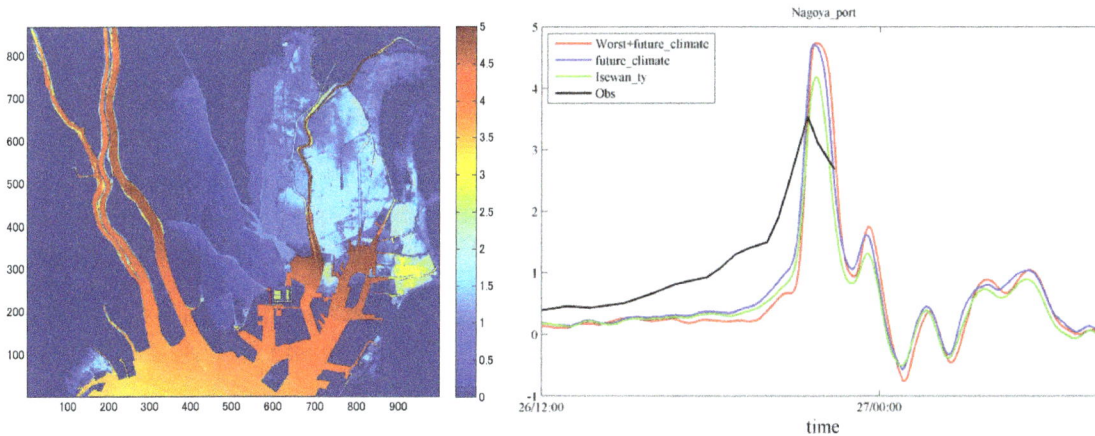

Fig. 9. Maximum storm surge height and inundation area of worst case scenario of typhoon Vera in Nagoya and Ise Bay considering global warming (*Left*: maximum storm surge height; *Right*: time series of storm surge height at Nagoya Port, black line: historical run, blue line: pseudo global warming, red line: pseudo global warming with worst case track) (Shibutani *et al.* 2014).

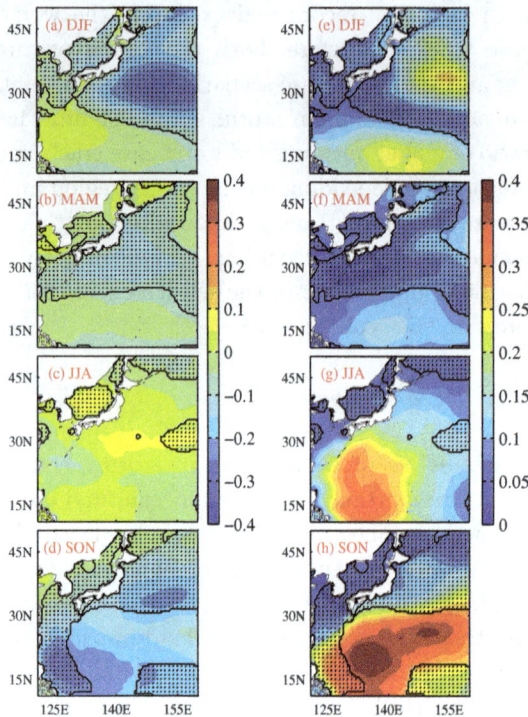

Fig. 10. Future change of in the mean seasonal wave height [m]. Dotted areas represent sea areas whose change width is indicated by the same sign over the four experiments (Shimura *et al.* 2015).

wave projection forced by different future condition with changes in sea surface temperature (SST) is presented based on our recent results (Shimura *et al.* 2015). The dynamic wave projection was conducted by the spectral wave model WaveWatch III version 3.14 (Tolman 2009). The forcing from sea surface winds are the results from a 60-km AGCM at the Meteorological Research Institute, as in the previous sections. This is a time-slice experiment forced by SRES A1B scenario targeting the period 1979 to 2009 for the present climate and 2075 to 2099 for the future climate.

Figure 10 shows the future change of the seasonal mean wave height in East Asia. The mean wave height changes caused by warmer atmospheric condition with the four patterns of SST show positive values in almost all regions of the world. The seasonal mean wave

height changes were evaluated in four seasons, December-January-February (DJF), March-April-May (MAM), June-July-August (JJA), and September-October-November (SON), respectively. The average changes and their width vary significantly by season and sea area. The seasonal characteristics can be summarized below.

DJF: The mean wave height significantly increases (up to approximately 0.4 m) in the center of the Northern Pacific, but decreases near 30°N in the western region.

SON: There is a significant decrease in the mean wave height in the Western North Pacific. All experiments projected the same trend in the changes in width for 55.2% of the total sea areas; the change in wave height is approximately 0 to 0.4 m.

In summary, the ensemble wave projection indicates the same trend in the width change for 62.0% of the sea areas over the globe. However, the uncertainty of projections is large in regional scales over the northwestern Pacific. The change width is approximately 7.5% of the mean wave height in the present condition. The width changes in the wave height off southern Japan and East Asia is significant are JJA and SON, but the projections in the sea area have a large uncertainty due to SST.

4. Water Resources

Changes of precipitation characteristics in Japan give significant impact on water resources. The Whole Japan Area Water Resource Model is a distributed model made up of five submodules (rice growth, land surface, irrigation, stream flow, and dam operation) (Fig. 11). This model is designed to simulate hydrological circulation considering not only the natural systems (e.g., runoff, stream flow, and rice growth) but also the human activities (e.g., irrigation and dam operations) using meteorological forcing data and land surface parameters. The land surface, irrigation,

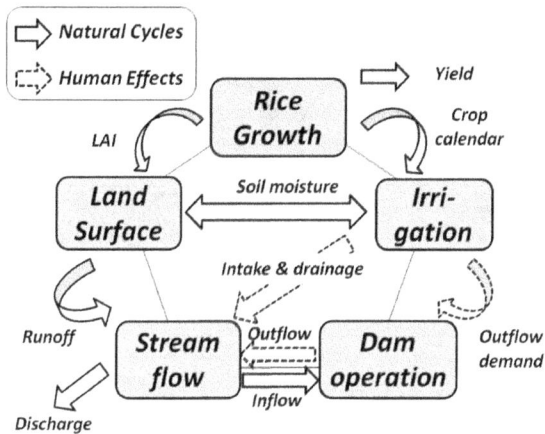

Fig. 11.   Structure of the water resource model (Kotsuki et al. 2013a).

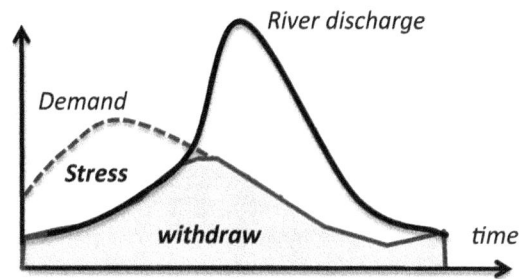

Fig. 12.   Schematic image of water stress calculation upon considering seasonality based on the CWD index. (Kotsuki *et al.* 2013a).

and rice growth modules deal with water balance and carbon dioxide assimilation, which are vertical to the land surface, while the stream flow and dam operation modules analyze horizontal water circulation.

The rice growth module is based on the model proposed by Iizumi *et al.* (2009) and follows the formula proposed by Nakagawa *et al.* (1995) for the development index (DVI). For the land surface and irrigation modules, SiBUC (Tanaka 2004) was used. To analyze stream flows, the kinematic wave method was used. The dam operation module was applied to 1,231 dams with storage capacities of $1,000,000$ m^3 or more. Since the objective here is to model the water resources in all of Japan to identify areas vulnerable to climate change, generalized operation rules based on the purpose of dams were applied instead of the operation rules specific to each dam. The dam operation module with the operation purposes divided into flood control, water utilization, and multi-purpose categories, was used to reduce the peak flow and release the water demanded in the downstream area. The outflow demand was determined by the irrigation module for agricultural water use or based on statistical data for domestic and industrial water use.

Based on the cumulative withdrawal to demand ratio (CWD) proposed by Hanasaki *et al.* (2008), a water stress evaluation was conducted on a weekly basis by taking into consideration the seasonality of river discharge (Fig. 12). Since Japan is expressed as a set of approximately 16,000 river basins in this model, the water stress was calculated for all the basins. The water supply in the river basin was set to the discharge at the river mouth. The water demand was calculated by adding the demand for irrigation and padding water to the industrial and domestic demands derived from statistical data. The industrial and domestic demands were assumed to remain constant throughout the year. However, the demands for irrigation and padding water had seasonality. As the infiltrated water was assumed to return to the water circulation system, infiltration during the ponding period and the water transport efficiency were not added to the agricultural water demand. The CWD was calculated from the weekly water balance in order to account for the time it takes for infiltrated water to return to the river. Although the model calculated the timing of padding, which requires plenty of water, it was considered to be the same day for each prefecture, when in reality it varies by a few days. This is another reason to evaluate CWD using a weekly time step.

A river discharge calculation was conducted under three climatic periods totaling 75 years:

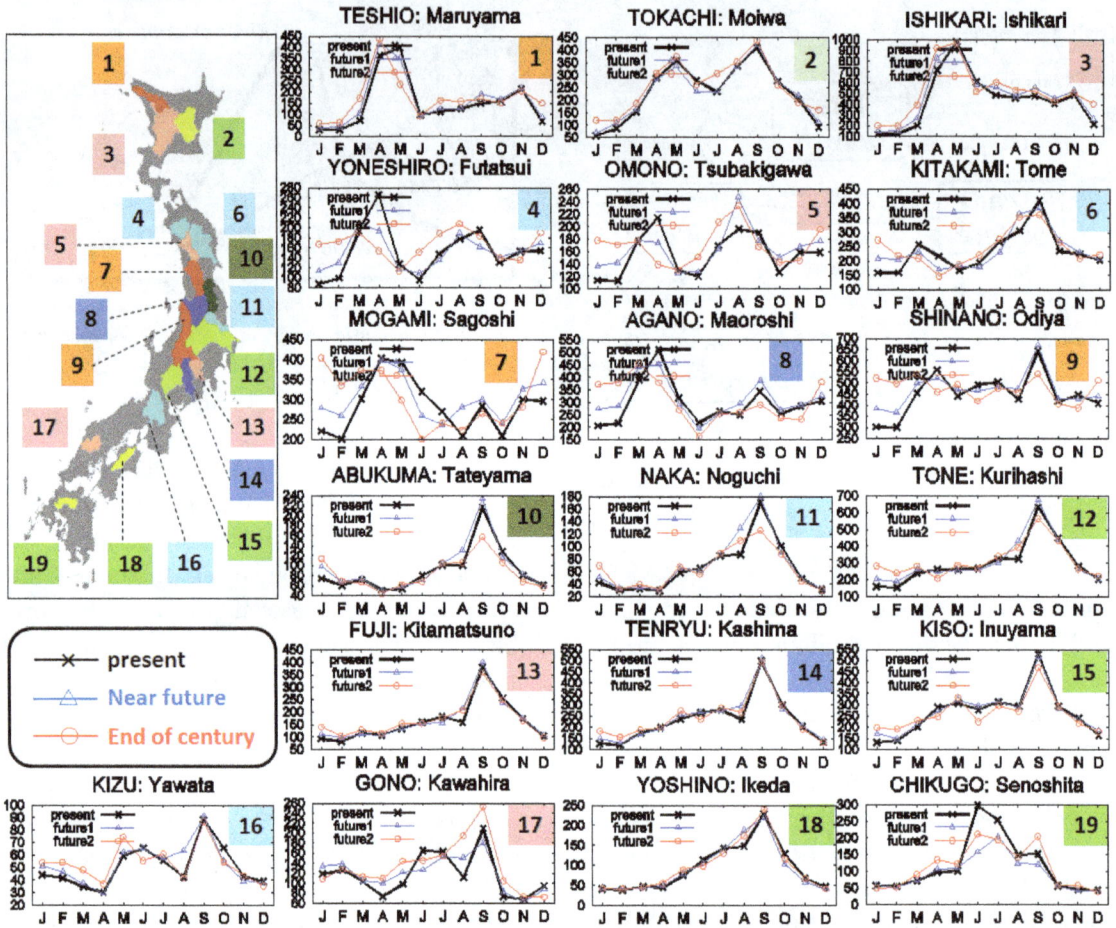

Fig. 13. Climatology of monthly river discharge of 19 first class rivers in Japan (m³/s). Black, blue, and red lines represent the river discharge under the present climate (1979 to 2003), the near-future climate (2015 to 2039), and the end-of-the-21st century climate (2075 to 2099), respectively. (Kotsuki *et al.* 2013a).

current, near future, and the end of the 21st century. Figure 13 shows the climatology of the monthly river discharge of 19 first class rivers in Japan for the three periods. The seasonality of river discharge changes noticeably in the Mogami River, Omono River, Kitakami River, Yoneshiro River, Shinano River, and Agano River. These rivers are all situated in heavy snowfall areas on the Japan Sea side of northern Honshu. Comparing the discharge for these rivers under the present climate (black line) with the end-of-the-21st century climate (red line), the discharge is projected to increase from

December to March and decrease from April to May. December to March is considered to be the snowfall season under the present climate, and the change from snowfall to rainfall and the increase of precipitation due to the air temperature rise lead to an increase in the river discharge in this season. It also affects the river discharge in the current snow melt season (from April to May) due to the decrease in the amount of accumulated snow.

The Nash coefficient was used to detect the degree of change in the seasonality of the river discharge by comparing the river discharge

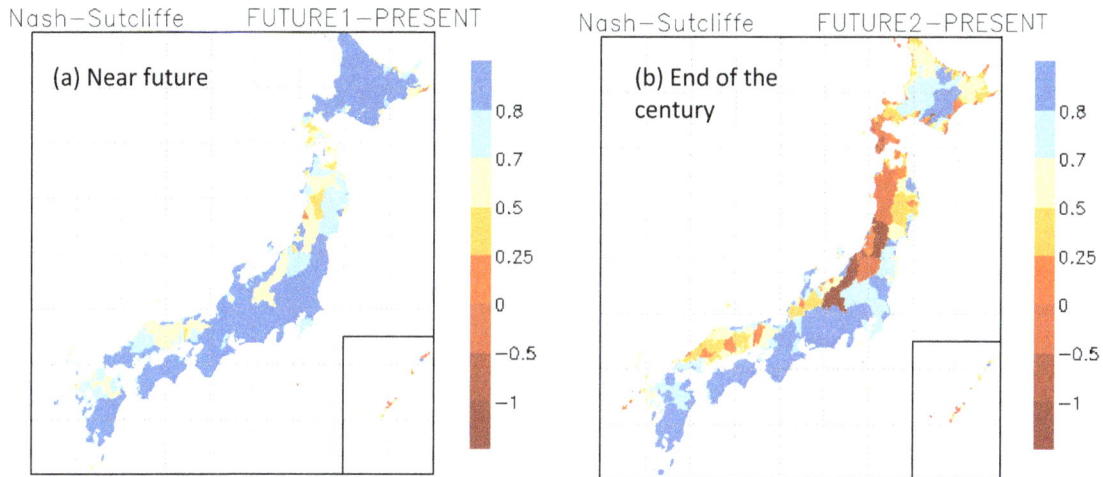

Fig. 14.   Changes in river flow regime due to climate change. Map (a) shows the Nash coefficient derived from the climatology of the monthly river discharge under the near-future climate and the present climate. Map (b) shows the Nash coefficient derived from the climatology of the monthly river discharge under the end-of-the-21st century climate and the present climate. Warmer colors indicate a more significant change in the river flow regime. (Kotsuki *et al.* 2013b).

under the near-future/end-of-the-century climate with the river discharge under the present climate. A Nash coefficient greater than 0.8 indicates a minor change, and a smaller Nash coefficient indicates a larger change. Figure 14 shows the Nash coefficients calculated using the climatology of river discharge for each river basin at the outlet. The seasonality of discharge changes prominently, especially for the end-of-the-21st century climate. The change is noticeable in snowy regions under the present climate, such as Hokkaido, Tohoku, Niigata, and the Chugoku regions on the Japan Sea side. This decrease in snowfall can be associated with climate change.

From the viewpoint of water resource utilization, the increase in precipitation associated with the accelerated water circulation has the benefit of an increase in available water resources. However, climate change, which causes an increase of evapotranspiration, also increases the agricultural water demand. Thus, water stress needs to be evaluated by taking into consideration the effects of benefits and drawbacks. Under the present, near-future, and end-of-the-21st century climates, the CWD's

for water stress and the mean annual water resources (precipitation minus evapotranspiration) in each river basin were calculated. Figure 15 compares (a) the near-future climate and (b) the end-of-the-21st century climate to the present climate. Maps (a.1) and (b.1) show the changes in water resources, while maps (a.2) and (b.2) depict the changes in water stress. Charts (a.3) and (b.3) plot the data from 601 river basins with a catchment area of 50 km^2 or more where the horizontal axis represents the change in water resources and the vertical axis represents the change in water stress. The water resources decrease significantly and the water stress increases toward the near-future climate in the vicinity of Kyushu. On the other hand, in regions from Hokkaido to Chubu on the Japan Sea side, the water resources increase and are not accompanied by effects such as water stress mitigation. In many river basins, the water resources decrease toward the end-of-the-21st century climate. In particular, this trend of an increase in water resources is a characteristics of heavy snowfall regions such as the Hokkaido, Tohoku, and Chugoku regions on the

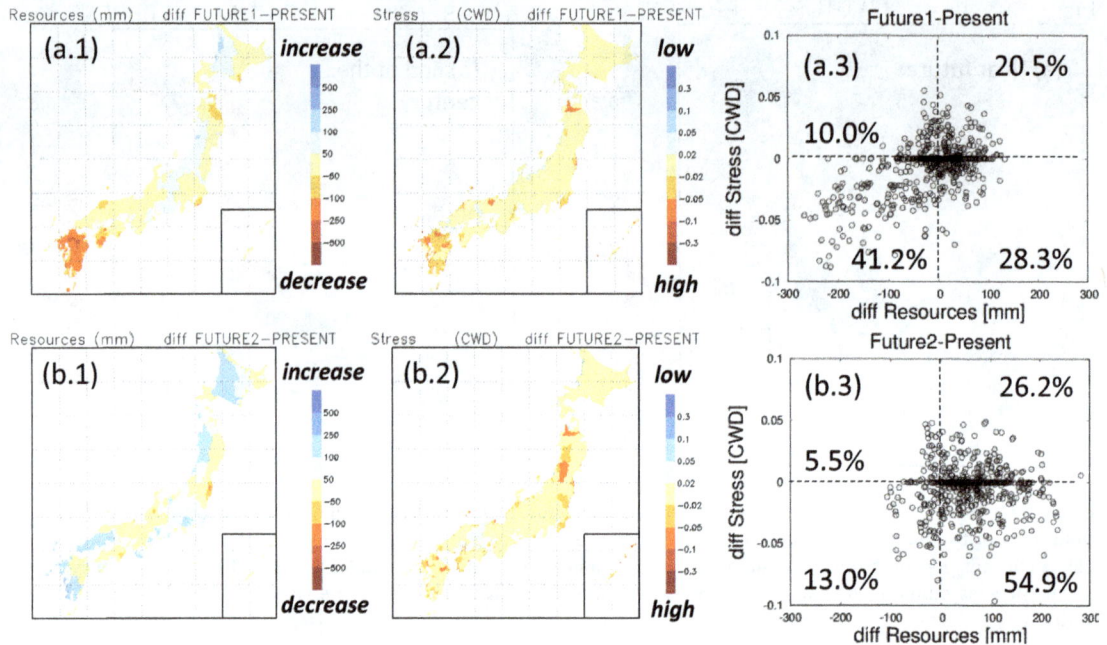

Fig. 15. Changes in the water resources (a.1, b.1) and water stress (a.2, b.2) caused by climate change. Relationship between the change in water resources and water stress in each river basin is shown in charts (a.3, b.3) where the value in each quadrant indicates the percentage of basins belonging to that quadrant. Difference between the near-future climate and the present climate (a.1, a.2, a.3) and the difference between the end-of-the-21st century climate and the present climate (b.1, b.2, b.3) are shown (Kotsuki *et al.* 2013b).

Japan Sea side. On the other hand, there are a small number of river basins where water stress decreases and, in the Tohoku region, water stress increases despite an increase in water resources. Another trend is that the water stress increases in rivers with a small basin area, which can be attributed primarily to the increase in the rainfall intensity associated with global warning.

Without dam, runoff water generated in the mountains area reaches the sea quickly and has little chance to be utilized. The lack of dams makes it more difficult to manage the water resources in the small basins. It should be also noted that the correspondence between the change in water resources and the change in water stress differ between the near-future climate and the end-of-the-21st century climate. The percentage of river basins where the increase/decrease of water resources matches the increase/decrease of water stress (i.e., the

river basins belonging to the first quadrant and the third quadrants) is 61.7% under the near-future climate, but is only 39.2% under the end-of-the-21st century climate.

5. Conclusions

The monsoon and tropical cyclone, typhoon are major sources of natural hazards for water resources in East Asia and Southeast Asia. Thus, it is important to know the changes of these phenomena and the related impacts on natural hazard intensity and water supply. Future changes of monsoon and tropical cyclone characteristics due to climate change is one of key issue of impact assessment in the East Asia and Southeast Asia regions. This chapter summarizes the latest findings on climate change and its effects on natural disasters and water resources in East Asia, Southeast Asia and Japan based on the latest high resolution 20-km and 60-km GCMs developed by the Japan Metrological Research Institute.

Future change in the frequency of precipitation will be higher in the western part of Japan and the areas along the Pacific coast. In general, a moist tongue and a low-level jet are observed on the south side of the Baiu frontal zone and are accompanied by heavy rain where the moisture flux is significant. The amount of precipitable water is greater than 35% in the region from southern China to off the southern coast of Honshu and Kyushu, and the moisture inflow due to southwesterly winds converges in Kyushu and in areas along the Pacific coast.

A statistically significant increase in the annual maximum daily discharge for the Irrawaddy River in East Asia is projected by the MRI-AGCM and MIROC5 models. These climate models indicate increasing trends in the upper reaches of the Salween and Mekong Rivers. However, in the central regions of Vietnam, the annual minimum daily discharge show a decrease.

The storm surge in the western part of Japan was projected based on direct use of high resolution downscaling results. The long-term return period of storm surge in the western part of Japan will be increased associated with extreme typhoons in the future climate. The worst-case scenario of storm surge in Ise bay of Japan was estimated for the historical storm surge of Typhoon Vera (1959) using a pseudo global warming approach. Both the global warming and the worst-case scenarios have significant impacts on maximum storm surge height.

The ensemble wave projection was conducted on the global scale. The future mean wave height indicates the decrease trend in the width change for 62.0% of the sea areas. The projections in low latitude sea areas in the northwestern Pacific and high latitude sea areas in the South Pacific vary by model. The change width is approximately 7.5% of the mean wave height at present condition. The similar magnitude of negative changes can be expected in the wave height off southern Japan.

A river discharge calculation was conducted under three climatic periods totaling 75 years. The river discharge in Japan is projected to increase from December to March and decrease from April to May. December to March is considered to be the snowfall season under the present climate, and the change from snowfall to rainfall and the increase of precipitation due the air temperature rise lead to an increase in the river discharge in this season. Also affected is the river discharge in the current snow melt season (from April to May) due to the decrease in the amount of accumulated snow. From the viewpoint of water resource utilization, the increase in precipitation associated with the accelerated water circulation has the benefit of an increase in available water resources. However, climate change, which causes an increase of evapotranspiration, also increases the agricultural water demand. Thus, water stress needs to be evaluated by taking both the effects of benefits and drawbacks into consideration. The percentage of river basins where the increase/decrease of water resources matches the increase/decrease of water stress is 61.7% under the near-future climate, but is only 39.2% under the end-of-the-21st century climate.

References

Duong, D. T., Y. Tachikawa, and K. Yorozu, 2014: Changes in river discharge in the Indochina Peninsula region projected using MRI-AGCM and MIROC5 datasets. *Journal of Japan Society of Civil Engineers, Ser. B1 (Hydraulic Engineering)*, **70**(4), I_115–I_120.

Hemer, M. A., Y. Fan, N. Mori, A. Semedo, and X. L. Wang, 2013: Projected changes in wave climate from a multi-model ensemble. *Nat. Clim. Change*, 6p. doi:10.1038/nclimate1791.

Iizumi, T., M. Yokozawa, and M. Nishimori, 2009: Parameter estimation and uncertainly analysis of large-scale crop model for paddy rice: Application of a Bayesian approach. *Agri. For. Meteorol.*, **149**, 333–348.

IPCC WGI and II, 2013: *Intergovernmental Panel on Climate Change Fifth Assessment Report*, WGI and II.

Hanasaki, N., S. Kanae, T. Oki, K. Masuda, K. Motoya, N. Shirakawa, Y. Shen, and K. Tanaka, 2008: An integrated model for the assessment of global water resources-Part 2: applications and assessments. *Hydrol. Earth Syst. Sci.*, **12**, 1027–1037.

Kamiguchi, K., O. Arakawa, A. Kitoh, A. Hamada, and N. Yasutomi, 2010: Development of APHRO_JP, the first Japanese high-resolution daily precipitation product for more than 100 years. *Hydrol. Res. Lett.*, **4**, 60–64.

Kim, S. Y., T. Yasuda, and H. Mase, 2008: Numerical analysis of effects of tidal variations on storm surges and waves. *Appl. Ocean Res.*, **30**(4), 311–322.

Kitoh, A., T. Ose, K. Kurihara, S. Kusunoki, and M. Sugi, 2009: Projection of changes in future weather extremes using super-high-resolution global and regional atmospheric models in the KAKUSHIN Program: Results of preliminary experiments. *Hydrol. Res. Lett.*, **3**, 49–53.

Kobayashi, S., Y. Ota, Y. Harada, A. Ebita, M. Moriya, H. Onoda, K. Onogi, H. Kamahori, C. Kobayashi, H. Endo, K. Miyaoka, and K. Takahashi, 2015: The JRA-55 Reanalysis: General specifications and basic characteristics. *J. Meteorol. Soc. Jpn.*, **93**, 5–48.

Kotsuki, S., K. Tanaka, T. Kojiri, and T. Hamaguchi, 2012: Parameter identification of distributed runoff model using the particle swarm optimization method. *Annual Journal of Hydraulic Engineering*, **56**, 523–528. (in Japanese)

Kotsuki, S., K. Tanaka, and T. Kojiri, 2013a: Estimation of climate change impact on Japanese water resources part 1: The development of a Japanese water resource model. *J. Jpn. Soc. Hydrol. Water Resour.*, **26**(3), 133–142. (in Japanese)

Kotsuki, S., K. Tanaka, and T. Kojiri, 2013b: Estimation of climate change impact on Japanese water resources part 2: water demand-supply balance, rice yield changes, and an adaptation plan. *J. Jpn Soc. Hydrol. Water Resour.*, **26**(3), 143–152. (in Japanese)

Mori, N., T. Yasuda, H. Mase, T. Tom, and Y. Oku, 2010: Projection of extreme wave climate change under the global warming, Special Collections of Weather extreme event projections and their impact assessments. *Hydrol. Res. Lett.*, **4**, 15–19.

Mori, N., M. Kato, S. Kim, H. Mase, Y. Shibutani, T. Takemi, K. Tsuboki, and T. Yasuda, 2014: Local amplification of storm surge by Super Typhoon Haiyan in Leyte Gulf. *Geophys. Res. Lett.*, **41**(14), 5106–5113. doi:10.1002/2014GL060689.

Mori, N. and T. Takemi, 2015: Impact assessment of coastal hazards due to future changes of tropical cyclones in the North Pacific Ocean. *Weather and Climate Extremes*, **11**, 53–69.

Nakagawa, H. and T. Horie, 1995: Modelling and prediction of developmental process in rice : II. A model for simulating panicle development based on daily photoperiod and temperature. *Jpn. J. Crop Sci.*, **64**(1), 33–42. (in Japanese)

Nakakita, E., H. Kusano, and S. Kim, 2015: Prediction on appearance frequency of atmospheric characteristics causing localized heavy rainfall during Baiu season under climate change. *Journal of Japan Society of Civil Engineers, Ser. B1 (Hydraulic Engineering)*, **71**(4), I_373–I_378. (in Japanese)

Okada, Y. and K. Yamazaki, 2012: Climatological evolution of the Okinawa Baiu and differences in large-scale features during May and June. *J. Climate*, **25**, 6287–6303.

Sampe, T. and S.-P. Xie, 2010: Large-scale dynamics of the Meiyu-Baiu rainband: Environmental forcing by the westerly jet. *J. Climate*, **23**, 113–134.

Shimura T., N. Mori, and H. Mase, 2015: Future projection of ocean wave climate: analysis of SST impacts on wave climate changes in the Western North Pacific. *J. Climate*, **28**, 3171–3190.

Shimura T., N. Mori, and H. Mase, 2015: Future projections of extreme ocean wave climates and the relation to tropical cyclones. *J. Climate*, doi:10.1175/JCLI-D-14-00711.1.

Shibutani, Y., S. Y. Kim, T. Yasuda, N. Mori, and H. Mase, 2014: Sensitivity of future tropical cyclone changes to storm surge and inundation: Case study in Ise Bay, Japan. *Coast. Eng. Proceedings*, **1**(34), 27.

Takayabu, I., K. Hibino, H. Sasaki, H. Shiogama, N. Mori, Y. Shibutani, and T. Takemi, 2015: Climate change effects on the worst-case storm surge: a case study of Typhoon Haiyan. *Environ. Res. Lett.*, **10**, 064011.

Takemi, T., 2012: Projected regional-scale changes in atmospheric stability condition for the development of summertime convective precipitation in the Tokyo metropolitan area under global warming. *Hydrol. Res. Lett.*, **6**, 17–22.

Takemi, T., S. Nomura, Y. Oku, and H. Ishikawa, 2012: A regional-scale evaluation of changes in environmental stability for summertime afternoon precipitation under global warming from super-high-resolution GCM simulations: A study for the case in the Kanto Plain. *J. Meteorol. Soc. Jpn.*, **90A**, 189–212.

Tanaka, K., 2004: Development of the New Land Surface Scheme SiBUC Commonly Applicable to Basin Water Management and Numerical Weather Prediction Model. Doctoral Dissertation, Graduate School of Engineering, Kyoto University, Kyoto, 289p.

Tolman, H. L., 2009: User manual and system documentation of WAVEWATCH III TM version 3.14. Technical note, MMAB Contribution, 276.

Yasuda, T., N. Katahira, N. Mori, and H Mase, 2015: Development of statistical bias correction method for climate model typhoons and ensemble future storm surge projection due to climate change. *Journal of Japan Society of Civil Engineers, Ser. B2 (Coastal Engineering)*, in press (in Japanese).

Chapter 8

Climate Change and Stream Temperature in the Willamette River Basin: Implications for Fish Habitat

Heejun Chang* and Eric Watson

Department of Geography, Portland State University,
Portland, OR 97201, USA
*changh@pdx.edu
www.pdx.edu/geography

Angela Strecker

Department of Environmental Science and Management,
Portland State University,
Portland, OR 97201, USA

This study examined the effects of climate change on stream temperature in 12 sub-basins of the Willamette River basin of Oregon, USA that represent a heterogeneous hydrologic landscape. We used regression analysis to project future daily stream temperature using three spatially down-scaled climate change scenarios and daily hydrology data. Daily maximum stream temperature was best explained by a combination of the 15 day moving average of daily air temperature and daily streamflow. Together with a reduction in summer streamflow, rising air temperature may increase stream temperature 1–4°C by the 2080s. Stream temperature rises modestly in groundwater-fed streams in high elevations, illustrating that groundwater-fed streams could be more resilient to climate warming and reduced surface flow. In contrast, lowland surfacewater-fed streams that lack riparian vegetation will be most vulnerable to climate change since stream temperature is projected to increase most (up to 4°C) under the highest climate change scenario. Such changes will have negative consequences for cold-water fishes that are already listed as endangered species. The exceedance probability of threshold temperature (17.8°C and 21°C) increased substantially, particularly in the lowland areas, suggesting future management practices for threatened and endangered fish species should focus on improving thermal conditions in those reaches.

1. Introduction

Stream temperature is a barometer of aquatic ecosystem health, affecting habitat quality for diverse fish species (Caissie 2006; Whitehead *et al.* 2009). Warmer stream temperature can accelerate instream biogeochemical cycles (Webb *et al.* 2008), increasing the concentrations of pollutants and their toxicity to fish (e.g., Patra *et al.* 2015). In the Pacific Northwest (PNW) region of the USA, salmonids are a cultural icon; however, several species, including Chinook salmon (*Oncorhynchus*

tshawytscha), coho salmon (*Oncorhynchus kisutch*), chum salmon (*Oncorhynchus keta*), sockeye salmon (*Oncorhynchus nerka*), and steelhead (*Oncorhynchus mykiss*) are listed as threatened or endangered on the US Endangered Species Act (USFWS 2016). The occurrence of salmonid species is highly dependent on the presence of cold water habitat (Richter and Kolmes 2005; Isaak *et al.* 2016).

The generalized life cycle of salmonids includes the hatching and rearing of juvenile fish in freshwaters prior to migrating to the ocean, where they mature and grow until they return

Bridging Science and Policy Implication for Managing Climate Extremes
Edited by Hong-Sang Jung and Bin Wang

to freshwater habitats to spawn (Quinn 2005). Temperature can affect all parts of this life cycle. As juveniles, temperature can influence the process of smoltification (preparing the fish for the transition from freshwater to seawater), growth, the timing of migration to the marine environment, and can induce thermal stress (McCullough *et al.* 2001; Richter and Kolmes 2005). High water temperatures can slow swimming and impede adult migration (Schreck *et al.* 1994, McCullough *et al.* 2001). Spawning may also be negatively affected by high temperatures, including reduced fertilization rates and embryo survival, increased egg mortality, and slower egg maturation (Richter and Kolmes 2005).

The thermal criteria for streams in the state of Oregon is 17.8°C (Oregon Department of Environmental Quality 2010), while for the spawning season, salmonids require temperatures <13°C. The upper optimal limit for juvenile rearing is 16°C (Richter and Kolmes 2005), and the upper optimal limit for survival is 21°C (Hicks 2000). Thus, the accurate simulation of stream temperature is important for sustainable water resources and fish management.

According to recent climate change simulation results in the Willamette River basin (WRB) that used 13 GCMs, some sub-basins of the WRB are projected to have reduced summer precipitation and streamflow more than 20% by the end of the 21st century (2070–2099) (Chang and Jung 2010). Stream temperature has increased approximately 0.1–0.2°C/decade during the last 30–40 years in the Columbia River basin (Isaak *et al.* 2012).

Together with a rise in air temperature, reduction in summer streamflow is likely to increase summer stream temperature. In a study of the Columbia River basin, Ficklin *et al.* (2014) projected an increase in summer stream temperature up to 7°C in an interior Columbia River basin in the 2080s under the RCP 8.5 scenario, while Hill *et al.* (2014) estimated an increase in annual stream temperatures of 3°C in the Cascade Mountains of Oregon. In a study of a small urbanizing basin in Oregon, Chang and Lawler (2011) projected an increase in the occurrence of threshold critical temperatures in the 2050s and 2080s using synthetic climate change scenario of the warming of 3°C in air temperature and 10% streamflow reduction. Together with a decline in summer streamflow, rising air temperatures are projected to increase stream temperatures by 2°C by the end of the 21st century. However, in a heterogeneous river basin such as the Willamette River, the response of each stream to changes in a range of future hydroclimate has not been fully investigated. Our current analysis seeks to fill in the gap by assessing the effects of air temperature and streamflow on stream temperature in selected representative streams under different warming scenarios.

2. Data and Methods

2.1. *Study area*

The 29,700 km^2 Willamette River basin (WRB) is a relatively water-rich basin located in the Pacific Northwest of the USA. The basin receives approximately 970 mm of annual precipitation with distinct dry summer and wet winter seasons. The WRB exhibits diverse hydrological landscape regions from the Coast Range in the west to the Cascades in the east, where the Willamette valley is located between the two physiographic regions (Fig. 1). The valley is an elongated structural and erosional lowland filled with flows of Columbia River Basalt and younger unconsolidated sediment (Laenen and Risley 1997). The Coast Range, composed of tertiary marine sandstone, shale, and mudstone interbedded with basalt flows and volcanic debris, occupies approximately 20 percent of the basin. The Cascade Range accounts for more than 50 percent of the basin area and is further divided into two regions. The High Cascades, located in the far eastern part of the basin, have relatively young volcanic rocks such as Quaternary basaltic and andesitic lava flows

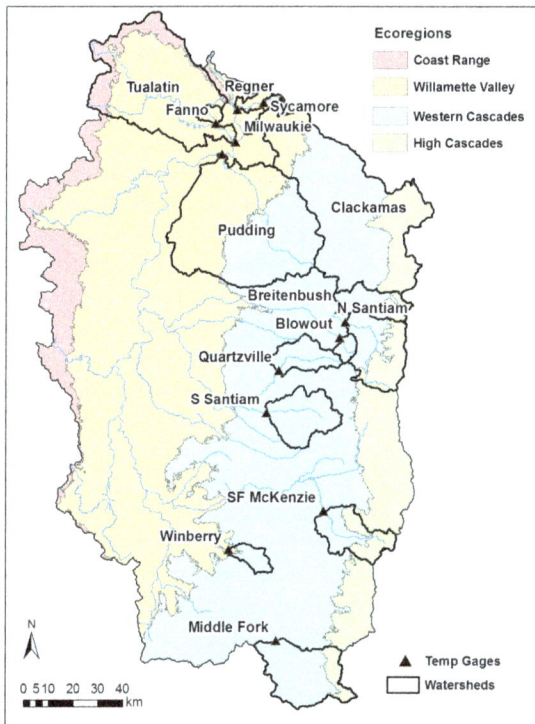

Fig. 1.   Study basins in the Willamette River basin.

and associated deposits. The Western Cascades have relatively old dissected Tertiary basaltic and andesitic rocks and associated volcanic debris.

Considering its diverse physiographic regions with different geological settings that affect the hydrologic regime of each sub-basin (Chang and Jung 2010), the WRB makes an excellent laboratory to investigate each sub-basin's stream temperature response to changes in climate change.

We selected 12 sub-basins located in three physiographic regions that represent three ecoregions. These sub-basins were chosen because they have at least three years of continuous daily stream temperature and streamflow data and have not been affected by dams or excessive human diversions of water. They also represent different elevation and land cover gradients being located in the three physiographic regions.

2.2. Data

Both observed and future climate data were used for our analysis. Gridded daily maximum and minimum air temperature and daily precipitation data were obtained from the University of Idaho Interactive Numeric and Spatial Information Data Engine (INSIDE) (Abatzoglou 2013). Based on observed weather data and topography, these gridded data were spatially interpolated at a resolution of 4 km^2. Future daily maximum and minimum temperatures and daily precipitation data were also obtained from the INSIDE. We used three GCMs — GFDL-ESM2M RCP 4.5, MIROC5 RCP 8.5, and HadGEM2-ES RCP 8.5. The three scenarios project different trajectories for global climate and they differ based on a set of assumptions for the future of human resource use and land management practices. MIROC5 was used as a reference scenario, which assumes little change in the rates and methods in resource management, while GFDL and HadGEM represent lower and higher intensity trajectories, respectively (see Fig. 2). Both MIROC5 and GFDL-ESM2M GCMs project slight increases in annual precipitation, while HadGEM2-ES projects slight decreases in annual precipitation. Mean annual temperatures are projected to increase in all three GCMs (0.8–3.2°C).

For historical hydrology data, mean daily streamflow data were obtained from the United States Geological Survey (USGS 2015). Future daily streamflow data were gathered from the simulated streamflow by the Willamette Hydrologic Model (WHM). The WHM, developed as part of the Willamette Water 2100 project (http://water.oregonstate.edu/ww2100/) is based on a conceptual model HBV-light (Seibert and Vis 2012) and has shown reasonable simulations for tracking daily streamflow in selected sub-basins of the Willamette River basin (R^2 ranging from 0.25 to 0.49), including the majority of the stations used in the current analysis (Jaeger et al. 2017).

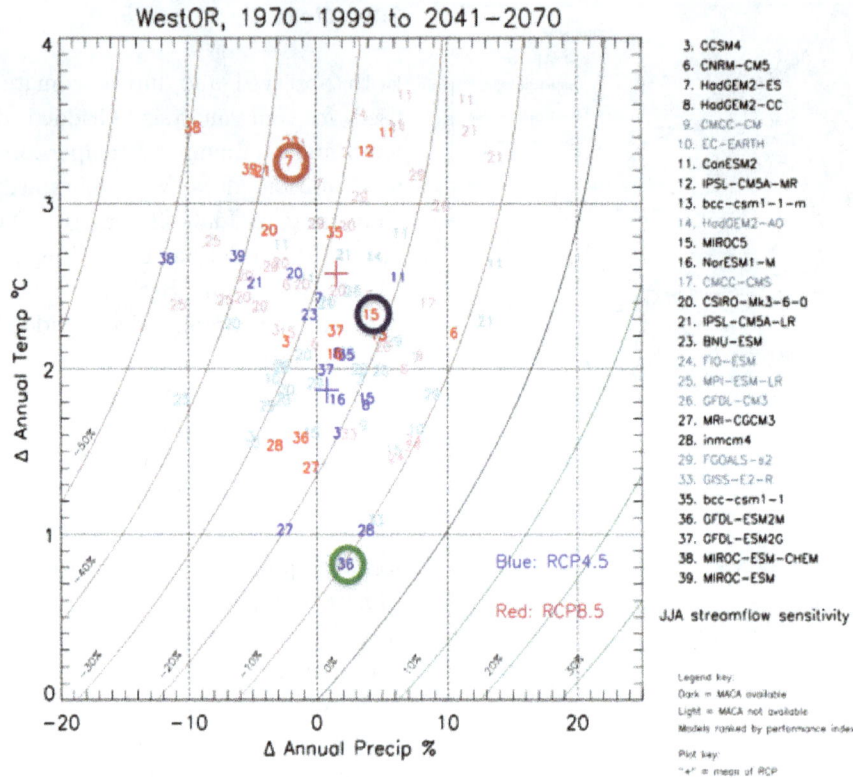

Fig. 2. Change in annual precipitation and temperature in Western Oregon. Circles represent the three climate change scenarios used in this paper.

Daily stream temperature data were obtained from the same USGS gaging stations that have daily flow records. Only summer months' daily maximum and minimum temperature data were used (Table 1).

2.3. Methods

Daily air temperature and streamflow data were used to estimate daily maximum stream temperature for each site. We used time series multiple regression analysis by taking into account the effects of temporal lags in streamflow and air temperature (e.g., 15-day average of daily maximum air temperature, 15-day average of daily streamflow) on stream temperature. Regression models were generated in R using the multivariate regression function, which automates the process of independent variable selection.

The coefficients of multiple regression equations were then used to project possible future changes in daily stream temperature from 2010 to 2199 (Table 1). Daily streamflow output from the Willamette Hydrologic Model runs for these three scenarios was used alongside the climate data to project future daily stream temperatures.

Like any statistical analysis, projecting future stream temperatures beyond the historical period has inherent uncertainty because the stationarity assumption may be invalid. However, given that other future meteorological and landscape drivers (e.g., solar radiation and land covers) that affect stream temperature are largely unknown, regression models based on empirical relationships between air temperature and stream temperature have been successfully applied for projecting future stream temperature

Table 1.   Time frames for available historical stream temperature gauging stations and regression model coefficients.

USGS ID	Location description	Years (# of samples)	Coefficients Intercept	$T_{max\ 15}$	T_{min}	Q15
14180300	Blowout	1998–2011 (5110)	1.82	0.38	0.23	−0.00207
14179000	Breitenbush	1997–2011 (5110)	3.46	0.30	0.18	−0.00078
14206950	Fanno	2002–2011 (3650)	0.92	0.53	0.33	0.00528
14211550	Johnson-Milwaukie	1998–2011 (5110)	2.16	0.50	0.26	−0.00087
14211400	Johnson-Regner	1999–2011 (4745)	0.47	0.50	0.35	0.01609
14211500	Johnson-Sycamore	1998–2011 (5110)	0.17	0.53	0.36	0.00865
14144800	Middle Fork Willamette	1979–1987 (3285)	2.16	0.36	0.21	−0.00043
14178000	North Santiam	1998–2011 (5110)	2.92	0.30	0.18	−0.00067
14185900	Quartzville	2008–2011 (1460)	1.08	0.52	0.22	−0.00061
14159500	South Fork McKenzie	2004–2011 (2920)	3.39	0.22	0.15	−0.00058
14185000	South Santiam	2009–2011 (1460)	1.61	0.45	0.32	−0.00026
14150800	Winberry	2008–2011 (1460)	0.50	0.49	0.26	—

in previous studies (e.g., Mohseni *et al.* 1999; Wu *et al.* 2012; van Vlict *et al.* 2013; Calwell 2014).

Analysis of the future stream temperatures focused mainly on the projected maximum 7-day average daily maximum stream temperature (7DAD T_{max}) for each year from 2010 to 2099. 7DAD T_{max} is calculated by averaging daily maximum stream temperatures three days before and after the day of the event. Thermal sensitivity is estimated based on the ratio of stream temperature increase to a rise in air temperature. This is derived from the slope of linear regression. Thermal sensitivity has been used in the literature to investigate the response of individual streams to warming (Chang and Psaris 2013).

ArcGIS 10.2 was used to derive percent land cover of each sub-basin in 2011 using USGS NLCD data. A ten meter resolution digital elevation model was used to derive other topographic parameters such as mean elevation and slope. We estimated baseflow index (BFI), a measure of relative contribution of groundwater to streams, for each monitoring station based on recursive digital filter method (Eckhardt 2004).

2.4. *Model goodness of fit*

The regression models perform well as measured by high NSE values for daily stream

temperatures ranging from 0.83 to 0.94 (Table 2). The regression models tend to underestimate peak stream temperature for some summer months as indicated by higher than 1°C RMSE for some sub-basins. Hence, our model estimates are somewhat conservative in estimating summer maximum stream temperatures. Because aquatic species are sensitive to subtle stream temperature changes, an over- or under-estimation of stream temperatures will have implications for fishery management.

$$\text{NSE} = 1 - \frac{\sum (O - S)^2}{\sum (O - \bar{O})^2} \qquad (1)$$

$$\text{RMSE} = \sqrt{\frac{1}{n} \sum (O - S)^2} \qquad (2)$$

$$\text{MAE} = \frac{1}{n} \sum |S - O| \qquad (3)$$

Root mean square error (RMSE)
Nash–Sutcliffe efficiency (NSE)

where O is observed flow, $\bar{O}$ is mean observed flow, and S is simulated flow.

3. Results

3.1. *Stream temperature and basin characteristics*

There was a negative relationship between elevation and 7DAD T_{max} (Fig. 3a). Half of the

Table 2. NSE, RMSE, and ME values for the 12 stream temperature models.

Sub-basin	NSE	RMSE	MAE	% High Cascade	% Urban	% riparian vegetation
Blowout	0.89	1.4	1.20	57.03	0.49	82.64
Breitenbush	0.90	1.0	0.84	51.84	0.56	82.69
Fanno	0.94	1.3	0.99	0.00	89.62	3.10
Johnson-Milwaukee	0.93	1.3	1.00	0.00	66.89	0.00
Johnson-Regner	0.92	1.3	1.00	0.00	33.77	32.08
Johnson-Sycamore	0.92	1.4	1.10	0.00	43.70	5.16
Middle Fork Willamette	0.83	1.5	1.20	64.01	0.04	93.28
North Santiam	0.92	1.0	0.76	2.86	1.17	22.50
Quartzville	0.89	1.7	1.40	7.41	0.81	56.87
South Santiam	0.89	1.6	1.20	0.03	0.86	53.33
South Fork McKenzie	0.90	0.81	0.65	45.16	0.37	80.33
Winberry	0.91	1.5	1.30	0.00	0.03	59.69

[a]NSE = Nash Sutcliff Efficiency.
[b]RMSE = Root Mean Square Error.
[c]MAE = Mean Absolute Error.

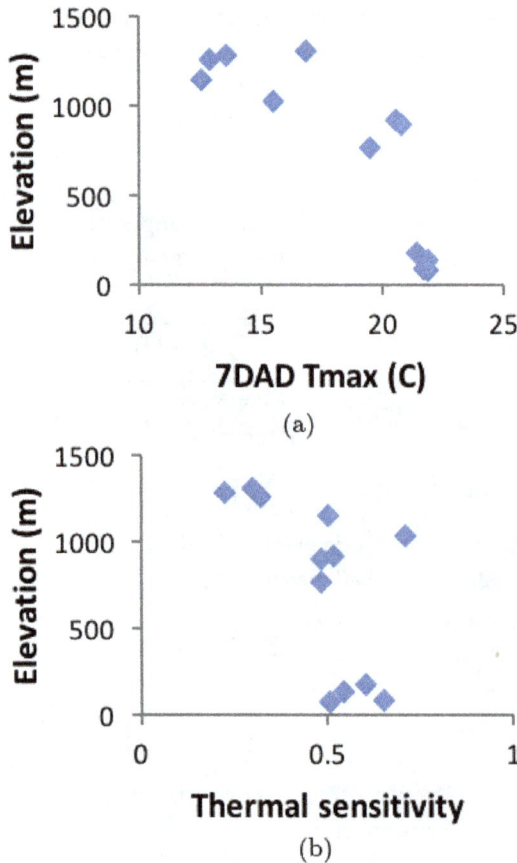

Fig. 3. Relation between (a) elevation between 7DAD T_{max} and (b) elevation and thermal sensitivity.

study basins were above the threshold value for salmonids of 7DAD T_{max} 17.8°C. These basins were located in either low valley areas or the Western Cascades, mostly lain on either agricultural or urban land covers. For sub-basins that were located >1000m elevation, 7DAD T_{max} is always lower than 17.8°C.

Thermal sensitivity was generally lower in High Cascade streams, while it was higher in valley urban streams (Fig. 3b). Western Cascade streams exhibit a wide range of thermal sensitivity. Streams primarily fed by groundwater (high BFI) had lower thermal sensitivity, while surface-water dominated streams (low BFI) had high thermal sensitivity.

3.2. Change in daily maximum air temperature by month

Changes in annual air temperature for three representative basins in three physiographic regions of the Willamette River basin are highlighted in Fig. 4. Daily maximum air temperature was projected to increase with different ranges under different climate change scenarios (Fig. 4). The HadGEM model projected a 3–6°C increase in the 2080s (2070–2099) compared to the 2020s

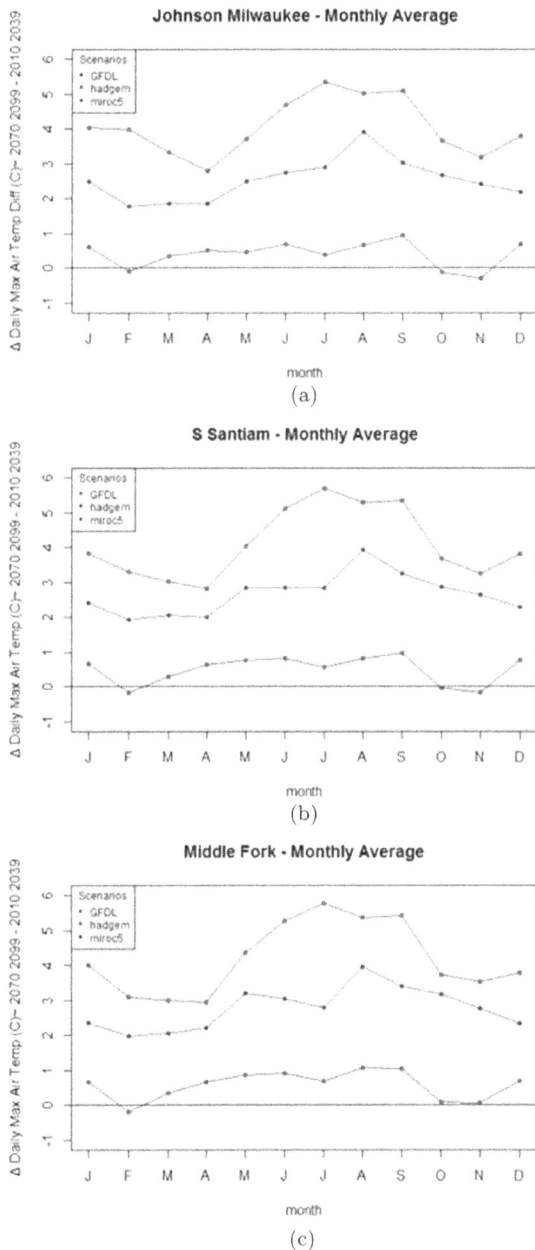

Fig. 4.   Change in daily maximum air temperature by month.

months under the Hadley model. The GFDL model projected the slightest decline in daily maximum air temperature in February and November. The Middle Fork Willamette basin, located at the highest elevation among the three, appeared to be warming faster than the other two basins, particularly in the summer months.

3.3. Change in daily streamflow by month

While the three GCMs do not necessarily agree with each other in terms of the magnitude of changes in monthly streamflow, they tend to agree that streamflow in summer months (June to September) is likely to decline (two of the three GCMs agree) (Fig. 5). Additionally, streamflow in winter months (December to March) was projected to increase under at least two of the three GCMs used in the analysis. The HadGEM projected decreases in streamflow in December and January, while both MIROC5 and GFDL GCMs projected increases in these early winter months. All three models agree in that they projected increases in March streamflow and declines in November streamflow.

3.4. Change in daily maximum stream temperature by month

Stream temperature changes largely mirror changes in maximum air temperature over monthly time scales (Fig. 6). The HadGEM GCM projected the largest increases (2–4°C) in maximum stream temperature, while GFDL projected the smallest increase (less than 0.5°C) in stream temperature. All models projected the largest increases in summer stream temperature compared to other months. Stream temperature increases in summer months were 2–3× larger than those in other non-summer months. Under the HadGEM, stream temperature is projected to increase least in April when air temperature

(2010–2039), while the GFDL model projected the slightest increase (less than 1°C) in daily maximum air temperature. Across the sites, April and November temperatures were projected to be lowest compared to the summer

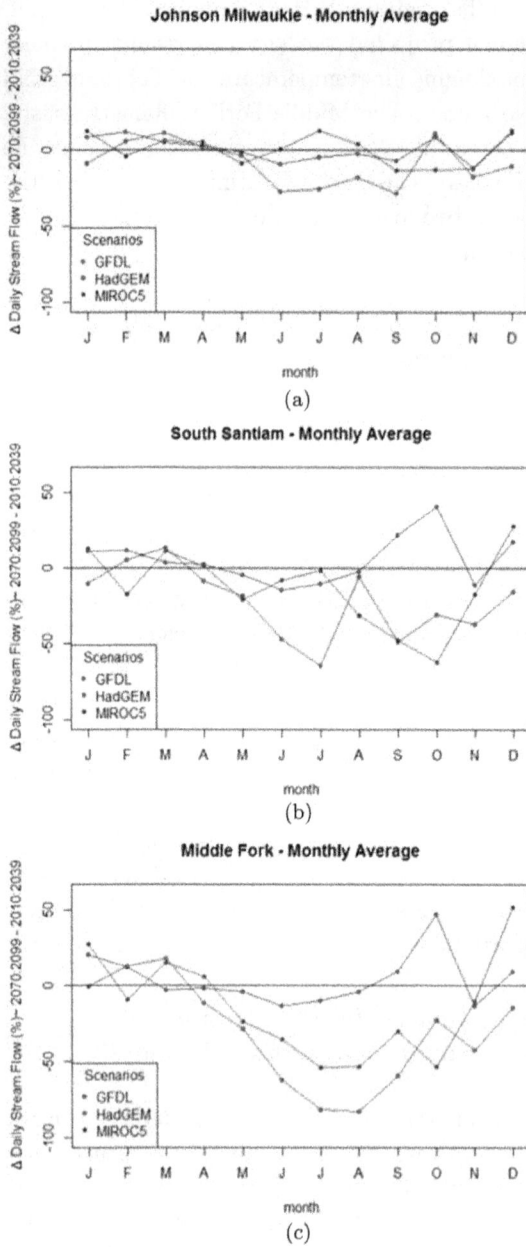

Fig. 5. Change in streamflow for three basins by month.

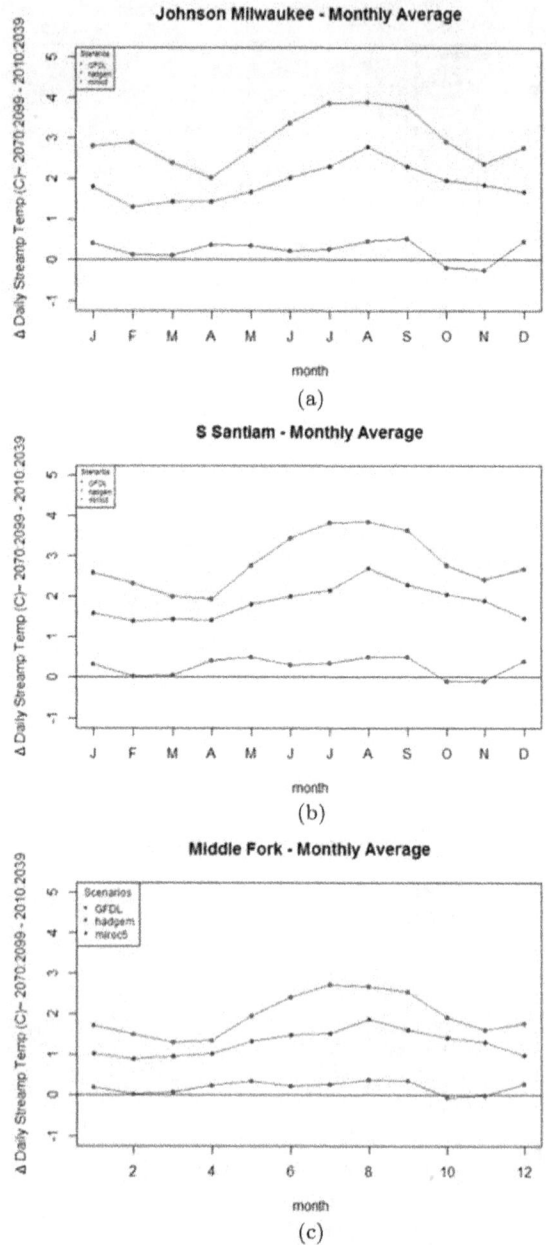

Fig. 6. Change in stream temperature for three basins by month.

is projected to increase least but streamflow was projected to increase somewhat compared to other months.

Across the three basins, stream temperature was projected to increase least in the Middle Fork Willamette basin that is located

in the highest elevation. Stream temperatures in summer months were projected to increase less than 3°C, while stream temperature in the other two basins was projected in increase close to 4°C.

3.5. *Change in probability of exceedance stream temperature*

The probability of threshold temperature of 7DAD T_{max} all exceeded under the three GCMs in the future (Fig. 7). For the low-lying basins in the Willamette Valley, the models predicted that more than 25 days will violate the critical threshold temperature of 17.8°C. For the basins that are partially in the High Cascades, 7DAD T_{max} exceeded the critical threshold temperature fewer than 5 days in a given year. Similar spatial patterns are shown for the probability of exceeding the threshold temperature of 21°C (i.e., upper optimal limit for survival).

4. Discussion

4.1. *Sub-basin response to change in hydroclimate*

Our results show that each sub-basin's response to changes in air temperature and streamflow varies depending on where it is located. Sub-basins in the low elevation valleys typically exhibit higher sensitivity to climate change than those in the upper elevations. This is likely to be associated with the fact that these more urban sub-basins have higher thermal sensitivities compared to forested sub-basins (Chang and Psaris 2013). Such a diverse range of thermal sensitivity values is related to two primary factors. First, the degree of urban impervious surfaces plays a role in regulating stream temperature (Nelson and Palmer 2007; Kaushal *et al.* 2010). Sub-basins in the lower elevations are more covered by impervious surfaces. These impervious surfaces absorb and release solar radiation more quickly than vegetated pervious surfaces, increasing air temperature of the surrounding environment. Additionally, lack of vegetated surfaces along the riparian corridor contributes to the direct exposure to sunlight and added heat to the water body (Johnson 2004; Simmons *et al.* 2014). The broad distribution of

urban impervious surface areas also contributes to the reduced infiltration of rainwater and thus baseflow during the dry period (Chang 2007), reducing the buffering effect of thermal regulation.

Second, thermal sensitivity values are highly affected by geology. Streams where groundwater contributions are high year around typically exhibit reduced variations in stream temperature compared to those in surface-water fed streams (Tague *et al.* 2007; Chang and Psaris 2013). For example, those sub-basins that have the lowest thermal sensitivity values were all located in the High Cascades where relatively young volcanic rocks store and release water gradually. This is consistent with other studies that have demonstrated lower thermal sensitivity in cold water streams (Luce *et al.* 2014). As shown in previous studies (Tauge and Grant 2009; Chang and Jung 2010), future streamflow in these High Cascades sub-basins will not be reduced substantially compared to those in Western Cascades sub-basins. For example, the South Fork McKenzie sub-basin has the lowest thermal sensitivity (0.22) amongst the 12 sub-basins studied. Indeed, the McKenzie sub-basin provides approximately a third of summer flow to the Willamette River while it only occupies 13% of the whole basin. The sub-basins located in the Western Cascades show a wide range of thermal sensitivities, suggesting that factors other than vegetation and groundwater may control thermal dynamics of these sub-basins.

4.2. *Utility of regression models for predicting future stream temperatures*

While simple, our time series regression models effectively predict daily maximum stream temperatures. Unlike complex process-based simulation models such as SWAT-TEMP (Ficklin *et al.* 2012), our models require less input data for simulating daily stream temperatures. Many

(a)

(b)

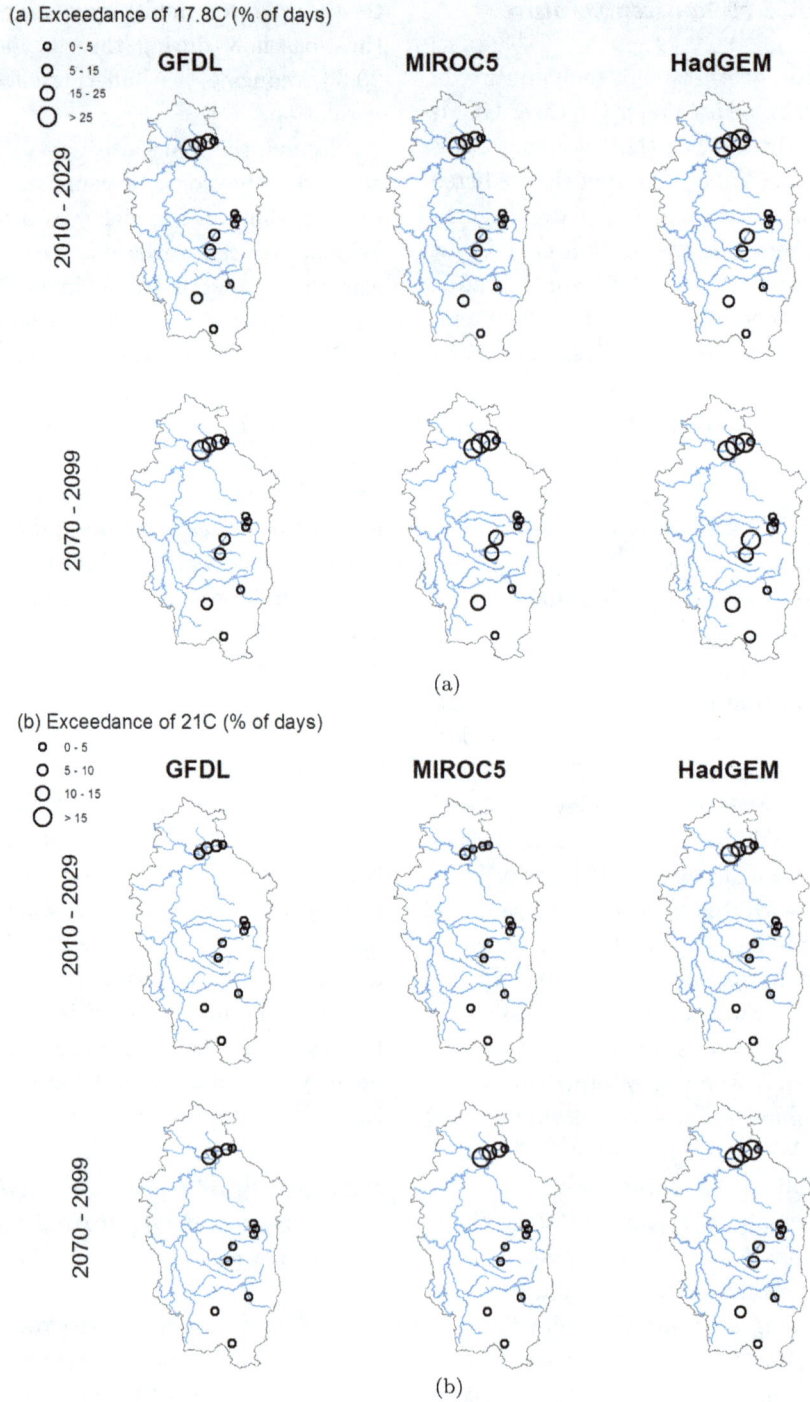

Fig. 7. Change in the probability of threshold temperature (a) exceedance of 17.8°C and (b) exceedance of 21°C.

process-based models require intensive empirical data for model parameterization, potentially introducing uncertainty in estimating model parameters when data are sparse. In contrast, our models required only few input data to derive future stream temperature conditions and reduce uncertainty associated with future stream temperature predictions. In other words, given that daily air temperature and stream temperature data are widely available, once empirical relationships between stream temperature and hydroclimate variables are developed, one can use this relationship for projecting future stream temperatures if simulated streamflow values are available. Considering the historical rate of change in stream temperature, our estimates (2–4°C in stream temperature) are also reasonable and within the range of the values reported in the literature. In contrast, some simulated models project unlikely increases in stream temperatures (up to 8°C) in some parts of the Columbia River basin in the 21st century. Such overestimation could be attributed to the simulation model used in the analysis (personal communication, Daniel Issak, US Forest Service). However, simple regression models built only on air temperature and streamflow are good predictors of stream temperature that affect the aquatic species in specific areas. As such, the site specific results should not be used for generalizing the habitat suitability of salmonid species in the entire basin.

4.3. *Implications for fish management*

Together with changes in air temperature, changes in streamflow have important impacts on the optimal stream temperature conditions for salmonid species. These species have evolved over millions of years to adapt to changing environmental conditions (Kovach *et al.* 2015). If climate change occurs more rapidly than these species can evolve, their entire suite of life-history traits may be threatened (Waples

et al. 2008). As shown in our results, given the probability of temperature increases over the 21st century that exceed physiological thresholds, salmonids will likely be both directly and directly impaired. First, they can be physically impaired, leading to excessive metabolic activity and lethargy. More subtle indirect effects would be slow growth, more frequent disease, predation or fatigue (Quinn 2005). If refuge is not found in a given stretch of a river in a timely manner in a warming environment, mortality might be inevitable. Thus, it might be necessary to close recreational fishing temporarily when such moments occur. Additionally, to be more proactive, it is necessary to create adequate space for refuge habitat by placing large woody debris and shading. Such adaptive management requires spatially explicit analysis of changes in the probability of maximum stream temperature by each sub-basin. Our current study attempts to provide such an answer. Future research should incorporate more fine-scaled variation in land use, geology, and riparian cover to understand the presence and persistence of cold-water refugia for salmonids. For instance, Lawrence *et al.* (2014) constructed detailed future water temperature predictions for two sub-watersheds of the John Day River, USA and paired them with fish-habitat models and remotely-sensed riparian land use to provide guidance on where to restore vegetation to provide thermal refugia. We consider our study an important first step in identifying, at broad spatial scales, the sub-basins that are mostly likely to be thermally impaired in the future. This research may serve as a springboard for future, more detailed studies on within-stream variation in thermal conditions.

5. Conclusions

Using available continuous daily climate and hydrology data, this study developed regression models to estimate daily maximum stream temperatures in the 12 sub-basins of the

Willamette River basin, USA. We then projected future stream temperatures in these sub-basins throughout the 21st century based on the combination of simulated future daily climate and streamflow data obtained by INSIDE and WHM, respectively. The changes in the probability of critical temperatures (17.8°C and 21°C) were then used to offer insights on future fish management under climate change scenarios.

Our study suggests that even though changes in air temperature are similar across the study sub-basins, sub-basin characteristics such as vegetation and geology can control the sensitivity of each sub-basin to rise in air temperature in the future. While sub-basins in the High Cascades may be less affected by rising air temperature primarily due to constant groundwater input during low flow seasons, fish may not be able to reach these cold stream reaches if lower sections of the stream warm faster than the upper sections. Thus, future fish management practices should focus on these lowland areas that serve as migratory pathways for salmonid species. Creating refuge areas by adding riparian vegetation or groundwater augmentation can help mitigate future warming.

Acknowledgments

This material is based upon work supported by the National Science Foundation NSF-BES Grant #1026629, NSF-WSC #1038925, and NSF-IGERT Program Grant #0966376. We appreciate Norman Buccola at the U.S. Geological Survey for providing R codes for estimating daily maximum stream temperature that were used for our analysis. Thanks also go to Mike Psaris who initially helped write a script to organize the climate data. An earlier draft of this paper was presented at the U.S. Geological Survey seminar in November 2015. Views expressed are our own and do not necessarily reflect those of our sponsoring agencies.

References

Abatzoglou, J. T., 2013: Development of gridded surface meteorological data for ecological applications and modelling. *Int. J. Climatol.*, **33**, 121–131.

Caissie, D., 2006: The thermal regime of rivers: a review. *Freshwater Biol.*, **51**, 1389–1406.

Caldwell, P., C. Segura, S. G. Laird, G. Sun, S. G. McNulty, M. Sandercock, J. Boggs, and J. M. Vose, 2014: Short-term stream water temperature observations permit rapid assessment of potential climate change impacts. *Hydrol. Process.*, **29**, 2196–2211.

Chang, H. and I.-W. Jung, 2010: Spatial and temporal changes in runoff caused by climate change in a complex large river basin in Oregon. *J. Hydrol.*, **388**(3–4), 186–207.

Chang, H. and K. Lawler, 2011: Impacts of climate variability and change on water temperature in an urbanizing Oregon basin. Water Quality: Current Trends and Expected Climate Change Impacts. *IAHS Publication*, **348**, 123–128.

Chang, H. and M. Psaris, 2013: Local landscape predictors of maximum stream temperature and thermal sensitivity in the Columbia River basin, USA. *Sci. Total Environ.*, **461–462**, 587–600.

Ficklin, D. L., L. Yuzhou, I. T. Stewart, and E. P. Maurer, 2012: Development and application of a hydroclimatological stream temperature model within the soil and water assessment tool. *Water Resour. Res.*, **48**, W01511.

Ficklin, D. L, B. L. Barnhart, J. H. Knouft, I. T. Stewart, E. P. Maurer, S. L. Letsinger, and G. W. Whittaker, 2014: Climate change and stream temperature projections in the Columbia River basin: habitat implications of spatial variation in hydrologic drives. *Hydrol. Earth Syst. Sc.*, **18**, 4897–4912.

Hicks, M., 2000: Evaluating standards for protecting aquatic life in Washington's surface water quality standards. Draft discussion paper and literature summary. Revised 2002. Washington State Department of Ecology, Olympia, WA, 197p.

Hills, R. A., C. P. Hawkins, and J. Jin, 2014: Predicting thermal vulnerability of stream and river ecosystems to climate change. *Clim. Change*, **125**, 399–412.

Isaak, D. J., M. K. Young, C. Luce, S. W. Hostetler, S. J. Wnger, E. E. Peterson, J. M. Ver Hoef, M. C. Groce, D. L. Horan, and D. E. Nagel, 2016: Slow climate velocities of mountain streams portend their role as refugia for cold-water biodiversity. *PNAS*, **113**(16), 4374–4379.

Isaak, D. J., S. Wollrab, D. Horan, and G. Chandler, 2012: Climate change effects on stream and river temperatures across the northwest U.S. from 1980–2009 and implications for salmonid fishes. *Clim. Change*, **113**, 499–524.

Jaeger, K. W., A. Amos, D. P. Bigelow, H. Chang, D. R. Conklin, R. Haggerty, C. Langpap, K. Moore, P. Mote, A. Nolin, A. J. Plantinga, C. Schwartz, D. Tullos, and D. T. Turner, 2017: Finding water scarcity amid abundance using human-natural system models. *Proceedings of the National Academy of Sciences*, **111**(45): 11884–11889.

Johnson, S. L., 2004: Factors influencing stream temperatures in small streams: substrate effects and a shading experiment. *Can. J. Fish. Aquat. Sci.*, **61**, 913–923.

Kaushal, S. S., G. E. Likens, N. A. Jaworski, M. L. Pace, A. M. Sides, D. Seekell, K. T. Belt, D. H. Secor, and R. L. Wingate, 2010: Rising stream and river temperatures in the United States. *Front. Ecol. Environ.*, **8**, 461–466.

Kovach, R. P., C. C. Muhlfeld, A. A. Wade, B. K. Hand, D. C. Whited, P. W. DeHaan, R. Al-Chokhachy, and G. Luikart, 2015: Genetic diversity is related to climatic variation and vulnerability in threatened bull trout. *Glob. Change Biol.*, **21**, 2510–2524.

Laenen, A. and J. C. Risley, 1997: Precipitation-runoff and streamflow-routing models for the Willamette River Basin, Oregon. US Geological Survey Water-Resources Investigations Report 95–4284.

Lawrence, D. J., B. Stewart-Koster, J. D. Olden, A. S. Ruesch, C. E. Torgersen, J. J. Lawler, D. P. Butcher, and J. K. Crown, 2014: The interactive effects of climate change, riparian management, and a nonnative predator on stream rearing salmon. *Ecol. Appl.*, **24**(4), 895–912.

Luce, C., B. Staab, M. Kramer, S. Wenger, D. Isaak, and C. McConnell, 2014: Sensitivity of summer stream temperatures to climate variability in the Pacific Northwest. *Water Resour. Res.*, **50**(4), 3428.

McCullough, D. A., S. Spalding, D. Sturdevant, and M. Hicks, 2001: *Issue Paper 5. Summary of technical literature examining the physiological effects of temperature on salmonids*. EPA-910-D-01-005, prepared as part of U.S. EPA Region 10 Temperature Water Quality Criteria Guidance Development Project. Available at http://yosemite.epa. gov/R10/WATER.NSF/6cb1a1df2c49e49688256 88200712cb7/5eb9e547ee9e111f88256a03005bd66 5/\$FILE/Paper%205-Literature%20Temp.pdf.

Mohseni, O., T. R. Erickson, and H. G. Stefan, 1999: Sensitivity of stream temperatures in the United States to air temperatures projected under a global warming scenario. *Water Resour. Res.*, **35**(12), 3723–3733.

Nelson, K. C. and M. A. Palmer, 2007: Stream temperature surges under urbanization and climate change: Data, models, and responses. *J. Am. Water Res. Assoc.*, **43**(2), 440–452.

Oregon Department of Environmental Quality, 2010.

Patra, R. W., J. C. Chapman, R. P. Lim, P. C. Gehrke, and R. M. Sunderam, 2015: Interactions between water temperature and contaminant toxicity to freshwater fish. *Environ. Toxicol. Chem.*, **34**, 1809–1817.

Quinn, T. 2005: The behavior and ecology of Pacific salmon and trout. University of Washington Press: Seattle, Washington.

Richter, A. and S. Kolmes, 2005: Maximum temperature limits for Chinook, coho, and chum salmon, and steelhead trout in the Pacific Northwest. *Rev. Fish. Sci.*, **13**, 23–49.

Schreck, C. B., J. C. Snelling, R. E. Ewing, C. S. Bradford, L. E. Davis, and C. H. Slater, 1994: *Migratory behavior of adult spring chinook salmon in the Willamette River and its tributaries*. Oregon Cooperative Fishery Research Unit, Oregon State University, Corvallis, Oregon. Project Number 88-160-3, Prepared for Bonneville Power Administration, Portland, OR.

Seibert, J. and M. J. P. Vis, 2012: Teaching hydrological modeling with a user-friendly catchment-runoff-model software package. *Hydrol. Earth Syst. Sci.*, **16**, 3315–3325.

Simmons, J. A., M. Anderson, W. Dress, C. Hanna, D. J. Hornbach, A. Janmaat, F. Kuserk, J. G. March, T. Murray, J. Niedzwiecki, D. Panvini, B. Pohlad, C. Thomas, and L. Vasseur, 2015: A comparison of the temperature regime of short stream segments under forested and non-forested riparian zones at eleven sites across North America. *River Res. Appl.*, **31**, 964–984. doi:10.1002/rra.2796.

Tague, C. and G. Grant, 2009: Groundwater dynamics mediate low flow response to global warming in snow-dominated alpine regions. *Water Resour. Res.*, **45**, W07421. doi:10.1029/2008WR007179.

Tague, C., M. Farrell, and G. Grant, 2007: Hydrogeologic controls on summer stream temperatures in the McKenzie River basin, Oregon. *Hydrol. Process.*, **21**(24), 3288–3300.

Waples, R. S., G. R. Pess, and T. Beechie, 2008: Evolutionary history of Pacific salmon in dynamic environments. *Evol. Appl.*, **1**, 189–206.

Webb, B. W., D. M. Hannah, R. D. Moore, L. E. Brown, and F. Nobilis, 2008: Recent advances in stream and river temperature research. *Hydrol. Process.*, **22**(7), 902–918.

Whitehead, P. G., R. L. Wilby, R. Battarbee, M. Kernan, and A. Wade, 2009: A review of the potential impacts of climate change on surface water quality. *Hydrolog. Sci. J.*, **54**, 101–123.

Wu, H., J. S. Kimball, M. M. Elsner, N. Mantua, and R. F. Adler, 2012: Projected climate change impacts on the hydrology and temperature of Pacific Northwest rivers. *Water Resour. Res.*, **48**, W11530, doi:10.1029/2012WR012082.

United States Fish and Wildlife Service. 2016: Endangered species. Accessed 1 February 2016 at http://www.fws.gov/endangered/?ref=topbar.

United States Geological Survey. 2015: USGS Water-Quality Data for Oregon. Available at http://waterdata.usgs.gov/or/nwis/qw. Accessed July 20, 2015.

Van Vliet, M. T. H., W. H. P. Franssen, J. R. Yearsley, F. Ludwig, and I. Haddeland, 2013: Global river discharge and water temperature under climate change. *Global Environ. Chang.*, **23**, 450–464.

Chapter 9

An Integrated Approach for Flood Inundation Modeling on Large Scales

Venkatesh Merwade*, Mohammad Adnan Rajib and Zhu Liu

Lyles School of Civil Engineering, Purdue University,
West Lafayette, IN 47907, USA
*vmerwade@purdue.edu
http://web.ics.purdue.edu/~vmerwade/

Contemporary flood forecasting frameworks are typically based on point discharge measurements at discrete locations; hence they are unable to provide spatio-temporal information of flood inundation extents over large scales and at high spatial resolution. The use of hydrologic/hydraulic models has recently appeared to be the most suitable option to bridge this gap. This chapter features coupling of a spatially distributed hydrologic model, Soil and Water Assessment Tool (SWAT), with a 2D hydraulic model LISFLOOD-FP, for flood inundation modeling over the Ohio River Basin, USA. Simulated 100-year return period flow outputs from SWAT are set at multiple input boundary locations of LISFLOOD-FP and routed along the streamlines to generate corresponding flood maps. Model evaluation shows that simulated daily streamflow from SWAT has Kling-Gupta and Nash-Sutcliffe Efficiency scores in the range of 0.4–0.72 when compared against observed records across the basin, and the modeled inundation area from LISFLOOD-FP has more than 70% agreement with the corresponding 100-year reference flood maps along the main river channels. While these results are promising, this chapter also reveals several avenues which can lead towards an integrated flood modeling framework having enhanced predictability.

1. Introduction

The changes in climate and land use are affecting the occurrence of high magnitude low frequency floods in many parts of the world (AWDR 2006; Chen and Yu 2015; Kundzewicz *et al.* 2014; Peduzzi *et al.* 2009; Syvitski and Brakenridge 2013). To deal with this rising threat of high floods and associated socio-economic damages, there is a growing interest to develop advanced flood warning and inundation mapping systems (Gaughan and Waylen 2012; Kauffeldt *et al.* 2016; Schumann *et al.* 2013).

Conventionally, the data recorded at streamflow gauging stations are used to generate information on historical flooding and future prediction by applying various statistical/stochastic methods (Bourdin *et al.* 2012). While these gauging stations provide critical information on floods at locations where the data are being collected, observations along many small rivers or streams are too short, of too poor a quality, or even non-existent (Sivapalan 2003). Existing gauge density is continuously declining in many parts across the globe (Swain *et al.* 2015). Thus, it has become increasingly difficult to obtain ready-to-use information on flood occurrence or inundation extent at ungauged sites. Recognizing this challenging problem, International Association of Hydrological Sciences (IAHS) launched an initiative aiming at enhancement in the ability to make Predictions in Ungauged Basins (PUB) (Hrachowitz *et al.* 2013; Sivapalan 2003). A well-established way is to transfer hydrologic information from one gauged watershed (donor) to an ungauged one (receptor), which is "hydrologically similar" with the donor (Razavi and Coulibaly 2013).

This approach, referred as the Regionalization, typically involves regression analysis of flood statistics (statistical moments of streamflow time series) of a gauged watershed against watershed characteristics (e.g., topography, climate, land use, soil, and drainage area.), then apply the statistical relationship in a similar ungauged watershed (He *et al.* 2011; Loukas and Vasiliades 2014). However, uncertainty related to the regionalization procedure that stems from the arbitrary choices of statistical fitting, length of observed data, watershed characteristics as well as definition/measures of hydrologic similarity can have a significant effect on the resultant prediction in the ungauged locations (Loukas and Vasiliades 2014; Pruski *et al.* 2013; Sellami *et al.* 2014). Moreover, such statistical regressions based on historical records may not comply with climate/land use changes in the long run.

One way to make the use of sparsely observed data to get information at ungauged sites is the use of computational models for simulating hydrologic and hydraulic processes. With the advancements in numerical weather prediction, simulation techniques for surface-subsurface hydrologic processes and high performance computational resources, there are some ongoing initiatives to design and implement large scale flood modeling frameworks in the United States, including the Coastal and Inland Flooding Observation and Warning project (CI-FLOW; NOAA 2014) and the National Flood Interoperability Experiment in USA (NFIE; Maidment 2015). Similarly, studies by Alfieri *et al.* (2013), Paiva *et al.* (2013, 2011), Pappenberger *et al.* (2012), Thielen *et al.* (2009), Winsemius *et al.* (2013) highlight some other recent attempts on other parts of the world, specifically focusing on Europe, parts of Africa and South America.

Numerous hydrologic/hydraulic models exist in literature, the majority of these models are developed for research purposes as off-line stand-alone applications on local computers.

Using an established model for large scale applications within a cyber environment requires the model to be publicly available, easily deployable on the internet, and widely evaluated under different geographic and climate settings. Besides the suitability of an individual model for large scale application in the web, individual model's representation of physical processes for runoff generation, streamflow routing and flood propagation, should also be taken into account. All these factors lead to a need for testing different model combinations which serve the purpose of flood modeling and mapping at large scales with accuracy and efficiency. In an attempt to address such need, this chapter proposes an integrated flood inundation modeling framework in which the Soil and Water Assessment Tool (SWAT) hydrologic model is loosely coupled with the LISFLOOD-FP hydrodynamic model.

To demonstrate the capabilities of the proposed framework, the Ohio River Basin (ORB), with a total drainage area of 490,000 km^2, is chosen as the study area that drains through eleven states within the contiguous United States (Fig. 1). Despite the frequent history of significant flood events in ORB, especially in the recent years of 2008, 2009, 2011 and 2014, limited studies exist related to flood modeling for the entire basin. Voisin *et al.* (2011) applied a 10-day ensemble weather forecast into the Variable Infiltration Capacity (VIC) model to evaluate a potential streamflow forecasting scheme for the Ohio Basin. Most studies related to flooding in this region have focused on relatively smaller scales (e.g., Goodwell *et al.* 2014; O'Donnell *et al.* 2000; Whitehead and Ostheimer 2009; Whitehead and Ostheimer 2014).

The proposed SWAT and LISFLOOD-FP coupling presented here is a prototype developed only for the major river reaches within the ORB, and applies historic climate information obtained from the available weather stations to force the hydrologic model component (SWAT). Eventually, the models could be

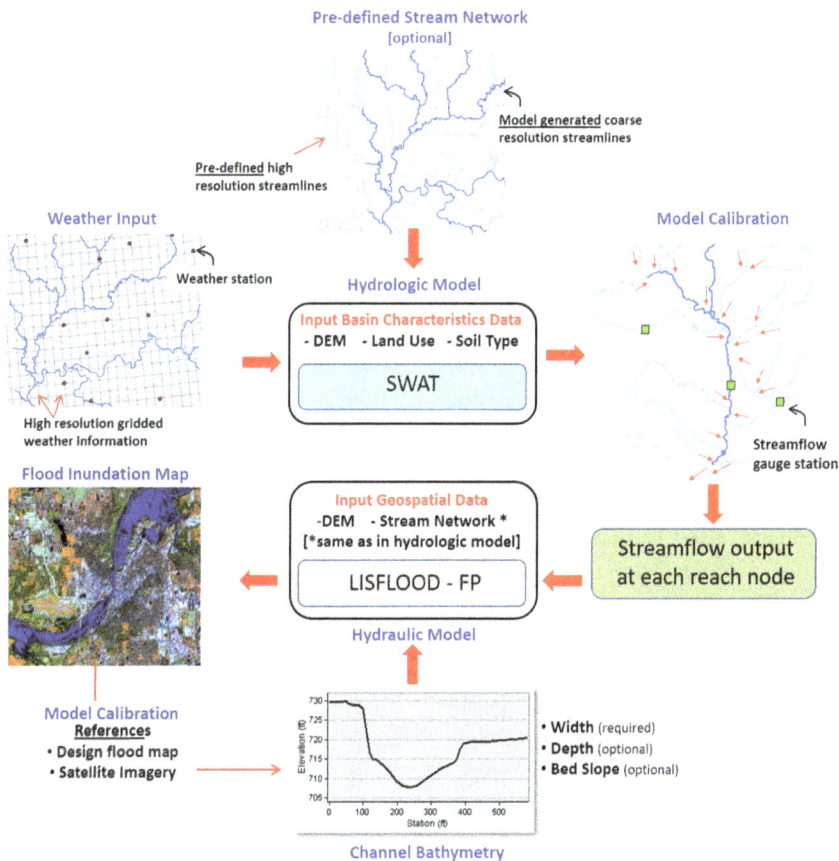

Fig. 1.   A generic workflow diagram of SWAT and LISFLOOD-FP flood modeling framework.

re-set with higher resolution stream network (e.g., Maidment 2015) and forced with gridded weather forecast products (e.g., Schumann *et al.* 2013).

As the way forward, this chapter provides several directions through which the proposed framework or similar ones can be executed with a holistic integrated design, including the application of remote sensing observations for spatial calibration of hydrologic/hydraulic models, accounting for topographic accuracy and channel bathymetry data, as well as the ensemble prediction from multi-model outputs. The integrated approach presented here will provide valuable insights for implementing an operational flood forecasting framework over large scales.

2. Coupling of Hydrologic and Hydraulic Model for Flood Inundation Mapping

The proposed flood inundation modeling framework is shown in Fig. 1. The overall vision for the proposed integrated hydrologic-hydraulic modeling framework involves the ingestion of weather input from multiple sources to feed into multiple hydrologic models to create an ensemble of streamflow output. These streamflow ensembles are then used to drive multiple hydraulic models to get an ensemble of water surface profiles along a stream. The ensemble of water surface profiles can then be used to create probabilistic inundation maps. The hydrologic and hydraulic models used in this integrated framework will also assimilate near real-time data from field sensors and satellite images to update model parameters as the hydrologic conditions change with time. The proposed framework is expected to provide probabilistic flood inundation maps for reaches ranging from first order streams to major rivers at the continental scale.

In an attempt to realize the vision presented above, a prototype modeling framework is developed by loosely coupling SWAT hydrologic model with LISFLOOD-FP hydrodynamic model for ORB. The SWAT model is calibrated by using historical flow data observed at selected streamflow gauging stations. The weather data that are used to drive the SWAT model come from field weather stations. The output discharge from SWAT is used to drive the LISFLOOD-FP hydrodynamic model for a coarser resolution stream network. LISFLOOD-FP takes the flow as input and provides water surface elevation and inundation extent for each reach in the stream network, after being calibrated for surface and channel roughness parameters. Both SWAT and LISFLOOD-FP are currently loosely coupled in the sense that the output information from SWAT is not automatically transferred to LISFLOOD-FP. A brief description of each model and its development is presented in the following section.

3. Model Description

3.1. *SWAT*

SWAT (Neitsch *et al.* 2011; Arnold *et al.* 2012) is a conceptual, semi-distributed watershed model widely used to assess the impact of land management practices and climatic change on long-term water, sediment and pollutant yields. Arnold *et al.* (1999) were the first to apply SWAT for the entire USA under the auspices of Natural Resources Conservation Service's HUMUS (Hydrologic Unit Model for the United States) project, simulating river discharges at around

6,000 gauging stations. SWAT is a widely used model in various national level governmental projects in the USA, such as the U.S. Department of Agriculture's CEAP (Conservation Effects Assessment Project) initiative, as well as the U.S. Environmental Protection Agency's BASINS (Better Assessment Science Integrating Point and Non-point Sources) project. Across the globe, SWAT has been applied in a number of applications ranging from catchment to continental scales, including India (Gosain *et al.* 2006), Africa (Schuol *et al.* 2008; Faramarzi *et al.* 2013), China (Zang and Liu 2013) and Europe (Abbaspour *et al.* 2015). While large scale application of SWAT is common, building a finer resolution large scale SWAT model has become possible only in recent times due to the increased data handling capacity of the GIS interface for SWAT model setup (e.g. Arc-SWAT). Moreover, recent development of cyber-enabled high performance simulation platforms such as SWATShare (Rajib *et al.* 2016), has made calibration of large scale SWAT models automatic, computationally economical and easier. All these factors make SWAT a suitable tool for modeling floods over large scales.

Based on topography and stream network, SWAT model divides a basin into a number of smaller sub-basins where flow routing is simulated. The sub-basins can be considered as the spatial units of model simulation or can be further discretized into Hydrologic Response Units (HRUs) consisting of uniform land use, soil type, topography (slope) and management practices. The HRU is a basic unit in SWAT where fundamental processes such as surface runoff generation, soil water dynamics and evapotranspiration are simulated. Depending on the mode of spatial discretization, a sub-basin can also function as an HRU.

Total SWAT stream flow is calculated as

$$Q = Q_{surf} + Q_{lat} + Q_{gw}$$

where Q is total streamflow of the day, Q_{surf} is surface runoff, Q_{lat} is subsurface lateral flow and

Q_{gw} is groundwater contribution to the stream. Surface runoff, lateral flow and groundwater flow are generated from each HRU and aggregated at the main channel of each sub-basin, then routed to obtain the total streamflow for the watershed.

Surface runoff (Q_{surf}) is estimated using the Soil Conservative Service (SCS) curve number procedure:

$$Q_{surf} = \frac{(R_{day} - 0.2S)^2}{R_{day} + 0.8S}$$

where R_{day} is daily precipitation (mm H$_2$O) and S is a retention parameter defined as:

$$S = 25.4\left(\frac{1000}{CN} - 10\right)$$

where CN is the curve number for the day, which is a function of the input parameter CN2 (initial curve number at average moisture condition) and initial soil profile water content of the day (mm H$_2$O) excluding the amount of water held at wilting point. SWAT also gives the option to calculate surface runoff using Green-Ampt infiltration method.

Subsurface lateral flow occurs when a soil layer is saturated in a sloped HRU:

$$Q_{lat} = \sum_{ly}^{n}\left(\frac{0.048s(SW_{ly} - FC_{ly})K_{sat}^{ly}}{(\varphi_{ly} - \theta_{FC}^{ly})L_s}\right)$$

where Q_{lat} is the total lateral flow from an HRU for the day (mm H$_2$O); SW_{ly} is the soil water content of the layer above wilting point and FC_{ly} is the water content at field capacity (both in mm H$_2$O units); K_{sat}^{ly} is the saturated hydraulic conductivity (mm/h), s is the slope steepness (m/m), φ_{ly} is the total porosity of the soil and θ_{FC}^{ly} is the volumetric moisture content at field capacity, L_s is the slope length (m). The index ly indicates a layer-specific parameter and n is the total number of soil layers in the profile.

When the amount of water stored in the shallow aquifer is greater than a defined threshold limit, groundwater flow is generated:

$$Q_{gw}^i = (Q_{gw}^{i-1})exp[-\lambda_{gw}] + w_{rch}(1 - exp[-\lambda_{gw}])$$

where Q_{gw}^i and Q_{gw}^{i-1} designate groundwater contribution to the main channel on day i (current time-step) and $i - 1$ respectively (previous time-step) in mm H$_2$O unit, λ_{gw} is the baseflow recession constant, and w_{rch} is the amount of recharge entering the shallow aquifer on day i (mm H$_2$O).

SWAT provides two options for in-stream routing, including the Muskingum K-X and the Variable Storage Area method (Neitsch et al. 2011).

3.2. LISFLOOD-FP

LISFLOOD-FP is a 1D/2D coupled hydrodynamic model to simulate spatial flood inundation extents (Bates and De Roo 2000; Horritt and Bates 2001; Bates et al. 2013). Inundation is calculated using an intelligent volume-filling process based on the raster grid; flow in the channel is unidirectional but on the floodplain it is integrated over two dimensions.

LISFLOOD-FP has two major assumptions in its flood process simulation, including gradually varied flow and no exchange of momentum between main channel flow and that in the floodplain. Flood wave propagating downstream through the channel is simulated in the model using 1D continuity and momentum equations. This method is evolved from the most basic dynamic wave routing scheme with a simplification of the full one-dimensional St. Venant equation, obtained by eliminating local acceleration, convective acceleration and pressure terms in the momentum equation.

$$\partial Q/\partial x + \partial A/\partial t = q$$

$$S_0 - \frac{n^2Q^2P^{4/3}}{A^{10/3}} - \left[\frac{\partial h}{\partial x}\right] = 0$$

where Q is the volumetric flow rate in the channel, x is the distance between cross sections, A is the cross sectional area, S_0 is the slope of the channel bed, n is the Manning's roughness coefficient, P is the wetted perimeter and h is the flow depth.

When the bankfull flow depth is reached, water ceases to be contained solely in the main river channel, hence spills onto adjacent shallow gradient floodplains. These floodplains act either as temporary storage for this water or additional routes for flow conveyance.

The model uses a "Diffusive Solver" for the treatment of floodplain flows, being discretized over a grid of square cells. More simply, flow between two cells in the floodplain is taken as a function of the free surface height difference between those cells based on 2D continuity and momentum equations (Horritt and Bates 2001):

$$\frac{dh^{i,j}}{dt} = \frac{Q_x^{i-1,j} - Q_x^{i,j} + Q_y^{i,j-1} - Q_y^{i,j}}{\Delta x \Delta y}$$

$$Q_x^{i,j} = \frac{h_f^{5/3} \left[\frac{h^{i-1,j}-h^{i,j}}{\Delta x}\right]^{1/2} \Delta y}{n}$$

where $h^{i,j}$ is the free water surface height at node (i, j), Δx and Δy are the cell dimensions, n is Manning's roughness coefficient, Q_x and Q_y are the volumetric flow rate between cells respectively in x and y direction, h_f is defined as the elevation difference between the highest free water surface in the two cells and the highest bed elevation.

The second option for the simulation of floodplain flow is to use a "Sub-grid Solver". This is relatively a new addition in the model adding the mass/inertial term in the St. Venant equation to simulate the mass exchange among floodplain cells or between floodplain and river channel. Further technical details can be found in Bates *et al.* (2010) and Bates *et al.* (2013).

Regardless of the solver method used in the model, the final output is a series of gridded water depth file for the study area. These water depth files can be viewed in ArcGIS interface and processed in order to get a spatial flood inundation map.

4. SWAT and LISFLOOD-FP Coupling for the Ohio River Basin

Creation of a SWAT model using ArcSWAT interface requires topography, soil texture, land use and climate data. Digital elevation model (DEM) at 30 m spatial resolution is obtained from the United States Geological Survey's (USGS) National Elevation Dataset (USGS-NED 2015). In this study, the 1:250,000 scale State Soil Geographic Data (STATSGO) is used from the SWAT 2012 database. The land use dataset at 30 m horizontal resolution is obtained from the USGS's National Land Cover Database of year 2011 (USGS-NLCD 2015). The climate data are obtained from the National Climatic Data Center (NCDC) for 112 weather stations that are uniformly distributed within the basin (Fig. 2). From each station's record, total daily precipitation, average daily maximum and minimum temperature records covering the 80-year period of simulation (1935–2014) are used to drive the SWAT model. All other related climatic components, e.g., solar radiation and relative humidity, are computed using the internal weather generator within ArcSWAT. Penman-Monteith equation is selected for computing potential evapotranspiration (PET). Observed daily streamflow time series for model calibration and validation are obtained from USGS.

The spatial heterogeneity is represented by dividing the entire basin into 125 sub-basins using 0.5% of the basin area (2,500 km^2) as the critical source area threshold. Considering the elevation differences across the basin, slope in the land surface is divided into three classes (0–4%, 4–10%, and >10%) while defining HRUs. Dominant land use, soil type and slope class within a sub-basin is chosen as the basis for HRU definition, resulting into one HRU in each sub-basin. This leads to the assumption that the simulated values for a hydrologic flux/state variable over a particular sub-basin is representative of the average condition of the whole sub-basin. The SCS Curve Number and Variable Storage

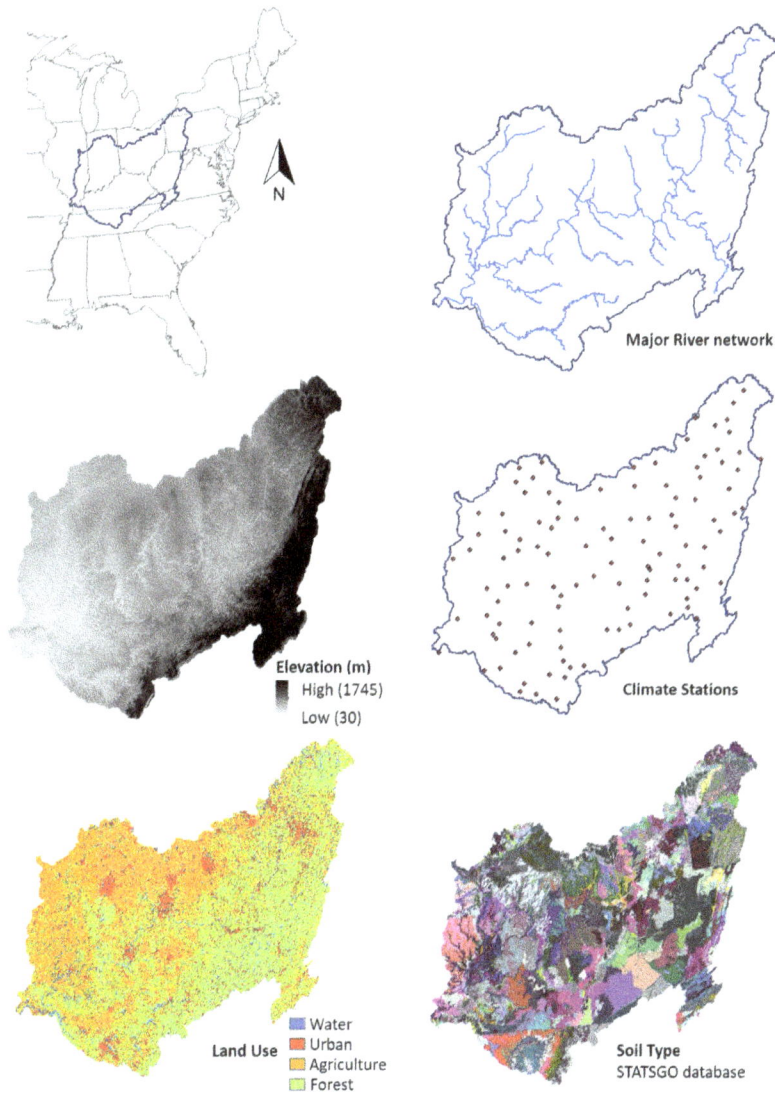

Fig. 2.   Geospatial data layers and weather stations used in SWAT model creation.

methods (Neitsch *et al.* 2011) are selected for surface runoff generation and channel routing process, respectively. Nearly 30% of ORB has artificial subsurface drainage (tile drainage) (Jaynes and James 2007). Due to the absence of a database listing the actual spatial coverage of subsurface drainage, their depth, spacing and date of installation, all the sub-basins having "poorly-drained" and "very poorly-drained" soil

are simulated by assigning tile drainage parameters (Neitsch *et al.* 2011).

The SWAT model is calibrated against daily streamflow observations for 80 years (1935–2014) from 9 USGS gauging stations (Fig. 3a), with the first 5 years as warm-up period (1935–1939). A long period of simulation is necessary in this case to calculate 100-year "design" flood magnitude along the ungauged reaches.

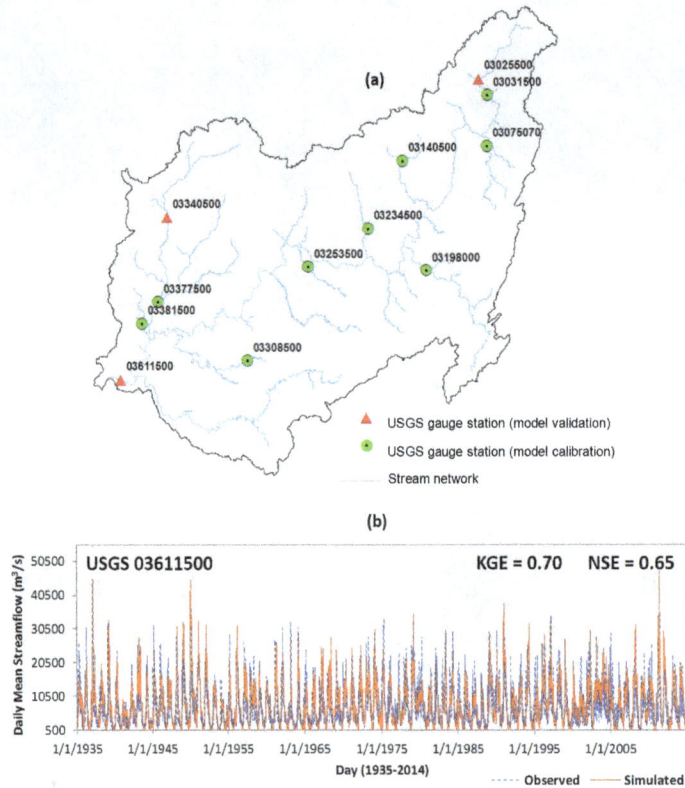

Fig. 3.   (a) Model calibration and validation locations (numbers correspond to respective USGS station IDs); (b) evaluation of model performance at a near-outlet gauge station.

The selection of parameters for calibration and their initial ranges are based on the review of existing literature and prior knowledge of the study area (e.g. Kumar and Merwade 2009; Larose et al. 2007; Rajib and Merwade 2016), as well as suggestions from model developers (Neitsch et al. 2011; Abbaspour 2015). Calibration is conducted by using the Sequential Uncertainty Fitting alogorithm-version2 (SUFI-2), which is a semi-automated inverse modeling procedure available inside SWAT-CUP (Abbaspour 2015). Kling-Gupta Efficiency (KGE) (Gupta et al. 2009; Kling et al. 2012) is used as the goal/objective function to measure the association between simulated and observed streamflow hydrographs. In this multi-objective parameter optimization simultaneously using 9 sets of observations, the best estimates of the

parameters are sought through maximizing an aggregated value of the goal (i.e. KGE) following the approach shown by Abbaspour et al. (2015). After calibration, the model is re-run for the entire period of simulation (warm-up and calibration) with respective best parameter values to get continuous time-series of simulated streamflow for all reaches in the basin. The model is validated by comparing the simulated streamflow with observed data at three separate gauge stations which are not included in the calibration process (Fig. 3a). The goodness of fit scores (Kling-Gupta Efficiency (KGE) and Nash-Sutcliffe Efficiency (NSE)) are found in the range of 0.4–0.72 and 0.55–0.7 for the calibration and validation locations respectively. These fitness statistics are within acceptable range (e.g. Moriasi et al. 2007) considering the

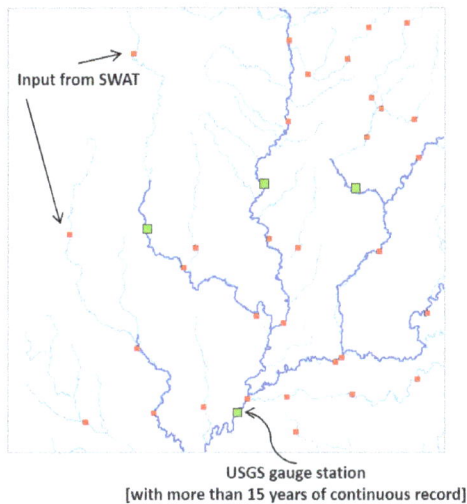

Fig. 4. Schematic presentation of SWAT input into LISFLOOD-FP.

"averaging effect" of multi-objective optimization being executed globally on such large scale.

The calibrated SWAT model provides daily streamflow data for 125 major reaches/rivers in the ORB. These hydrographs can be used to drive LISFLOOD-FP to get flood inundation at any time. For the prototype application, SWAT results are used to estimate 100-year design flow value for each reach by assuming the annual maximum series to follow the Log Pearson Type III distribution, as suggested in Bulletin 17B by USGS (IACWD 1982). Figure 4 schematically shows points where the data are transferred from SWAT results to LISFLOOD-FP for hydrodynamic simulation. It also shows the relative availability of observed streamflow data along major rivers, which is very sparse compared to the model generated streamflow locations. From the information published in the National Hydrography Dataset Plus-version 2 (NHDPlus-v2), ORB has only 700 USGS gauging stations spread over the 490,000 km^2 drainage area, out of which less than 50% of the stations have continuous record for recent 15 years. This leads to a gauge density of only one station per 700 km^2 of area on average, which is inadequate to produce flooding

information with local level details. The ORB test case included here validates the necessity for high resolution modeling in larger basins.

The channel/reach parameters required to run LISFLOOD-FP include the bankfull width, depth, bed slope and Manning's roughness coefficient (n value). In this case, a uniform depth is assumed for all the reaches (e.g., Horritt and Bates 2001). A simple approach is adopted to obtain width information, first by superimposing the stream network onto the DEM and then computing the approximate width at any cross section along a particular reach. Alternatively, the bankfull width and depth could have been extracted from the global database (http://gaia.geosci.unc.edu/rivers/) developed by Andreadis *et al.* (2013), although that is not applicable for smaller reaches. LISFLOOD-FP calculates a uniform bed slope from the DEM. The Manning's n is used as the calibration parameter as suggested by Horritt and Bates (2001).

The ORB is divided into six regions (Fig. 5) based on the clustering of the major river tributaries. Such sub-division helps better calibration of the model for the Manning's n value, considering that the land use and topography vary significantly across ORB. Even though the discretization of the ORB for simulating hydrodynamics is different from that for SWAT modeling, the loosely coupled nature between the two models still supports the exchange of information from each reach of SWAT to LISFLOOD-FP. The sub-division of ORB for hydrodynamic simulation also enables parallel execution using high performance computational resources. In this case, all six segments of ORB are simultaneously executed in Purdue University's Carter cluster, using the Diffusive Solver method and with different combinations of channel and floodplain n values (0.01–0.05 and 0.03–0.15, respectively) for each segment. These iterative simulations are continued until the model generated inundation map matched well with the 100-year reference flood map developed by the

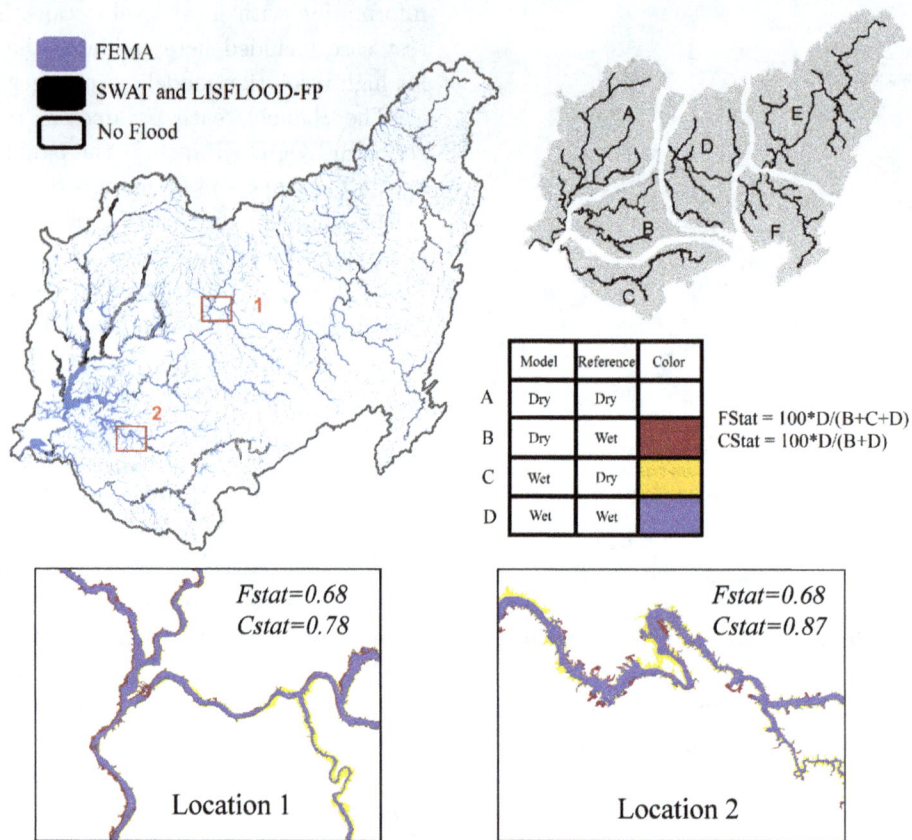

Fig. 5.   Basin discretization for parallel processing of LISFLOOD-FP and the evaluation of model generated 100-year flood inundation with corresponding FEMA reference map.

U.S. Federal Emergency Management Agency (FEMA, https://msc.fema.gov/portal) (Fig. 5).

The F and C statistics (e.g. Horritt and Bates 2002; Sangwan and Merwade 2015) are used as evaluation metrics to compare LISFLOOD-FP generated inundation maps with FEMA maps (Fig. 5). The F and C values range from 0 to 1, where 0 stands for no overlap between the model predicted inundation area and FEMA flood polygon; 1 stands for the case when both are identical by 100%. The scores for the two selected areas (denoted as Location 1 and 2 in Fig. 5) are found close to 70% or higher, thus suggesting reasonable performance by the SWAT and LISFLOOD-FP framework over such a large scale.

The illustrations in Fig. 5 are solely deterministic. Despite the acceptable performance both from SWAT (Fig. 3) and LISFLOOD-FP (Fig. 5), the model results may not be very accurate mainly due to the uncertainty induced from input climate forcing, models' physical conceptualizations and calibration equifinality. From this perspective, Figure 6 shows a more conservative prediction where the SWAT generated maximum and minimum 100-year flow (95% confidence interval) are both fed into the LISFLOOD-FP model in order to create a probabilistic flood inundation map. Figure 6 also compares the inundation extent if only the available USGS gauge station flow data is used in LISFLOOD-FP instead of SWAT outputs.

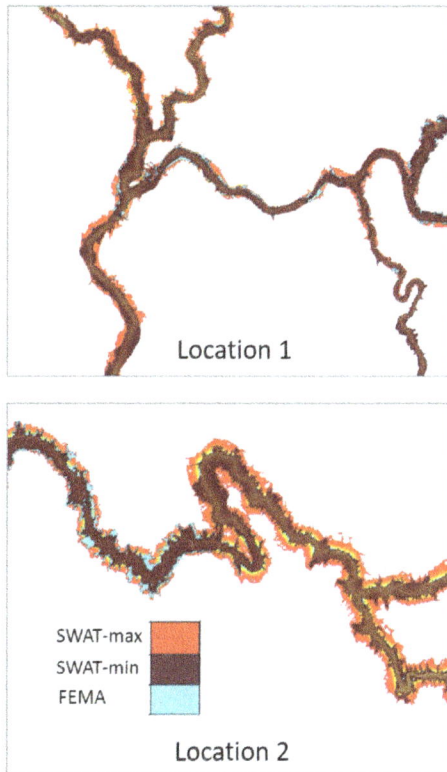

Fig. 6. Probabilistic flood mapping. See adjacent discussion for detail mechanism. Locations 1 and 2 are identical as in Fig. 5.

5. Realizing the Integrated Flood Inundation Modeling Framework Vision

The prototype application presented with the coupling of SWAT and LISFLOOD-FP shows the utility of such an approach for getting flood inundation information over large areas. However, this approach needs to be augmented with several functionalities, including model simulation at higher resolution, ingestion of multiple weather inputs, model calibration with spatially distributed remote sensing observations, execution of multiple models, and better representation of river geometry in hydrodynamic modeling, among others. How these functionalities can be incorporated into the integrated framework is briefly discussed below.

5.1. Representation of high resolution stream network

The prototype model setup shown in Fig. 2 simulates streamflow for only 125 reaches and sub-basins, thus providing the hydrograph at limited locations. Although this setup addresses the issue of relative data scarcity in terms of the availability of existing gauging stations, such coarse resolution setup is not capable of producing local level details for lower order streams. This is clearly revealed from Fig. 5, where the model is able to create flood maps only along the main rivers. Hence, to support an operational flood forecasting system, a large scale model should also include a high resolution stream network.

Figure 7 compares the current setup of SWAT and LISFLOOD-FP with a high resolution stream network derived from the National Hydrography Dataset Plus-version 2 (NHDPlus-v2). Compared to only 125 reaches of the current setup (Fig. 2), Fig. 7 has more than 30,000 reaches. The proposed SWAT and LISFLOOD-FP framework is capable of ingesting such high resolution stream network without significant re-development of the SWAT model. The ability to execute LISFLOOD-FP in parallel by using HPC resources will produce flood inundation for any number of reaches included in the hydrodynamic simulation.

5.2. Multi-model ensemble flood modeling

Outputs from any hydrologic or hydraulic model get uncertainty from multiple sources including the input data, model structure and process representation. Therefore, using a single model for simulating hydrology or hydrodynamics as presented in the SWAT and LISFOOD-FP coupling will create uncertain outputs. In an integrated streamflow and flood modeling framework, it is necessary to include multiple models so an ensemble of predictions can be produced to get

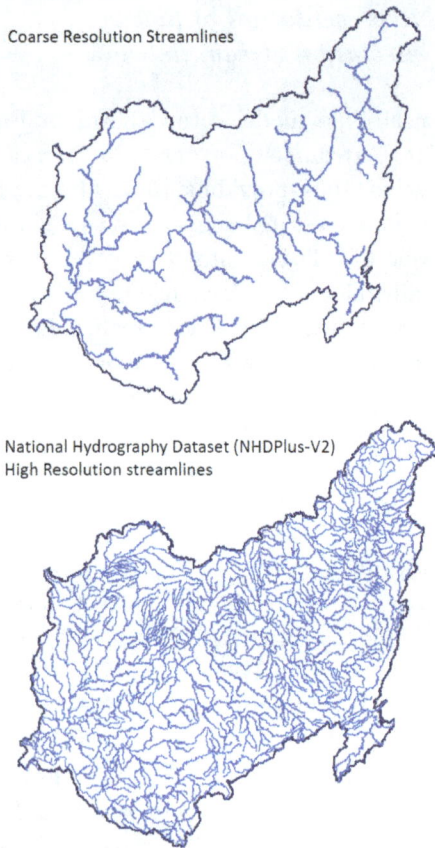

Fig. 7. Spatial comparison between a low and high resolution stream network.

probabilistic flood inundation maps. However, multi-model comparison over large scales and with high spatial resolution is yet to be fully explored.

In a related study, SWAT and VIC-RAPID modeling frameworks are compared for streamflow prediction skills over ORB. Gridded outputs of VIC generated surface/subsurface runoff from the North American Land Data Assimilation System, version 2 (NLDAS-v2; Xia *et al.* (2012), Cai *et al.* (2014)) are ingested into the high resolution stream network (Fig. 7), and then routed through the Routing Application for Parallel Computation of Discharge model (RAPID; David *et al.* (2011), Tavakoly *et al.* (2016)). To maintain a uniform comparison, SWAT is re-set with the same spatial resolution

and forced with NLDAS-v2 primary climate forcing (Rui and Mocko 2014). Both SWAT and RAPID are calibrated against streamflow observations at the same gauge stations as shown in Fig. 3. Figure 8a compares the simulated streamflow hydrographs from the two models along with the observed data for a flood event that occurred in February, 2008. The Flow-Duration Curves (FDCs) showing the exceedance probability of high flow conditions (80th percentile) over a period of 8 years (2000–2008) are also presented.

Figure 8b compares the flood inundation extents from two different hydraulic models, LISFLOOD-FP and HEC-RAS 2D (US-ACE, 2015), with 100-year FEMA reference maps for a short reach extent of St. Joseph river (USGS 04101000) in north-eastern Indiana, USA where both models are run with the same DEM and 100-year flood flow.

Clearly the simulated streamflow and flood inundation extent from the respective hydrologic/hydraulic models can be significantly different, but it is difficult to conclude whether one model is better or worse than the other. The differences in model outputs presented here support the notion of including an ensemble of models to account for prediction uncertainty in an integrated flood modeling framework.

5.3. *Ingesting remotely sensed observations in model evaluation*

With the rapid growth in instrumentation technology and internet, information on various hydro-climatic variables from field, air-borne and space sensors are becoming easily accessible to researchers and general public. Ingesting or assimilating this information into existing models can significantly improve model prediction. The following discussion briefly presents two approaches for evaluating the performance of hydrologic and hydraulic model by using remotely sensed data.

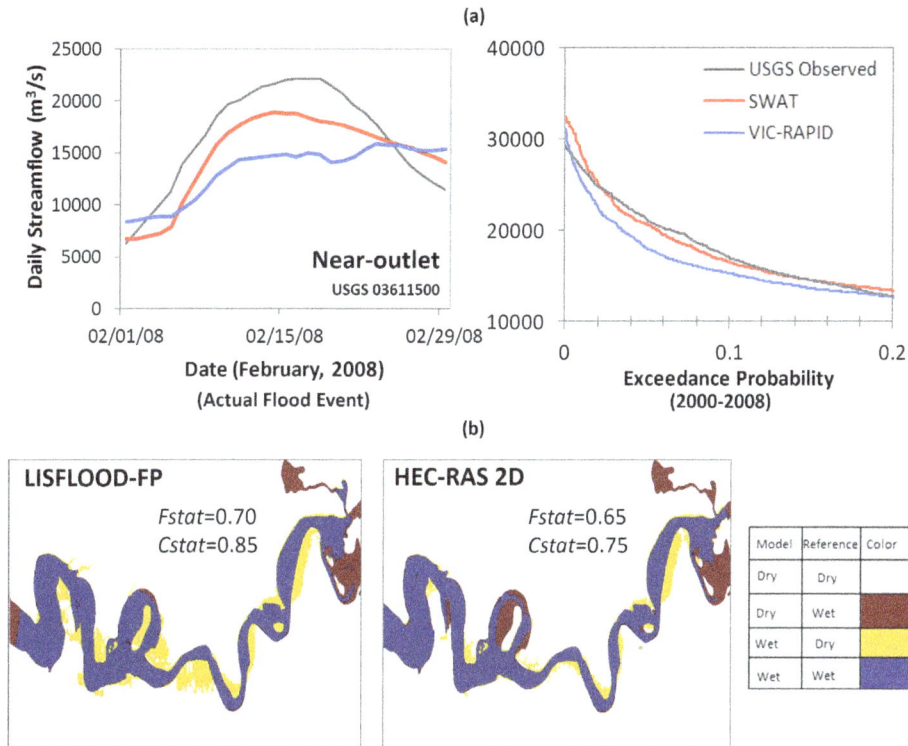

Fig. 8.   Comparison of multi-model prediction skills.

Generally a hydrologic model is evaluated or validated by comparing model generated streamflow hydrograph with observed hydrograph at one or more locations. Similarly, a hydraulic model is evaluated by comparing water surface elevations observed at discrete locations along a stream. In such point-based evaluation, it is not possible to evaluate how the model is simulating watershed's hydrologic responses related to runoff generation, soil moisture dynamics and evapotranspiration that vary in both space and time. As a result, a calibrated model can still contain significant uncertainty and equifinality (Beven 1993). Using remotely sensed spatially distributed surface/subsurface observations in model calibration, in addition to streamflow or stage height, can help to constrain the model's parameters to take into account spatial dynamics, thereby improving the model's overall performance.

Figure 9 shows the application of remotely sensed surface soil moisture ($\sim$1 cm top soil) estimates in calibrating the SWAT model simultaneously at individual sub-basins, along with observed streamflow data at particular stream location. The application is based on the Upper Wabash watershed (18,000 km^2, USGS 03335500) in the north-western part of ORB. Remote sensing data are obtained from the gridded level-3 land surface product (Njoku 2004) of the Advanced Microwave Scanning Radiometer — Earth Observing System (AMSR–E) and disaggregated into individual sub-basins using a python-based scaling algorithm. Figure 9 shows that a model calibrated by using only streamflow produces reasonable output in terms of the fitness coefficients such as the KGE and R^2. However, when the model was calibrated using a multi-objective approach by simultaneously using AMSR-E surface moisture

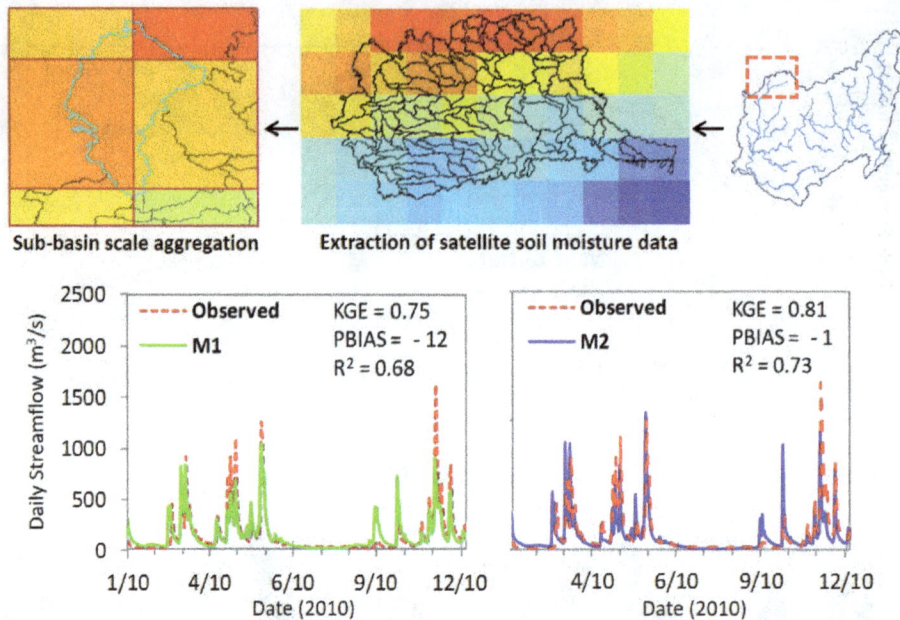

Fig. 9.   Calibration of SWAT model with remote sensing surface soil moisture data (Upper Wabash watershed, USGS 03335500); M1: Calibration only with streamflow data at watershed's outlet, M2: multi-objective calibration using sub-basin scale remote sensing data along with M1.

at each sub-basin along with the streamflow data at the outlet (M2), the model performance is improved. Specifically, the model's ability to capture the peak flows in the M2 configuration is quite distinctive. Besides soil moisture, remotely sensed evapotranspiration can also be used to constrain model parameterization by using a similar multi-objective approach. While the attempt to enhance streamflow prediction skills by calibrating hydrologic models with remotely sensed subsurface observations is relatively recent, many studies have already attempted the ingestion of remotely sensed/*in situ* measurements via data assimilation technique with limited successes (e.g. Alvarez-Garreton *et al.* 2015; Brocca *et al.* 2012; Lei *et al.* 2014; Pauwels *et al.* 2002).

Figure 9 showed how hydrologic model's evaluation and performance can be improved by making use of remotely sensed soil moisture information. Similarly, hydraulic modeling can also be improved by using remotely sensed

information. The evaluation of hydraulic model for simulated inundation extents shown in Fig. 5 and Fig.8 is based on standardized regulatory reference maps corresponding to a flood magnitude of 100-year return period, which is not an observed event. This was done primarily to compare how the 100-year simulation from a large scale hydraulic model compares with another model that is locally created through extensive data collection and calibration. Another approach that can be used for evaluating the model output is through comparison with actual historical flood extents.

Figure 10 shows two aerial images from the Landsat satellite, taken before and during the June 2008 flood in southern Indiana, USA. Due to very distinct water reflectance, it is possible to separate water surface from other land covers, applying various image classification techniques. Ancillary data such as topography and soil based floodplain information (e.g., Sangwan and Merwade 2015) can also augment accuracy

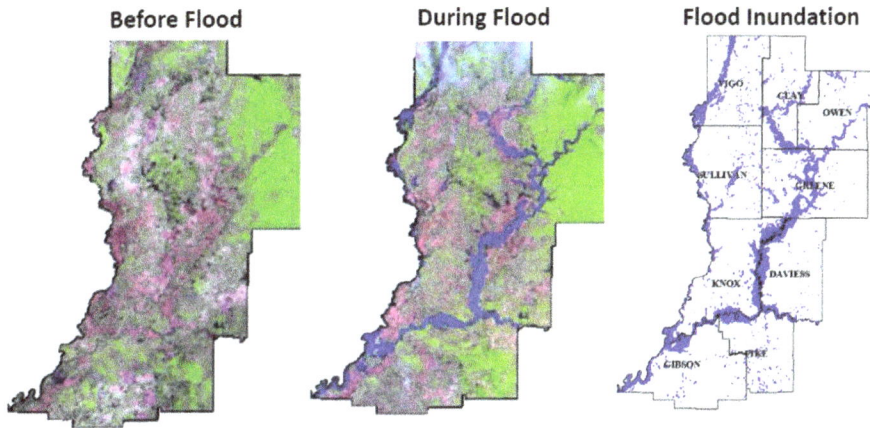

Fig. 10. Landsat images before and during June 2008 flood event in Southern Indiana, USA and the inundation extent map after applying image classification techniques (modified from Shan *et al.* 2009).

in the image classifications. Thus, using a classified set of sequential imagery in pre-, during and/or post- event time stamps (Fig. 10), it is possible to reconstruct a realistic propagation of inundation over the floodplain. Use of these sequential maps as a reference to calibrate hydraulic models is likely to produce best possible estimates of parameters which can reasonably represent different wetness conditions.

In general, imageries can be obtained from a variety of passive sensors (e.g., Landsat, MODIS and SPOT) (Marcus and Fonstad 2008) and radar based active sensors (e.g., ALOS PALSAR, Envisat ASAR, RADARSAT) (Schumann *et al.* 2012). Passive optical remote sensing is only possible under clear weather conditions; hence, images from these sources can be useful to map flood extent provided they are acquired at the time of the occurrence or immediately after the flood. This is a major limitation considering the likelihood that floods are most often associated with severe weather and cloudy conditions. In comparison, active remote sensing through radar based sensors can acquire images under any weather conditions during day and night. Fusion of multi-sensor images can also be used to create inundation maps, which can then be applied to parameterize hydraulic models for improved simulation of flood events.

5.4. Bathymetry representation in hydraulic/hydrodynamic modeling

Topography plays a major role in producing better hydraulic modeling results and flood inundation extents. Both the horizontal resolution of the data and the vertical accuracy directly affect the water surface elevation (WSE) estimate and the inundation extent. With the availability of high quality digital elevation models (DEMs) acquired through Light Detection and Ranging (LiDAR), the representation of floodplain has improved significantly compared to the data acquired through traditional surveys and other digital datasets. LiDAR DEMs can have horizontal resolution as low as 0.5 m and a vertical accuracy ranging from 0.15 to 0.25 m. However, many areas still lack LiDAR generated DEM due to the higher cost associated with the data collection. LiDAR data are available for less than 40 percent of the continental United States (Snyder 2012), although the coverage may have increased since 2012. Other DEMs, such as those derived from the Shuttle Radar Topography Mission (SRTM) and Interferometric Synthetic Aperture Radar (IfSAR), have very coarse horizontal resolution (greater than 30 meters) and low vertical accuracy in the range of 3–7 meters

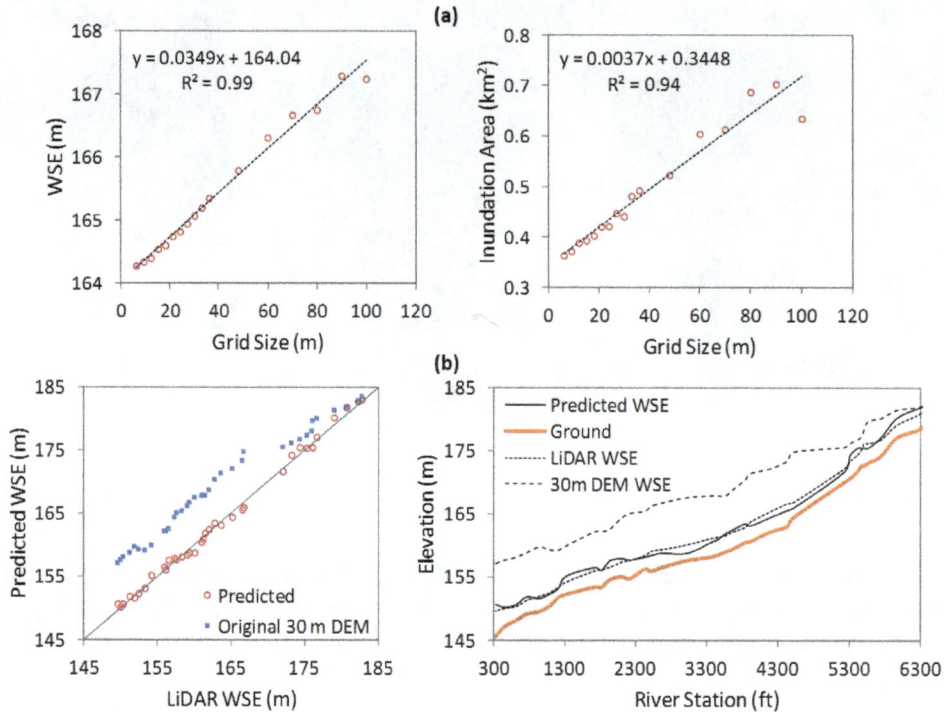

Fig. 11.   (a) Relationship between WSE and inundation area with DEM grid size for Strouds Creek in North Carolina. (b) Improvement in WSE when the linear relationship is applied to results from 30 m DEM. The water profile obtained from 30M DEM shifts to match with LiDAR WSE (modified from Saksena and Merwade 2015).

(Gesch *et al.* 2014). In areas with poor topography representation, especially in the developing countries, it is difficult to get reliable flood inundation extent. To address the issue of poor topography, Saksena and Merwade (2015) developed an approach to improve the hydraulic modeling results by applying a correction based on the horizontal resolution and vertical accuracy of the topography dataset. Figure 11 shows that the WSE estimate and the inundation extent are linearly related to DEM properties and this linear relationship can be extrapolated to get the WSE that would correspond to a higher resolution DEM with high vertical accuracy. In creating flood inundation extent at regional or continental scale, it may not be possible to have a homogenous DEM for the entire area — some areas will have a better DEM

compared to the other areas. The approach suggested by Saksena and Merwade (2015) could be used in such circumstances to improve the flood modeling results and get reliable inundation extents.

Better topographic representation in floodplain solves one issue associated with topography. Most LiDAR DEMs and other topography datasets still do not include river bathymetry information. Inadequate or no representation of river bathymetry can also have significant impact on hydraulic simulations and flood inundation extents (Cook and Merwade 2009). Most river bathymetry information is still collected in the form of traditional cross section surveys or by using a boat mounted single/multibeam echo-sounder, typically at a spacing of more than 100 meters along the river. As a

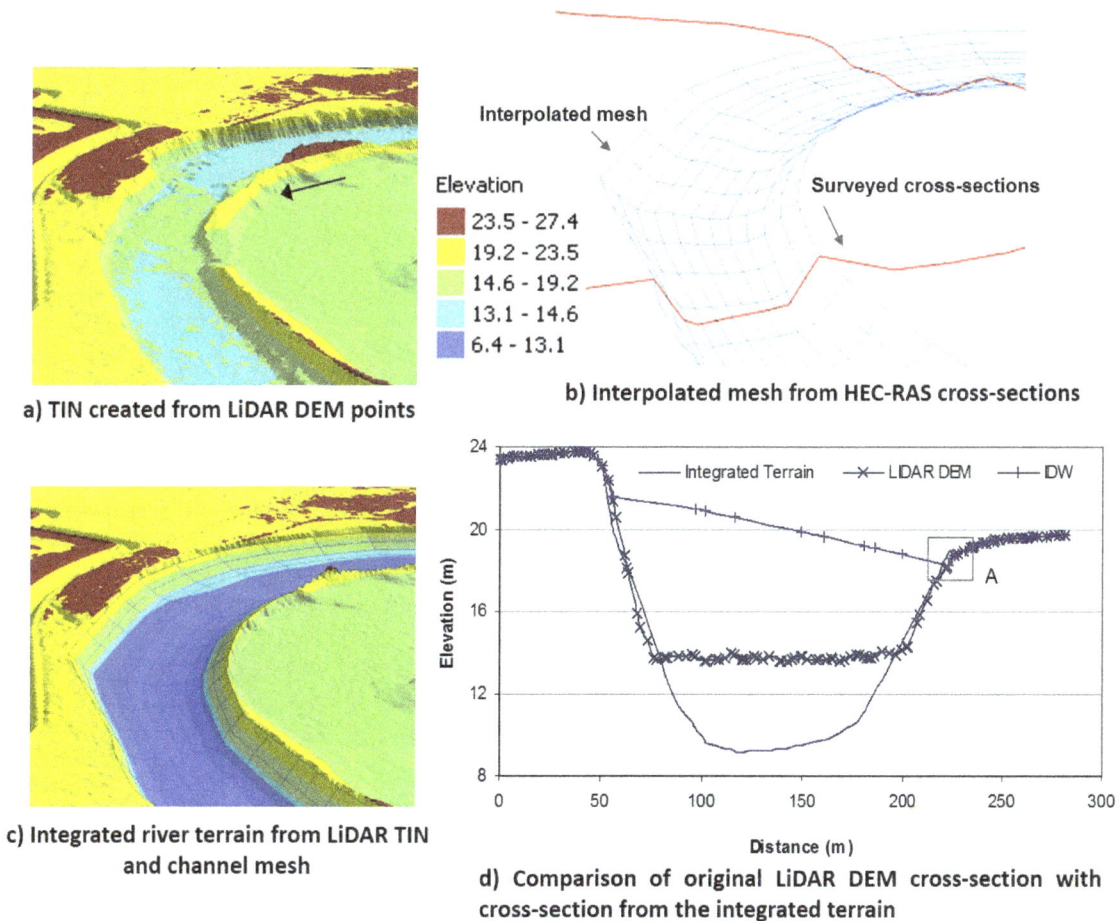

a) TIN created from LiDAR DEM points

b) Interpolated mesh from HEC-RAS cross-sections

c) Integrated river terrain from LiDAR TIN and channel mesh

d) Comparison of original LiDAR DEM cross-section with cross-section from the integrated terrain

Fig. 12.   Terrain model after integrating river bathymetry into DEM for Brazos river (USGS 08114000) in Texas, USA (modified from Merwade *et al.* 2008).

result, these discrete datasets, which are available as points or lines, do not necessarily match with the resolution of the DEM/LiDAR. Considering these inconsistency factors, Merwade *et al.* (2008) proposed a method for integrating cross-sections into DEMs to create an integrated elevation model that includes both river bathymetry and floodplain topography (Fig. 12).

Similarly, other techniques such as creating surfaces using empirical equations (e.g. Merwade and Maidment 2004) can be used to generate river bathymetry in data sparse rivers. Data for river geometry can also be collected from past studies. For example, Fig. 13 shows cross sections from HEC-RAS models that have been used in creating Flood Insurance Rate Maps (FIRMs) in the state of Alabama, USA. The cross sections cover approximately 40% of the streams in the state. It is possible that these data can be used in conjunction with other newly collected or surveyed as well as empirically derived datasets, enabling development of an integrated DEM and continuous river bathymetry for the entire state.

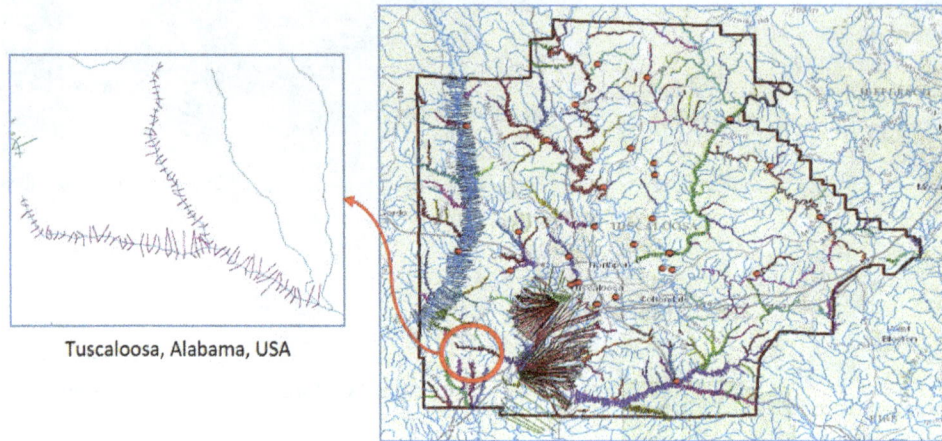

Fig. 13.   Map of river cross-sections from FEMA FIRMS for the state of Alabama, USA.

6. Concluding Remarks

This chapter presented a vision for an integrated hydrologic and hydraulic modeling approach to enable high resolution flood inundation mapping at regional to continental scale. The prototype presented here using SWAT and LISFLOOD-FP shows promise in creating flood inundation maps at much higher resolution than what is possible by using only gauged streamflow information on major river reaches. However, the results from this prototype have uncertainty that can be addressed by using multiple models to create an ensemble of results to give probabilistic flood inundation maps. Similarly, the modeling framework needs to incorporate as much information as possible from emerging field and remote sensing instruments. Specifically, remote sensing data deliver information on hydrologic variables, such as soil moisture and evapotranspiration, which have previously been either ignored or not fully incorporated in creating hydrologic models. Calibrating hydrologic models against point observations of streamflow can sometimes compromise the heterogeneous representation of various hydrologic processes within a model. Therefore, incorporating spatio-temporal information related to soil moisture and evapotranspiration provides another avenue to

reduce uncertainty in hydrologic predictions by improving models' representation of hydrologic processes.

An accurate hydrologic prediction can still yield poor flood inundation if the topography used in simulating the river hydrodynamics is of poor quality. Some new approaches show how hydrodynamic results obtained from a coarse resolution DEM can be improved by relating DEM properties such as grid size and vertical accuracy to flood modeling results. Similarly, availability of bathymetry data and the tools to incorporate such information into a DEM can significantly improve modeling of both high and low flows in a large scale hydrodynamic model.

While the models and their results can be improved by including more information, creating high resolution large scale models is still hindered by access to computational resources. The prototype presented in this chapter used distributed computing for LISFLOOD-FP, but SWAT was still executed as a single model. The execution of LISFLOOD-FP using distributed computing was still time intensive. Moreover, not many hydrologic and hydraulic models can be run using distributed computing. Some of these challenges provide opportunities to advance the computational side of hydrology and hydrodynamics to realize the true vision of integrated high resolution flood inundation modeling at large scales.

Acknowledgments

The authors are thankful to Kimberly Peterson and Deb Burrow from the Lyles School of Civil Engineering at Purdue University for proofreading this chapter.

References

Abbaspour, K. C., 2015: SWAT-CUP 2012: SWAT calibration and uncertainty programs — a user manual. Available online at: http://swat.tamu.edu/media/114860/usermanual_swatcup.pdf

Abbaspour, K. C., E. Rouholahnejad, S. Vaghefi, R. Srinivasan, H. Yang, and B. Kløve, 2015: A continental-scale hydrology and water quality model for Europe: Calibration and uncertainty of a high-resolution large-scale SWAT model. *J. Hydrol.*, **524**, 733–752. doi:10.1016/j.jhydrol.2015.03.027.

Alfieri, L., P. Burek, E. Dutra, B. Krzeminski, D. Muraro, J. Thielen, and F. Pappenberger, 2013: GloFAS — global ensemble streamflow forecasting and flood early warning. *Hydrol. Earth Syst. Sc.*, **17**, 1161–1175. doi:10.5194/hess-17-1161-2013.

Alvarez-Garreton, C., D. Ryu, A. W. Western, C.-H. Su, W. T. Crow, D. E. Robertson, and C. Leahy, 2015: Improving operational flood ensemble prediction by the assimilation of satellite soil moisture: comparison between lumped and semi-distributed schemes. *Hydrol. Earth Syst. Sc.*, **19**, 1659–1676. doi:10.5194/hess-19-1659-2015.

Andreadis, K. M., G. J.-P. Schumann, and T. Pavelsky, 2013: A simple global river bankfull width and depth database. *Water Resour. Res.*, **49**, 7164–7168. doi:10.1002/wrcr.20440

Arnold, J., D. Moriasi, P. Gassman, K. Abbaspour, M. White, R. Srinivasan, C. Santhi, R. D. Harmel, and A. Van Griensven, 2012: SWAT: Model use, calibration, and validation. *Trans. ASABE*, **55**, 1491–1508.

Arnold, J. G., R. Srinivasan, R. S. Muttiah, and P. M. Allen, 1999: Continental scale simulation of the hydrologic balance. *J. Am. Water Resour. Assoc.*, **35**, 1037–1051.

AWDR, 2006: Freshwater Resources in Africa: African Water Development Report, 380p.

Available online at: http://repository.uneca.org/handle/10855/22091 (last cited on February 14, 2016).

Bates, P., M. Trigg, J. Neal, and A. Dabrowa, 2013: LISFLOOD-FP user manual — version 5.9.6, University of Bristol, UK. Available online at: http://www.bristol.ac.uk/media-library/sites/geography/migrated/documents/lisflood-manual-v5.9.6.pdf (last cited on February 14, 2016).

Bates, P. D., M. S. Horritt, and T. J. Fewtrell, 2010: A simple inertial formulation of the shallow water equations for efficient two-dimensional flood inundation modelling. *J. Hydrol.*, **387**, 33–45. doi:10.1016/j.jhydrol.2010.03.027

Bates, P. D. and A. P. J. De Roo, 2000: A simple raster-based model for flood inundation simulation. *J. Hydrol.*, **236**, 54–77. doi:10.1016/S0022-1694(00)00278-X.

Beven, K., 1993: Prophecy, reality and uncertainty in distributed hydrological modelling. *Adv. Water Resour.*, **16**, 41–51. doi:10.1016/0309-1708(93)90028-E.

Bourdin, D. R., S. W. Fleming, and R. B. Stull, 2012: Streamflow modelling: A primer on applications, approaches and challenges. *Atmosphere-Ocean*, **50**, 507–536. doi:10.1080/07055900.2012.734276.

Brocca, L., T. Moramarco, F. Melone, W. Wagner, S. Hasenauer, and S. Hahn, 2012: Assimilation of surface- and root-zone ASCAT soil moisture products into rainfall-runoff modeling. *IEEE Trans. Geosci. Remote Sens.*, **50**, 2542–2555. doi:10.1109/TGRS.2011.2177468.

Cai, X., Z.-L. Yang, Y. Xia, M. Huang, H. Wei, L. R. Leung, and M. B. Ek, 2014: Assessment of simulated water balance from Noah, Noah-MP, CLM, and VIC over CONUS using the NLDAS test bed. *J. Geophys. Res. Atmos.*, **119**, 13,751–13,770.

Chen, Y.-R. and B. Yu, 2015: Impact assessment of climatic and land-use changes on flood runoff in southeast Queensland. *Hydrol. Sci. J.*, **60**. doi:10.1080/02626667.2014.945938.

Cook, A. and V. Merwade, 2009: Effect of topographic data, geometric configuration and modeling approach on flood inundation mapping. *J. Hydrol.*, **377**, 131–142. doi:10.1016/j.jhydrol.2009.08.015.

David, C. H., D. R. Maidment, G.-Y. Niu, Z.-L. Yang, F. Habets, and V. Eijkhout, 2011: River network routing on the NHDPlus dataset. *J. Hydrometeorol.*, **12**, 913–934. doi:10.1175/2011JHM1345.1.

Faramarzi, M., K. C. Abbaspour, S. Ashraf Vaghefi, M. R. Farzaneh, A. J. B. Zehnder, R, Srinivasan, and H. Yang, 2013: Modeling impacts of climate change on freshwater availability in Africa. *J. Hydrol.*, **480**, 85–101. doi:10.1016/j.jhydrol.2012.12.016.

Gaughan, A. E. and P. R. Waylen, 2012: Spatial and temporal precipitation variability in the Okavango–Kwando Zambezi catchment, Southern Africa. *J. Arid Environ.*, **82**, 19–30. doi:10.1016/j.jaridenv.2012.02.007.

Gesch, D. B., M. J. Oimoen, and G. A. Evans, 2014: Accuracy assessment of the U.S. geological survey national elevation dataset, and comparison with other large-area elevation datasets — SRTM and ASTER: U.S. Geological Survey Open-File Report 2014–1008, 10 p., http://dx.doi.org/10.3133/ofr20141008 (last cited on February 14, 2016).

Goodwell, A. E., Z. Zhu, D. Dutta, J. A. Greenberg, P. Kumar, M. H. Garcia, B. L. Rhoads, R. R. Holmes, G. Parker, D. P. Berretta, and R. B. Jacobson, 2014: Assessment of floodplain vulnerability during extreme Mississippi River flood 2011. *Environ. Sci. Technol.*, **48**, 2619–2625. doi:10.1021/es404760t.

Gosain, A. K., S. Rao, and D. Basuray, 2006: Climate change impact assessment on hydrology of Indian river basins. *Curr. Sci.*, **90**, 346–353.

Gupta, H. V., H. Kling, K. K. Yilmaz, and G. F. Martinez, 2009: Decomposition of the mean squared error and NSE performance criteria: Implications for improving hydrological modelling. *J. Hydrol.*, **377**, 80–91. doi:10.1016/j.jhydrol.2009.08.003.

He, Y., A. Bárdossy, and E. Zehe, 2011. A review of regionalisation for continuous streamflow simulation. *Hydrol. Earth Syst. Sc.*, **15**, 3539–3553. doi:10.5194/hess-15-3539-2011.

Horritt, M. S. and P. D. Bates, 2002: Evaluation of 1D and 2D numerical models for predicting river flood inundation. *J. Hydrol.*, **268**, 87–99. doi:10.1016/S0022-1694(02)00121-X

Horritt, M. S. and P. D. Bates, 2001: Predicting floodplain inundation: Raster-based modelling versus the finite-element approach. *Hydrol. Process.*, **15**, 825–842. doi:10.1002/hyp.188.

Hrachowitz, M., H. H. G. Savenije, G. Blöschl, J. J. McDonnell, M. Sivapalan, J. W. Pomeroy, B. Arheimer, T. Blume, M. P. Clark, U. Ehret, F. Fenicia, J. E. Freer, A. Gelfan, H. V. Gupta, D. A. Hughes, R. W. Hut, A. Montanari, S. Pande, D. Tetzlaff, P. A. Troch,

S. Uhlenbrook, T. Wagener, H. C. Winsemius, R. A. Woods, E. Zehe, and C. Cudennec, 2013: A decade of Predictions in Ungauged Basins (PUB) — a review. *Hydrol. Sci. J.*, **58**, 1198–1255. doi:10.1080/02626667.2013.803183.

Interagency Advisory Committee on Waster Data (IACWD), 1982: "Guidelines for determining flood flow frequency." *Bulletin 17B of the Hydrology Subcommittee*, Office of Water Data Coordination, U.S. Geological Survey, Reston, Va.

Jaynes, D. B. and D. E. James, 2007: The Extent of Farm Drainage in the United States. p. 50. In *Final Program and Abstracts, Soil and Water Conserv. Soc., 2007 Internat. Conf.*, 21–25 July, FL, USA. Available online at: http://www.ars. usda.gov/SP2UserFiles/Place/50301500/TheExt entofFarmDrainageintheUnitedStates.pdf (last cited on July 9, 2015).

Kauffeldt, A., F. Wetterhall, F. Pappenberger, P. Salamon, and J. Thielen, 2016: Technical review of large-scale hydrological models for implementation in operational flood forecasting schemes on continental level. *Environ. Model. Softw.*, **75**, 68–76. doi:10.1016/j.envsoft. 2015.09.009.

Kling, H., M. Fuchs, and M. Paulin, 2012: Runoff conditions in the upper Danube basin under an ensemble of climate change scenarios. *J. Hydrol.*, **424–425**, 264–277. doi:10.1016/j. jhyd rol.2012.01.011.

Kumar, S. and V. Merwade, 2009: Impact of watershed subdivision and soil data resolution on SWAT model calibration and parameter uncertainty. *J. Am. Water Resour. Assoc.*, **45**, 1179–1196. doi:10.1111/j.1752-1688.2009.00353.x.

Kundzewicz, Z. W., S. Kanae, S. I. Seneviratne, J. Handmer, N. Nicholls, P. Peduzzi, R. Mechler, L. M. Bouwer, N. Arnell, K. Mach, R. Muir-Wood, G. R. Brakenridge, W. Kron, G. Benito, Y. Honda, K. Takahashi, and B. Sherstyukov, 2014: Flood risk and climate change: Global and regional perspectives. *Hydrol. Sci. J.*, **59**. doi:10.1080/02626667.2013.857411.

Larose, M., G. C. Heathman, L. D. Norton, and B. Engel, 2007: Hydrologic and atrazine simulation of the Cedar Creek Watershed using the SWAT model. *J. Environ. Qual.*, **36**, 521–531. doi:10.2134/jeq2006.0154.

Lei, F., C. Huang, H. Shen, and X. Li, 2014: Improving the estimation of hydrological states in the SWAT model via the ensemble Kalman smoother: Synthetic experiments for the Heihe River Basin in northwest China. *Adv. Water*

Resour., **67**, 32–45. doi:10.1016/j.advwatres.2014.02.008.

Loukas, A. and L. Vasiliades, 2014: Streamflow simulation methods for ungauged and poorly gauged watersheds. *Nat. Hazards Earth Syst. Sci.*, **14**, 1641–1661. doi:10.5194/nhess-14-1641-2014.

Maidment, D. R., 2015: A conceptual framework for the national flood interoperability experiment, 22p. Available online at: https://www.cuahsi.org/Files/Pages/documents/13623/nfieconceptualfra mework_revised_feb_9.pdf (last cited on February 14, 2016).

Marcus, W. A. and M. A. Fonstad, 2008: Optical remote mapping of rivers at sub-meter resolutions and watershed extents. *Earth Surf. Process. Landforms*, **34**, 4–24. doi:10.1002/esp.

Merwade, V. and D. R. Maidment, 2004: Geospatial Description of River Channels in Three Dimensions. The University of Texas at Austin. Center for Research in Water Resources online report 04-8. Available at: http://www.crwr.utexas.edu/reports/2004/rpt04-8.shtml

Merwade, V., A. Cook, and J. Coonrod, 2008: GIS techniques for creating river terrain models for hydrodynamic modeling and flood inundation mapping. *Environ. Model. Softw.*, **23**, 1300–1311. doi:10.1016/j.envsoft.2008.03.005.

Moriasi, D. N., J. G. Arnold, M. W. Van Liew, R. L. Binger, R. D. Harmel, and T. L. Veith, 2007: Model evaluation guidelines for systematic quantification of accuracy in watershed simulations. *Trans. ASABE*, **50**, 885–900. doi:10.13031/2013.23153.

NOAA, 2014: CI-FLOW Fact Sheet: National Oceanic and Atmospheric Administration. Available online at: https://ciflow.nssl.noaa.gov/documents/ (last cited on February 14, 2016).

Neitsch, S. L., J. Arnold, J. Kiniry, and J. Williams, 2011: Soil & Water Assessment Tool theoretical documentation version 2009. Texas A&M University System, College Station, TX, USA.

Njoku, E. G., 2004: AMSR-E/Aqua Daily L3 Surface Soil Moisture, Interpretive Parameters, & QC EASE-Grids. Version 2. Boulder, Colorado USA: NASA National Snow and Ice Data Center Distributed Active Archive Center. doi:10.5067/AMSR-E/AE_LAND3.002.

O'Donnell, G. M., K. P. Czajkowski, R. O. Dubayah, and D. P. Lettenmaier, 2000: Macroscale hydrological modeling using remotely sensed inputs: Application to the Ohio River basin. *J. Geophys. Res.*, **105**, 12499–12516. doi: 10.1029/1999jd901193.

Paiva, R. C. D., W. Collischonn, and D. C. Buarque, 2013: Validation of a full hydrodynamic model for large-scale hydrologic modelling in the Amazon. *Hydrol. Process.*, **27**, 333–346. doi:10.1002/hyp.8425.

Paiva, R. C. D., W. Collischonn, and C. E. M. Tucci, 2011: Large scale hydrologic and hydrodynamic modeling using limited data and a GIS based approach. *J. Hydrol.*, **406**, 170–181. doi:10.1016/j.jhydrol.2011.06.007.

Pappenberger, F., E. Dutra, F. Wetterhall, and H. L. Cloke, 2012: Deriving global flood hazard maps of fluvial floods through a physical model cascade. *Hydrol. Earth Syst. Sci.*, **16**, 4143–4156. doi:10.5194/hess-16-4143-2012.

Pauwels, V. R. N., R. Hoeben, N. E. C. Verhoest, F. P. De Troch and P. A. Troch, 2002: Improvement of TOPLATS-based discharge predictions through assimilation of ERS-based remotely sensed soil moisture values. *Hydrol. Process.*, **16**, 995–1013. doi:10.1002/hyp.315.

Peduzzi, P., H. Dao, C. Herold, and F. Mouton, 2009: Assessing global exposure and vulnerability towards natural hazards: The Disaster Risk Index. *Nat. Hazards Earth Syst. Sci.*, **9**, 1149–1159. doi:10.5194/nhess-9-1149-2009.

Pruski, F. F., A. D. A. Nunes, P. L. Pruski, and R. D. G. Rodriguez, 2013: Improved regionalization of streamflow by use of the streamflow equivalent of precipitation as an explanatory variable. *J. Hydrol.*, **476**, 52–71. doi:10.1016/j.jhydrol.2012.10.005.

Rajib, M. A. and V. Merwade, 2016: Improving soil moisture accounting and streamflow prediction in SWAT by incorporating a modified time-dependent Curve Number method. *Hydrol. Process.*, **30**, 603–624. doi:10.1002/hyp.10639.

Rajib, M. A., V. Merwade, I. L. Kim, L. Zhao, C. Song, and S. Zhe, 2016: SWATShare — A web platform for collaborative research and education through online sharing, simulation and visualization of SWAT models. *Environ. Model. Softw.*, **75**, 498–512. doi:10.1016/j.envsoft.2015.10.032.

Razavi, T. and P. Coulibaly, 2013: Streamflow prediction in ungauged basins: Review of regionalization methods. *J. Hydrol. Eng.*, **18**, 958–975. doi:10.1061/(ASCE)HE.1943-5584.0000690.

Rui, H. and D. Mocko, 2014: North American Land Data Assimilation System — phase 2 products, Goddard Earth Sciences Data and Information Services Center. Available online at: ftp://hydro1.sci.gsfc.nasa.gov/data/s4pa/NLDAS/RE

ADME.NLDAS2.pdf (last cited on January 5, 2016).

Saksena, S. and V. Merwade, 2015: Incorporating the effect of DEM resolution and accuracy for improved flood inundation mapping. *J. Hydrol.*, **530**, 180–194. doi:10.1016/j.jhydrol.2015.09.069.

Sangwan, N. and V. Merwade, 2015: A faster and economical approach to floodplain mapping using soil information. *J. Am. Water Resour. Assoc.*, **51**, 1286–1304. doi:10.1111/1752-1688.12306.

Schumann, G. J.-P., Paul. D. Bates,G. Di Baldassarre, and D. C. Mason, 2012: The use of radar imagery in riverine flood inundation studies, in fluvial remote sensing for science and management (eds. P. E. Carbonneau and H. Piégay), John Wiley & Sons. doi:10.1002/9781119940791.ch6.

Schumann, G. J.-P., J. C. Neal, N. Voisin, K. M. Andreadis, F. Pappenberger, N. Phanthuwongpakdee, A. C. Hall, and P. D. Bates, 2013: A first large-scale flood inundation forecasting model. *Water Resour. Res.*, **49**, 6248–6257. doi:10.1002/wrcr.20521.

Schuol, J., K. C. Abbaspour, H. Yang, R. Srinivasan, and A. J. B. Zehnder, 2008: Modeling blue and green water availability in Africa. *Water Resour. Res.*, **44**, 1–18. doi:10.1029/2007WR006609.

Sellami, H., I. La Jeunesse, S. Benabdallah, N. Baghdadi, and M. Vanclooster, 2014: Uncertainty analysis in model parameters regionalization: A case study involving the SWAT model in Mediterranean catchments (Southern France). *Hydrol. Earth Syst. Sc.*, **18**, 2393–2413. doi:10.5194/hess-18-2393-2014.

Sivapalan, M., 2003: Prediction in ungauged basins: A grand challenge for theoretical hydrology. *Hydrol. Process.*, **17**, 3163–3170. doi:10.1002/hyp.5155.

Snyder, G. I., 2012: National Enhanced Elevation Assessment at a Glance. U.S. Geological Survey Fact Sheet 2012, 3088, http://pubs.usgs.gov/fs/2012/3088/.

Swain, J. B., R. Jha, and K. C. Patra, 2015: Stream flow prediction in a typical ungauged catchment using GIUH approach, in: *Aquatic Procedia.* **4**, 993–1000. doi:10.1016/j.aqpro.2015.02.125.

Syvitski, J. and G. R. Brakenridge, 2013: Causation and Avoidance of Catastrophic Flooding along the Indus. *GSA Today*, **23**. doi:10.1130/GSATG165A.1.

Tavakoly, A. A., D. R. Maidment, J. W. Mcclelland, T. Whiteaker, Z. L. Yang, C. Griffin, C. H. David, and L. Meyer, 2016: A GIS framework for regional modeling of riverine nitrogen transport: Case study, San Antonio and Guadalupe basins. *J. Am. Water Resour. Assoc.*, **52**, 1–15. doi:10.1111/1752-1688.12355.

Thielen, J., J. Bartholmes, M.-H. Ramos, and A. de Roo, 2009: The European flood alert system — Part 1: Concept and development. *Hydrol. Earth Syst. Sci. Discuss.*, **5**, 257–287. doi:10.5194/hessd-5-257-2008.

US-ACE. 2015: HEC-RAS River Analysis System 2D Modeling User's Manual — version 5.0, US Army Corps of Engineers report no. CPD-68A.

USGS-NED. 2015: National Elevation Dataset: United States Geological Survey National Map Viewer. Available at: http://viewer.nationalmap.gov/viewer/. Accessed 10 March, 2015.

USGS-NLCD. 2015: National Land Cover Data Set, 2006: United States Geological Survey National Map Viewer. Available at: http://viewer.nationalmap.gov/viewer/. Accessed 10 March, 2015.

Voisin, N., F. Pappenberger, D. P. Lettenmaier, R. Buizza, and J. C. Schaake, 2011: Application of a medium-range global hydrologic probabilistic forecast scheme to the Ohio River basin. *Weather Forecast.*, **26**, 425–446. doi:10.1175/WAF-D-10-05032.1.

Whitehead, M. T. and C. J. Ostheimer, 2009: Development of a flood-warning system and flood-inundation mapping for the Blanchard River in Findlay, Ohio: U.S. Geological Survey Scientific Investigations Report 2008–5234, 9p. Available online at: http://pubs.usgs.gov/sir/2008/5234/pdf/sir2008-5234.pdf (last cited on February 14, 2016).

Whitehead, M. T. and C. J. Ostheimer, 2014: Flood-inundation maps and updated components for a flood-warning system for the City of Marietta, Ohio and selected communities along the Lower Muskingum River and Ohio River: U.S. Geological Survey Scientific Investigations Report 2014–5195, 16p. http://dx.doi.org/10.3133/sir20145195 (last cited on February 14, 2016).

Winsemius, H. C., L. P. H. Van Beek, B. Jongman, P. J. Ward, and A. Bouwman, 2013: A framework for global river flood risk assessments. *Hydrol. Earth Syst. Sc.*, **17**, 1871–1892. doi:10.5194/hess-17-1871-2013.

Xia, Y., K. Mitchell, M. Ek, B. Cosgrove, J. Sheffield, L. Luo, C. Alonge, H. Wei, J. Meng, B. Livneh, Q. Duan, and D. Lohmann, 2012:

Continental-scale water and energy flux analysis and validation for North American Land Data Assimilation System project phase 2 (NLDAS-2): 2. Validation of model-simulated streamflow. *J. Geophys. Res.*, **117**, D03110. doi:10.1029/2011 JD016051.

Zang, C. and J. Liu, 2013: Trend analysis for the flows of green and blue water in the Heihe River basin, northwestern China. *J. Hydrol.*, **502**, 27–36. doi:10.1016/j.jhydrol.2013.08.022.

Chapter 10

Service and Research on Seasonal Streamflow Forecasting in Australia

P. M. Feikema*, Q. J. Wang‡, S. Zhou*, D. Shin*, D. E. Robertson†,
A. Schepen§, J. Lerat¶, J. C. Bennett†, N. K. Tuteja¶ and D. Jayasuriya*

*National Forecast Services, Bureau of Meteorology,
Melbourne, Victoria, Australia*

†*Land and Water, Commonwealth Science and Industrial Research Organisation,
Melbourne, Victoria, Australia*

‡*Department of Infrastructure Engineering, The University of Melbourne, Victoria, Australia*
quan.wang@unimelb.edu.au

§*Land and Water, Commonwealth Science and Industrial Research Organisation,
Brisbane, Queensland, Australia*

¶*National Forecast Services, Bureau of Meteorology, Canberra,
Australian Capital Territory, Australia*

Each month, the Australian Bureau of Meteorology issues operational seasonal streamflow forecasts of total water volumes for three months ahead for over 200 locations across Australia (http://www.bom.gov.au/water/ssf/). The forecasts are created using modelling approaches developed by the Commonwealth Scientific and Industrial Research Organisation, through a close and ongoing research partnership. Outcomes from the Seasonal Climate Research Program of the Bureau and hydrologic research conducted by the university sector in Australia have also directly contributed to the development and delivery of this service.

The publicly available forecasts are based on a statistical modelling approach that uses relationships from directly observed streamflow data and derived climate indices. Forecasts based on rainfall-runoff modelling forced by downscaled rainfall forecasts are currently being released to and evaluated by registered users. Work is underway to merge forecasts from these approaches to improve accuracy and reliability.

Delivery and development of the forecasting service requires cooperation with internal and external stakeholders, a comprehensive understanding of user needs, support for research and development, robust operational modelling tools and web-based service delivery systems, a communication and adoption strategy, and an implementation plan. This chapter discusses how these multiple elements are combined to provide a national seasonal streamflow forecasting service for Australia.

1. Introduction

For over 50 years, Australia's Bureau of Meteorology (the Bureau) has been providing routine flood forecasting services to assist emergency services mitigate the effects of flood and to support community safety. However, in 2010 the Bureau started to deliver a seasonal streamflow forecasting service[a] that issues timely, accurate and reliable water availability forecasts across the country (Tuteja 2015a; Vertessy 2013). This service allows water users and managers to plan operations by assessing the risks of different strategies for water use and management.

Several factors prompted the development of the seasonal forecasting service. One was the need to provide water allocation outlooks for water management agencies, and others were

---

[a]Available at http://www.bom.gov.au/water/ssf/.

Bridging Science and Policy Implication for Managing Climate Extremes
Edited by Hong-Sang Jung and Bin Wang

to inform water markets, to plan and manage water use and to manage drought. This need became more urgent during the drought that occurred between 1997 and 2009 (known as the millennium drought; van Dijk *et al.* 2013). Another factor was the *Water Act 2007*, which required the Bureau to provide forecasts of the availability of Australia's water resources over various time frames. Finally, there was the announcement in May 2008 of the *Water for the Future* program, which was backed by the *Water Act 2007* and key stakeholders, and included the *Improving Water Information Program*, administered by the Bureau.

As part of its seasonal forecasting service, at the start of each month the Bureau now provides probabilistic forecasts of total streamflow volume over the next three months for more than 200 locations across Australia.

The timescales for seasonal forecasting are typically from a few weeks up to one year ahead. Forecasting seasonal streamflow is a challenge, because of the low predictability of climate over such time frames. Conversely, the effects on streamflow of the initial conditions of soil moisture, groundwater and other water stores in a catchment can be relatively more predictable over a horizon of one to several months.

There are two broad approaches to seasonal streamflow forecasting. The *statistical approach* relies on past relationships between the observed streamflow and the antecedent climate and catchment conditions to make predictions. The *dynamic approach* is based on a model that simulates processes of climate and hydrological systems. State-of-the-art forecasting methods aim to:

- Capture information from both sources of streamflow predictability (climate and catchment conditions) to produce the most accurate forecasts possible
- Statistically represent the remaining predictive uncertainty in a reliable manner.

The science of seasonal streamflow forecasting has progressed thanks to the Water Information Research and Development Alliance (WIRADA), which brings together the Bureau and the Commonwealth Scientific and Industrial Research Organisation (CSIRO). The Bureau's publicly-available seasonal streamflow forecasting service uses an operational forecasting model — the Bayesian joint probability (BJP) statistical modelling method — developed by CSIRO (Wang *et al.* 2011a; Wang *et al.* 2009). Also, the Bureau, with support from universities and CSIRO, has developed a dynamic hydrologic forecasting approach. To take advantage of the different strengths of the two models and provide one official version of forecasts to the users, a method for merging these statistical and dynamic forecasts has been developed and will be used in the future.

This chapter describes the current seasonal streamflow forecasting service, the science and models that underpin it, and the processes involved in providing the service to end users, including data supply, systems and communication. It explains how the service has evolved to deliver forecasts across Australia, describes the collaborative process by which it was created, and how it is likely to develop.

2. Delivery of the Service

2.1. *Current forecasting service*

Provision of the seasonal streamflow forecasting service to the public started in December 2010. The Bureau provides probabilistic forecasts of total streamflow volumes over the next three months for locations across Australia, making the forecasts available early in each month. Some of the selected locations are gauging points along rivers; others represent flow volumes into water storages. Currently, there are over 200 forecast locations, from which 140 forecasts are made available to the public, and the remaining locations are made available to registered

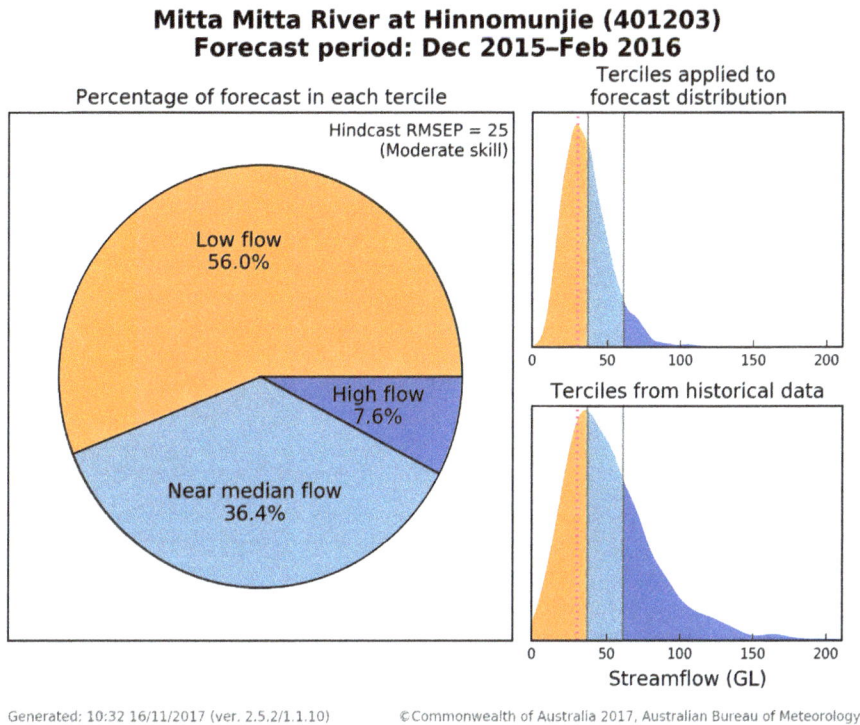

Fig. 1. A probabilistic forecast for the Mitta Mitta River at Hinnomunjie, Victoria, with historical data for the period 1970–2008 for reference.

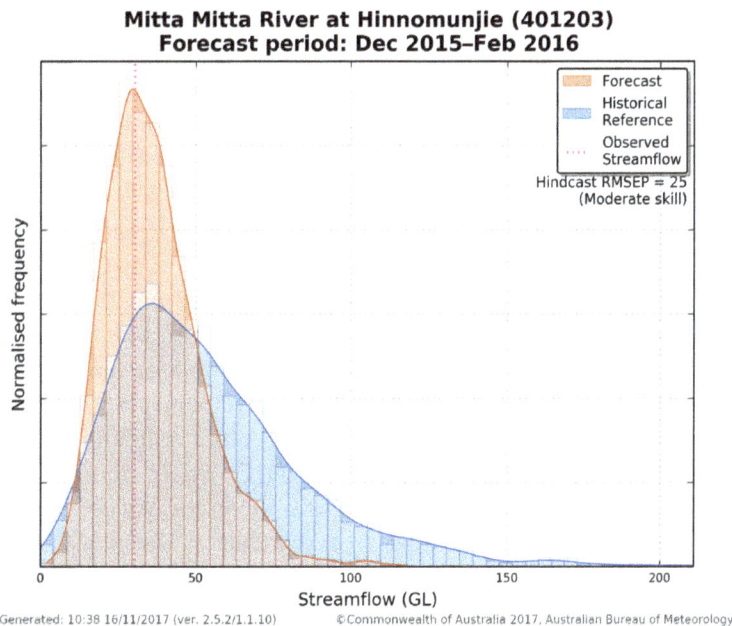

Fig. 2. The probability distributions for the forecast (orange) and historical reference (blue) for the site at Mitta Mitta River at Hinnomunjie. The forecast is a simulation from the Bayesian joint probability (BJP) model based on a 5,000-member ensemble. The historical reference is a probabilistic representation of the historical data.

Mitta Mitta River at Hinnomunjie (401203)
Forecast period: Dec 2015–Feb 2016

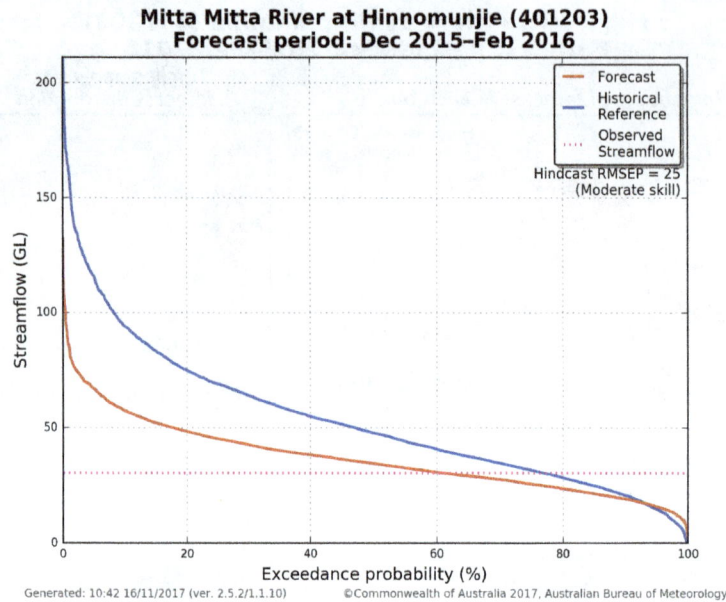

Fig. 3. The exceedance probabilities for the forecast (orange) and historical reference (blue) for the site at Mitta Mitta River at Hinnomunjie. This shows the probability that a given streamflow volume will be exceeded in the forecast period.

users (registered users are typically water management agencies, or similarly interested users). Forecast products include probabilistic forecasts (e.g., Fig. 1), probability distributions (e.g., Fig. 2), exceedance probabilities (e.g., Fig. 3) forecast performance metrics (e.g., Fig. 4) and observed data (e.g., Fig. 5) for each location. These downloadable products are accompanied by information such as frequently asked questions (FAQ), a glossary, and case studies, describing how forecasts are used in decision making. Products giving a national overview are included in a monthly email and in stakeholder briefings. Examples include a map showing the most likely flow category (Fig. 6).

The BJP statistical method empirically captures relationships between streamflow volume in a forecast period and the initial conditions of catchment and climate. Figure 7 shows the main steps the Bureau employs in using this approach. Observations of streamflow for the previous month are required, and must be available by the 4th working day of the month,

to allow time for the data to be checked for irregularities or missing data — a process that becomes more time consuming as more locations are included in the service. Each time series is inspected visually. Spurious peaks are removed and missing records are infilled by interpolating using nearby streamflow records. Next, the forecasts are generated. Text for the web front page and monthly email is drafted, summarising the forecasts and placing them in the context of recent observations of climate and streamflow. Finally, the forecasts are released to the public by the 7th working day of the month (Fig. 7).

The Bureau also produces forecasts using a conceptual rainfall-runoff model known as GR4J (Perrin *et al.* 2003) with climate forecasts as inputs. We refer to the forecast model as dynamic, to distinguish it from the statistical model; these forecasts are available to registered users only. The model relies on predictability from future climate conditions (particularly rainfall) and antecedent catchment conditions. This will eventually improve the performance of

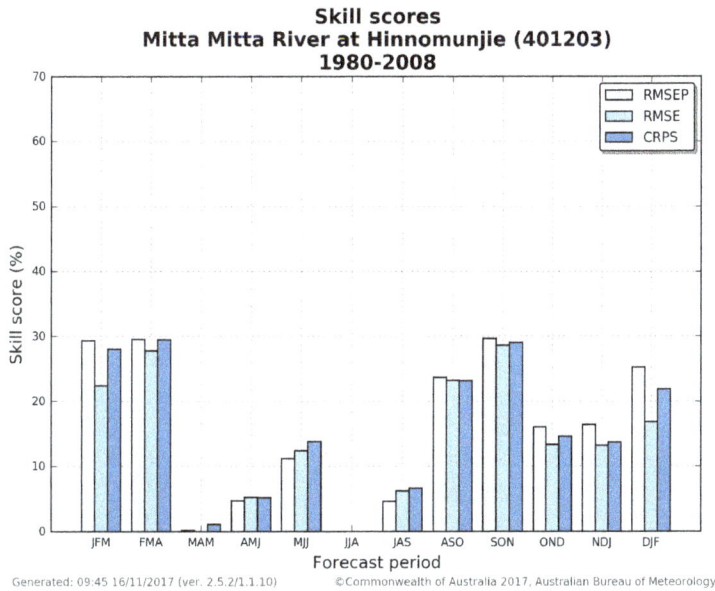

Fig. 4.   Chart showing skill scores (Root Mean Square Error in Probability (RMSEP), Root Mean Squared Error (RMSE) and Continuous Ranked Probability Score (CRPS)) for the site on the Mitta Mitta River at Hinnomunjie. These hindcasts are generated using leave-five-years-out cross validation with a reference period between 1980 and 2008. The RMSEP hindcast skill score is used to classify and communicate the skill of a particular forecast to end users. Zero is considered to be a forecast with no skill (equivalent skill to predicting using historical averages). Less than 5 is considered to be a forecast with very low skill. Between 5 and 15 is considered low skill. Between 15 and 30 is considered moderate skill, and higher than 30 is considered to be a forecast with high skill.

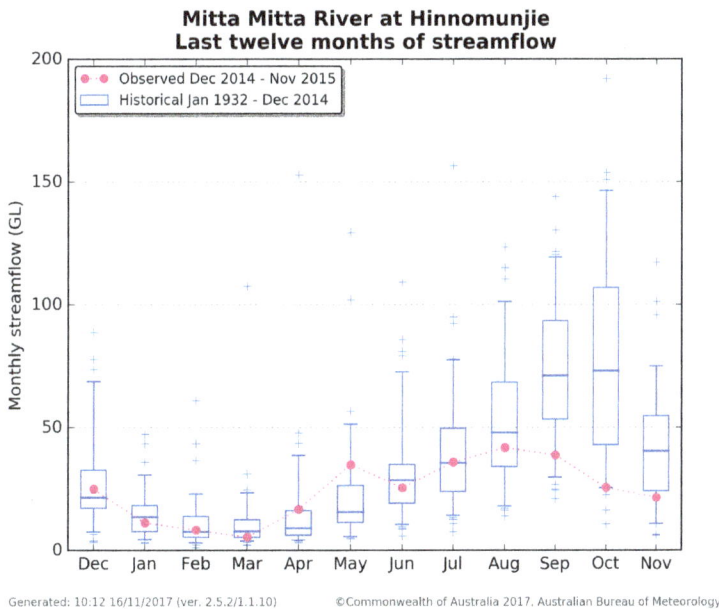

Fig. 5.   Monthly observed streamflow data for the past 12 months represented by pink dots and joined by a dotted line; the boxplot represents the distribution of historical monthly streamflow data. The box extends from the 25th percentile (lower quartile) to the 75th percentile (upper quartile), with a line at the median (50th percentile). Crosses show outliers. The time period of the historical data used is Jan. 1932–Dec. 2014.

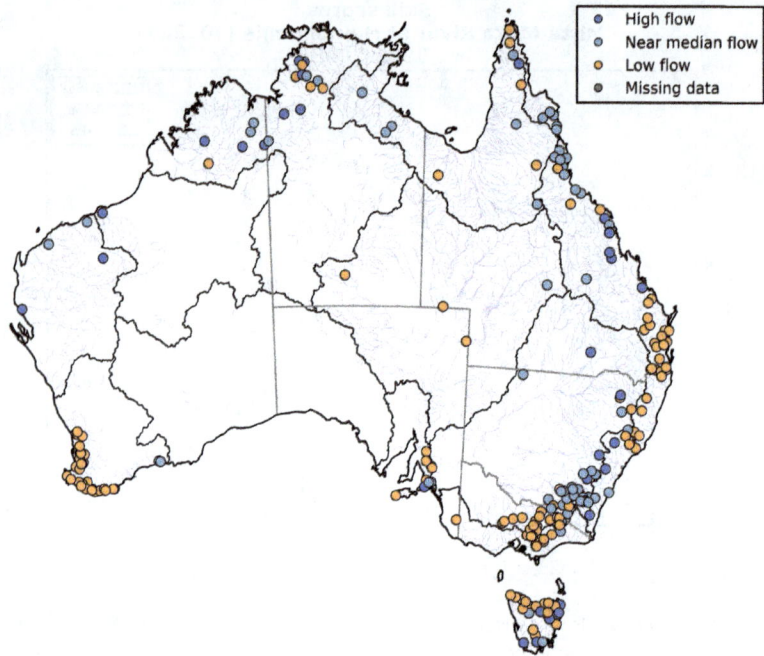

Fig. 6.   A forecast map of the most likely flow category for 227 locations for Dec. 2015–Jan. 2016.

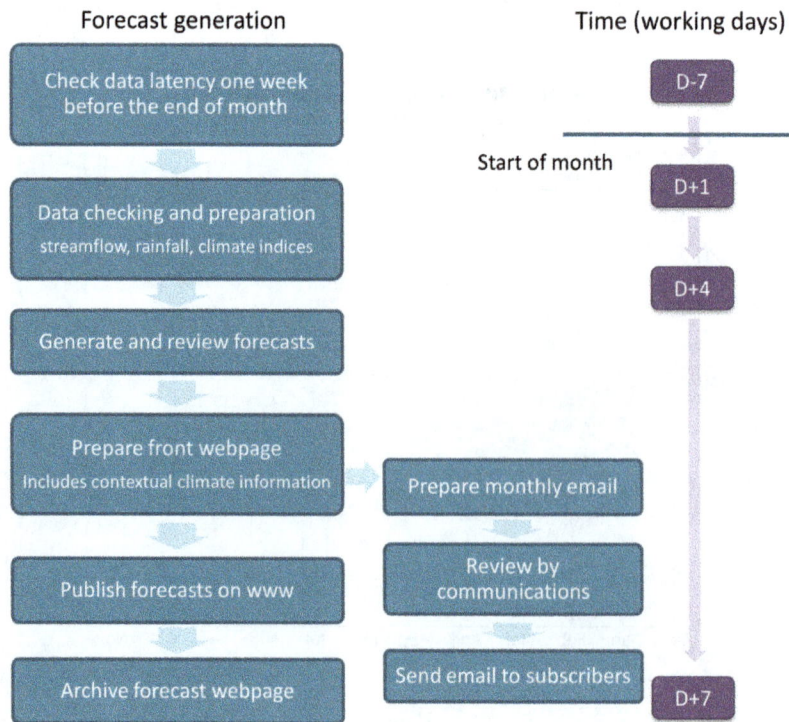

Fig. 7.   Flow chart showing the major steps in providing the forecasting service each month.

the forecasts by merging the forecasts generated from the statistical and dynamic approaches (see Sec. 1.4).

2.2. *WAFARi system*

Statistical and dynamic forecasts are produced by the Bureau using the Water Availability Forecasts of Australian Rivers (WAFARi) modelling system (Shin *et al.* 2011). Nearly all of the processes required to produce the monthly forecast have been automated within WAFARi. WAFARi integrates model kernels for various algorithms, including BJP, Bayesian total error analysis (BATEA) and rainfall-runoff models. The kernels were developed by research partners and were integrated using Python solutions.

Key features of WAFARi include a flexible and extendible modular structure, self-descriptive data files with fully annotated metadata, product customisation tools and a scriptable interface. The operation of BJP and the dynamic models requires a large amount of computing power, and this requirement will increase as the number of forecast locations increases. The system tools have therefore been designed to take advantage of parallel computing power in a computer cluster. Currently, the system runs the entire forecast verification procedure for more than 250 locations in parallel by spawning a child process for each location in the National Computational Infrastructure (NCI) cluster.

2.3. *Data supply*

The forecasting service requires near real-time observations of daily streamflow to be available at each forecast location. Therefore, efficient data delivery is a vital component of the service. The data are sourced from the Australian Water Resources Information System (AWRIS), which is able to receive, standardise, organise and interpret water data from across the continent. As stated above, data must be available by the 4th day of the month, so that they can be checked and quality-controlled before being used to generate a forecast on the 7th working day of the month.

Data checking is the most challenging part of the process. When working on large datasets, clear guidelines and processes are needed to reduce the risk of errors. The rapid expansion of locations from 85 to over 200 in the past 18 months has increased both the data checking workload and the risk that some of the observed data might not be available by the required date. To mitigate this situation, an internal service level agreement, initially developed in 2013 to provide data from AWRIS, was modified to include a list of locations covered by the agreement, and a protocol that must be followed when data are unavailable. The former is updated as new locations are added to the forecasting service.

The service level agreement requires the forecast team to monitor data supply and quality one week before the end of the month. This gives the data supply team time to act if there are any serious issues with data latency (i.e., the time it takes for data to be stored or retrieved). The team then investigates these issues, liaising with the data providers when required to ensure that, as far as possible, the necessary data are available for each location. The data team has proved to be a valuable link between Australia's water data suppliers and the forecasting service, greatly reducing the risk that the required streamflow observations will be unavailable.

The streamflow data feed from AWRIS is supplemented by estimates of total monthly inflow volumes, generally for water storages. Water managers submit these estimates via email in the first five working days of the month. These water managers include Icon Water, the Murray-Darling Basin Authority, Goulburn-Murray Water, Melbourne Water, WaterNSW and Water Corporation (in Western Australia).

Also available in the first five working days of each month are various climate indices — including those derived from sea surface temperature data that are published on the website of the National Oceanic and Atmospheric Administration (NOAA).

2.4. *Staff training and competency*

There are two broad types of tasks undertaken by the service team — the monthly forecast releases, and addition of forecast locations. In order to ensure successful delivery of the forecasting service each month, a forecast roster is prepared a month in advance to allocate team members to specific tasks. These tasks include (1) data preparation, (2) forecast generation, (3) web page editing, and (4) slide preparation for several internal external monthly briefings to communicate the forecasts and contextual information. Major steps for adding new locations to the WAFARi system are to (1) gather information on the new location, (2) configure the location to have observation data collected and ingested into the system database, (3) set up the new model in WAFARi, (4) run parameter sampling/calibration for the new model, and (5) commit the new model to the repository. Staffing has been amended to deal with the increased workload and with staff absences (planned or unplanned). This is particularly important in relation to data checking, which is labour intensive. New team members undergo training to ensure that they are competent in forecast-associated tasks. Such tasks include processing and checking of data, generation of forecasts and products, and compilation of text for the web and emails. Each month, a given team member is allocated to a different set of tasks, so they continue to gain experience and keep their skills up to date in all areas.

3. Science, Models and Methods

3.1. *Overview*

Robust science is fundamental to the service. The relationships between the Bureau and other partners — including CSIRO, universities and other research providers — have generated significant intellectual input into the service.

A particularly important partnership that has been of benefit to both parties is that of the Bureau and CSIRO through WIRADA. The partnership has increased the impact of CSIRO's research by focusing efforts on the Bureau's requirements for water forecasting. Also, it has allowed the research to be put into practice within the Bureau's operational services and systems.

In addition, the Bureau's own Research and Development group provided critical input on rainfall forecasts and downscaling methods, both of which are currently operational in the dynamic approach (Sec. 3.3).

Another beneficial partnership has been that between the Bureau and two universities: the University of Newcastle and the University of Adelaide. Research teams at these institutions have developed sophisticated methods for calibrating hydrological models. The methods consider uncertainty from different sources (e.g. errors in data measurement and in model structure). Again, the Bureau has been able to put this research into practice, with these calibration methods forming a key component of the fully automated WAFARi system, allowing the system to fully consider hydrological uncertainty.

Next we describe the various scientific models and methods that underpin the forecast service.

3.2. *Statistical approach*

The BJP modelling method (Robertson and Wang 2012, 2013; Robertson *et al.* 2013; Wang *et al.* 2011a; Wang *et al.* 2009) used in WIRADA was first developed through the South Eastern Australia Climate Initiative. The method empirically captures relationships between streamflow volume in a forecast period (predictand) and the initial conditions of catchment and climate (predictors). It has three components:

- model formulation, parameter inference and forecast generation;
- predictor selection; and
- probabilistic forecast verification.

Each of these components is discussed below.

3.2.1. *Model formulation, parameter inference and forecast generation*

In the BJP model (Wang *et al.* 2011a; Wang *et al.* 2009), a transformed multivariate normal distribution (rather than a regression) is used to model the joint distribution of the predictand and predictors. Using the joint probability model with a Bayesian inference approach makes it possible to use data even when records are non-concurrent, missing or censored. A transformation is employed to normalise variables and homogenise their variances. Initial versions of the BJP model exclusively used the Yeo-Johnson (Y-J) transformation (an extended form of the Box-Cox transformation). However, for variables such as streamflow volume and rainfall that have a physical lower bound of zero, the log-sinh transformation is used nowadays (Wang *et al.* 2012b). The Y-J transformation is still used for variables that can be positive or negative in value (e.g. climate indices).

When streamflow or rainfall is zero, the record is treated as censored data. In doing so, it is possible to use a continuous probability distribution rather than a less tractable mixed discrete and continuous distribution. Bayesian Markov chain Monte Carlo (MCMC) sampling is used to infer parameters and their uncertainty. A forecast can be generated by repeatedly sampling the posterior distribution of the predictand, which is conditional on given values of predictors. This produces an ensemble, which is a numerical representation of a probabilistic forecast.

3.2.2. *Predictor selection*

Antecedent streamflow and rainfall are candidate predictors used to represent the initial catchment condition, whereas various observed climate indices are candidate predictors used to represent the climate conditions. Climate indices are generally based on sea surface temperature (e.g., NINO3.4) or mean sea level pressure (e.g., SOI). Predictors are selected based on their ability to forecast independent events, rather than how well they fit with historical events. The CSIRO team developed and applied a stepwise predictor selection method based on a pseudo-Bayes factor (Robertson and Wang 2012). To separate effects of the initial catchment and climate conditions, the team selected climate predictors based on their ability to forecast rainfall rather than streamflow.

The model was later modified so that the output from a conceptual hydrological model was used to represent the influence of initial catchment condition (Robertson *et al.* 2013). Also, forecast rainfall and forecast climate indices from a dynamic climate model were considered as additional candidate predictors (Pokhrel *et al.* 2013). Finally, model averaging was considered as an alternative to selection of the best model (Schepen and Wang 2015; Wang *et al.* 2012a).

3.2.3. *Probabilistic forecast verification*

Performance of probabilistic forecasts is quantified by assessing their accuracy and reliability in uncertainty representation. A comprehensive suite of verification methods has been developed for this purpose (Wang *et al.* 2011a); it includes skill scores (e.g., Fig. 4), reliability plots and detailed displays for detecting error patterns. Verification can be done on a historical dataset using cross-validation, or on new forecast events.

As explained above, the Bureau issues monthly streamflow forecasts for the next three

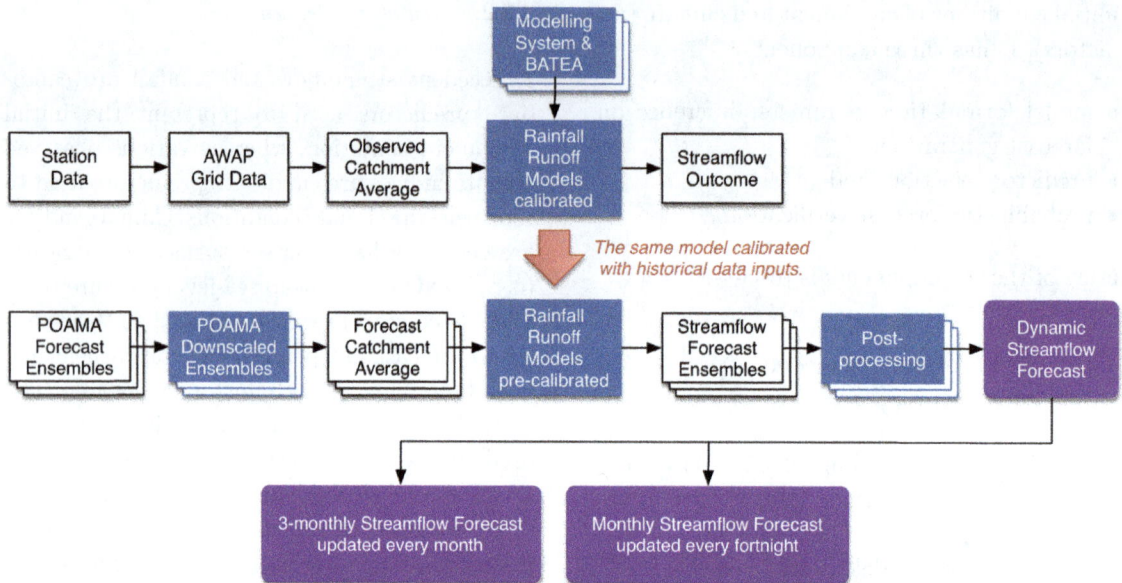

Fig. 8.   Dynamic modelling approach for seasonal streamflow forecasting.

months. It also makes available past forecasts and cross-validation forecast verification results (validation of the model on events independent of the set of events used for model calibration).

3.3. *Dynamic approach*

Registered users can now access a new seasonal streamflow forecasting service based on a dynamic modelling approach (Lerat *et al.* 2015; Tuteja 2015b). This service provides probabilistic forecasts for more than 100 sites every month to registered users, and further sites will be added over the next 12 months. The GR4J model used to generate these forecasts is a lumped conceptual hydrological model, forced by downscaled rainfall forecasts. The forecasts provide total flow volume for 1-month and 3-month periods. As shown in Fig. 8, there are three main parts to the modelling approach: downscaling of seasonal rainfall forecasts; rainfall-runoff calibration and modelling; and post-processing residual biases to adjust the ensemble spread.

Rainfall ensembles are used as inputs to generate streamflow ensembles from the GR4J rainfall-runoff model (Perrin *et al.* 2003). The rainfall ensembles are generated by post-processing output from the Predictive Ocean and Atmosphere Model of Australia (POAMA): a seasonal forecasting system based on a coupled ocean-atmosphere model (Hudson *et al.* 2013). POAMA rainfall forecast ensembles are downscaled from grid boxes approximately 250 km across to local-scale grid boxes about 5 km across, using an atmospheric analogue downscaling method (Charles *et al.* 2013; Shao and Li 2012). The GR4J model is calibrated using the BATEA statistical tool; this provides parameter and predictive uncertainty using MCMC sampling (Kavetski *et al.* 2006). BATEA is used to derive 40 parameter sets of the hydrologic model. In forecast mode, the model is first run over a warmup period of 5 years prior to the forecast date, using observed rainfall extracted from the AWAP dataset and each one of the 40 parameter sets. Then, for each of those 40 ensembles, each of the 150 members of the

POAMA downscaled rainfall ensembles is used to drive GR4J at a daily time step during the forecasting period. This produces an ensemble of daily forecast streamflow volume with 6,000 members. Each ensemble member can be aggregated in time to produce forecasts for one and three-month periods. An autoregressive moving average method is used to correct systematic biases in the aggregated streamflow volume.

Over the next two years, the Bureau will move to a new climate forecasting system, ACCESS-S, which is expected to greatly improve the quality of rainfall and streamflow forecasts. The improvements will result from an increase in spatial resolution (from 250 km to 60 km) and better assimilation of observed data to the climate model. The new seasonal rainfall forecasts from ACCESS-S will require changes in the seasonal streamflow forecasting service. In particular, it will be necessary to develop a rainfall post-processing method to replace the analogue downscaling method, and to modify catchment-scale hydrologic modelling to improve the forecasting of dynamic streamflow.

3.4. Merging statistical and dynamic forecasts

The streamflow forecasts produced by dynamic modelling complement those produced by statistical modelling (i.e. by BJP). This has been shown by cross-validation of results for the period 1980–2008 at 90 locations across Australia. Over many locations and in different seasons, the two approaches produced similar and consistent forecasts. However, in some situations, one approach produced more accurate forecasts than the other. Therefore, the forecasts produced by the two methods will be merged (using methods discussed below) to produce a single forecast that combines the advantages of both dynamic and statistical forecasting (Schepen and Wang 2015). The merged forecasts are expected to be available from 2017.

3.4.1. Merging methods

In merging statistical and dynamic probabilistic forecasts, it is important to ensure that the merged forecast accurately represents the original components. Two weighted averaging methods for merging ensemble forecasts have been developed through WIRADA:

- Bayesian model averaging (BMA), which averages forecast probability densities
- quantile model averaging (QMA), which averages forecast values across equal cumulative probabilities.

Both methods use cross-validation predictive performance to weight the components, and can make use of skewed and censored data (employing methods similar to those used in BJP modelling).

In cases where the original statistical and dynamic forecasts are identical, the two approaches — BMA and QMA — yield the same merged forecast. However, if the original forecasts differ significantly, the resulting merged forecasts can differ, as shown in Fig. 9. The statistical and dynamic forecasts are unimodal, as are QMA forecasts (with the ensemble spread appearing between the original forecasts). In contrast, the merged BMA forecasts can be bimodal and the ensemble spread can be wide.

3.4.2. Assessment

Applying BMA and QMA to more than 50 forecast locations has shown that the methods are similar in terms of overall skill scores and reliability plots. There are two reasons for this finding:

- the two original methods (statistical and dynamic) often produce concordant forecasts
- in situations where one of the original methods consistently produces far better forecasts, that method carries most weight in averaging.

It is rare that the statistical and dynamic approaches are given similar weight but produce very different forecasts. In such situations,

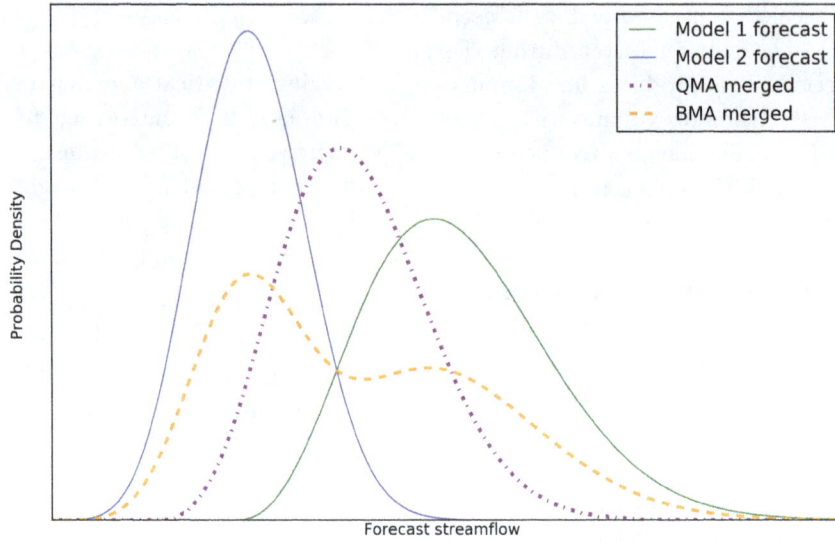

Fig. 9.   A comparison of merging two probabilistic forecasts using BMA and QMA based on a synthetic data set.

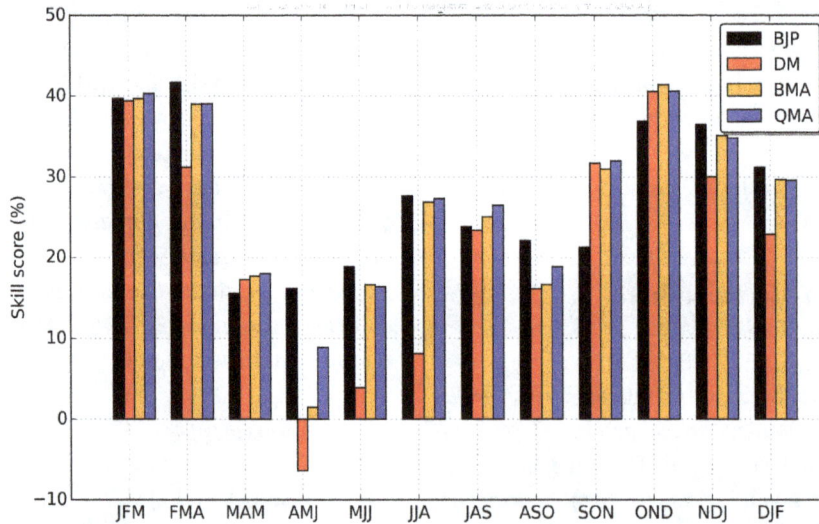

Fig. 10.   A comparison of continuous ranked probability score (CRPS) skill scores for Bayesian joint probability (BJP), dynamic (DM), Bayesian model averaging (BMA) and quantile model averaging (QMA) forecasts for Adelong Creek at Batlow Road in the Murrumbidgee region of southeastern Australia, demonstrating the advantage of forecast merging with BMA and QMA.

a QMA merged forecast is preferred for operational seasonal streamflow forecasting, because the result is unimodal and therefore more readily communicated to and interpreted by end users.

Figure 10 shows the improvement in skill attained by merging statistical and dynamic forecasts for a site in southeastern Australia. For September and October issued forecasts, forecasting skill is chiefly due to the dynamic approach. Forecasts issued in the other months are mainly skillful due to the statistical approach. The exceptions are January, March and July issues, when skill of the two approaches is very similar. Consequently, merging of forecasts takes advantage of whichever approach performs better in particular conditions of climate or hydrology. For more information on merging of forecasts, readers are referred to Schepen and Wang (2015).

3.5. *Extending forecast horizons to longer lead times*

Australian water agencies now routinely use the Bureau's monthly seasonal streamflow forecasting service that provides three-month forecasts of streamflow volumes. However, many agencies are interested in longer term forecasts of up to 12 months ahead. Forecasts for such extended periods are not expected to be skillful. Nevertheless, water agencies are interested in stochastic scenarios for long lead times, wishing to combine these with the more skillful shorter term forecasts for modelling water allocation for forward planning and operations.

In response to this interest, the CSIRO team has developed the Forecast-Guided Stochastic Scenarios (FoGSS) model, to produce 6 to 12-month streamflow ensemble forecasts. The model combines two elements:

- use of a calibration, bridging and merging (CBaM) method to post-process rainfall forecasts from coupled ocean–atmosphere dynamical model outputs (from POAMA); and
- use of hydrological and error models to convert the rainfall forecasts to 12-month streamflow forecasts.

These elements are discussed below.

3.5.1. *Calibration, bridging and merging of climate forecasts*

To derive monthly rainfall forecasts, the CBaM method is used to post-process outputs from POAMA (which produces 33-member ensemble forecasts of climate to a lead-time of 9 months).

Using the CBaM method has three advantages over simple POAMA forecasts, because the latter are:

- often biased (either too high or too low) and over-confident (too narrow in ensemble spread);
- generally of low skill; and
- too coarse for many catchment-scale applications.

The calibration part of CBaM deals with the first of these issues, as the BJP approach is used to statistically calibrate raw POAMA rainfall forecasts, producing unbiased and reliable probabilistic forecasts (Hawthorne *et al.* 2013; Schepen *et al.* 2012). The bridging part of CBaM deals with the second issue, by using the BJP method to produce rainfall forecasts, with predictors based on POAMA sea surface temperature (SST) forecasts. The scale issue is addressed by using observed monthly catchment rainfall as the predictand in the BJP model.

Bridging is necessary because the complexity of the coupling between the ocean and atmosphere in dynamic climate models such as POAMA means that the processes are not always well captured. Statistical models can sometimes be more effective than dynamic climate models at relating SST to rainfall (Peng *et al.* 2014; Schepen and Wang 2014; Schepen *et al.* 2014). Bridging takes advantage of the often good skill in forecasting SST by dynamic climate models.

Rainfall forecasts developed by calibration and bridging can offer supplementary skill; so they are merged using the BMA method (Wang *et al.* 2012a) in the merging part of CBaM. To form ensemble time series forecasts of monthly rainfall that are correlated temporally, CBaM is applied to each lead time individually, then the ensemble members of all lead times are connected using the Schaake shuffle (Clark *et al.* 2004). When POAMA climate forecasts are lacking for lead times of 10, 11 and 12 months, rainfall forecasts are produced by using

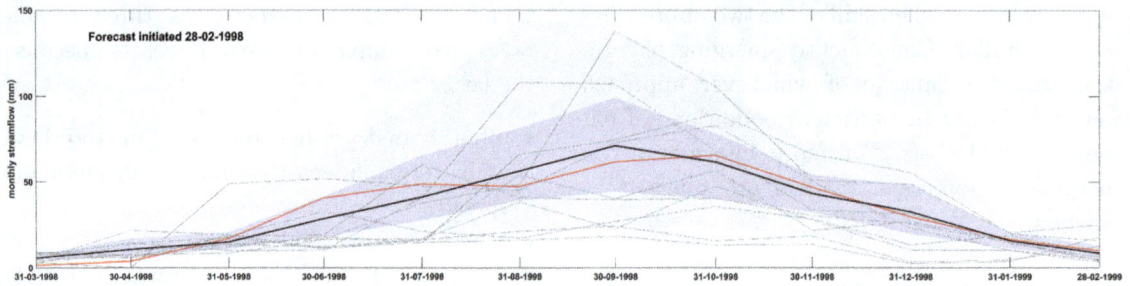

Fig. 11. Forecast generated with the forecast guided stochastic scenarios (FoGSS) forecasting system for the Daly River at Mount Nancar in the Northern Territory. Red line, observations; black line, forecast mean; shading, forecast confidence intervals (dark, 50%; light; 80%); grey lines, 10 randomly sampled members from a 1,000-member ensemble.

lagged statistical relationships to produce bridging rainfall forecasts.

3.5.2. *Forecast guided stochastic scenarios (FoGSS) of streamflow*

To generate a 12-month streamflow forecast, CBaM is used to generate a rainfall forecast, which is then input to the WAPABA monthly rainfall runoff model (Wang *et al.* 2011b). Next, a staged error model is applied to quantify hydrological uncertainty and propagate the uncertainty through the forecast lead times.

After several rounds of development and refinement, the FoGSS error model is now a three-stage model (Li *et al.* 2013; Li *et al.* 2015; Wang *et al.* 2014; Bennett *et al.* 2016). Stage 1 transforms streamflows with the log-sinh transformation, to normalise and homogenise the variance of errors. Stage 2 applies a bias-correction that improves forecast skill. Stage 3 applies an autoregressive error model to improve the forecast skill for shorter lead times.

Forecast skill varies with both catchment and season. Monthly streamflow forecasts are typically skillful to lead times of one to three months, whereas those for accumulated volumes are skillful to lead times of six to nine months. FoGSS streamflow forecasts are statistically reliable in uncertainty spread, for both individual monthly volumes and accumulated volumes over all lead times. When a streamflow forecast is not skillful, it transits seamlessly to stochastic scenarios. Each ensemble member is a temporally coherent 12-month streamflow time series. The forecast ensemble members (as shown by the example in Fig. 11) can be entered directly into water allocation models for planning and operations.

4. Stakeholder Engagement and Communication

Forecast information is intended to be used to inform decisions and improve social, economic and environmental prosperity (Plummer *et al.* 2009). However, end users need more from forecasts than just accuracy and reliability. Specifically, users need to be involved in the development and continued refinement of the service if it is to meet their needs. Hence, part of the Bureau's remit is to ensure it produces seasonal forecast information that is clear, appropriate for its audience and suitable for use in decision-making.

4.1. *Initial product and service design*

Engagement with stakeholders (both internal and external) was achieved initially through activities such as planning meetings, targeted workshops, surveys and visits to stakeholders. The aim was to gain a better understanding of the specific needs of potential users of the service. This allowed us to gauge the overall priority of particular forecast products. This informed the Bureau's research, data, system development and service requirements. The findings of these activities were then used in developing prototypes of the system and service which were trialed with end users. The prototypes elicited specific and detailed feedback, which was used in a process of ongoing improvement of services and products. This process was employed in the development of web pages and other products.

Many end users prefer deterministic and fairly certain forecasts. Hence, where the forecasts are probabilistic and inherently uncertain, it is important to clearly communicate this to the potential users. In response to feedback from stakeholders, the Bureau incorporated a range of product types. While the products are probabilistic in nature, some show continuous variables (e.g. Fig. 3), while others are categorical (e.g. Fig. 1).

To communicate the new service and demonstrate the benefits to users a communication and adoption strategy was developed. The strategy also aimed to promote uptake of the forecasting service and encourage collaboration between the Bureau and users to allow the forecasts to influence decision-making and become integrated with agency operating strategies.

Some groups within the Bureau had already established operational services in consultation with stakeholders, and it was useful to take this experience into consideration. For example, in developing the seasonal streamflow forecasting service, it was helpful to build on experience gained through developing the flood forecasting and warning service, and the seasonal climate outlook service.

It was possible to speed up the process of developing and testing prototypes by combining several mature third-party open-source software libraries. Both Python and Fortran programming languages were used, which made it possible to employ many features of Python in the modelling service while retaining a high level of overall computing performance.

4.2. Web survey

In 2013, a web-based survey of the monthly email recipients was undertaken to determine the performance of the service in terms of user satisfaction and adoption (Wilson *et al.* 2014). Also, visitor traffic to the public seasonal streamflow forecasts website was reviewed. The survey elicited valuable suggestions as to how the seasonal streamflow forecasts and the website could be improved.

Of the 87 stakeholders who completed the survey, 88% reported satisfaction with the existing service. Consequently, the service was exceeding its target of 85% user satisfaction and adoption. Responses were received from various organisations, ranging from government and industry to the general public, in sectors such as agriculture, engineering, mining and resources. This finding illustrated the depth of uptake of the forecasting service.

In some cases, users had changed their decisions based on the forecasts, to good effect. For example, the forecasts provide an indication on the potential for physical spills from water storages, which have affected water allocation and management decisions in the Murray Darling Basin. Furthermore, 9% of respondents were directly using the forecasts in their decision-making, and another 48% were using the forecasts to support decisions related to water allocation outlooks, water restrictions, scheduling irrigation, water markets, managing river operations and scheduling environmental watering.

4.3. Targeted workshops and meetings

In certain regions, forecast performance can be poor, due to regional characteristics. For example, in western Tasmania, catchment persistence is relatively low due to the combination of high rainfall and shallow soils. Low persistence is also seen in Western Australia, but in this case is due to the sudden transition from dry to wet periods.

Given that it is not possible to provide forecasts at all locations, it is important to involve stakeholders in the decisions about where best to add locations to the service. The Bureau used targeted workshops and meetings to involve potential end users in the expansion of the service in 2014–2015, particularly in Tasmania and Western Australia. The main aim was to ensure that the additional locations offered the best value, and to communicate with end users about the limitations of the forecasts.

Another opportunity for the Bureau to engage with external stakeholders is the monthly national climate and water briefings, held in Canberra. The Bureau uses these briefings to present accessible information on climate and water observations, current conditions, and seasonal outlooks.

Finally, the Bureau makes use of the streamflow forecasts when developing its monthly climate outlooks. The information is used in background documents, and in creating clear and concise materials for forecasters and other Bureau staff who deal with the media. The national map is included, to put the information on climate and water into perspective.

4.4. *Case studies*

Case studies that describe how individuals use the forecasts in decision-making are a powerful way to increase uptake of the service. The studies, which are summarised in a document usually about two pages long, show how the forecasts provide more accurate information than historical data alone. For example, one study explains how Icon Water used the forecasts in their decision to remove water restrictions after the millennium drought.[b]

Other case studies are planned in various regions across Australia. They will show how forecasts are being used in making decisions about allocating water for irrigation, hydropower or the environment, and about water storage management, in both rural and urban areas.

4.5. *Videos and social media*

Recently, the Bureau started using social media (e.g., Facebook, Twitter and YouTube) to engage stakeholders and reach a much wider audience. For example, since 2014 it has been producing a monthly four-minute video that describes recent conditions, and seasonal climate and water outlooks, but also provides contextual information (including current water storage levels and modelled soil moisture information). These videos allow the Bureau to highlight its expertise in areas ranging from climate and water monitoring to prediction, in a format that suits many consumers.

The videos are usually produced in the last week of the month, and are released via social media and the Australian Broadcasting Company (ABC) program, Landline. Via this initiative, the streamflow forecasts reach more than 500,000 viewers. The Bureau complements the monthly videos with educational videos about topics such as seasonal streamflow forecasting.

5. Future Plans and Summary

The Bureau wishes to expand the streamflow forecasting service to include additional locations, particularly in regions that are not currently included in the service. Emphasis will be placed on locations where there is need, and where forecast performance exceeds the service benchmark. As shown in Fig. 12, it is expected that about 200 locations will be added to the service, many of which will be in regions

---

[b]Available at http://www.bom.gov.au/water/about/.

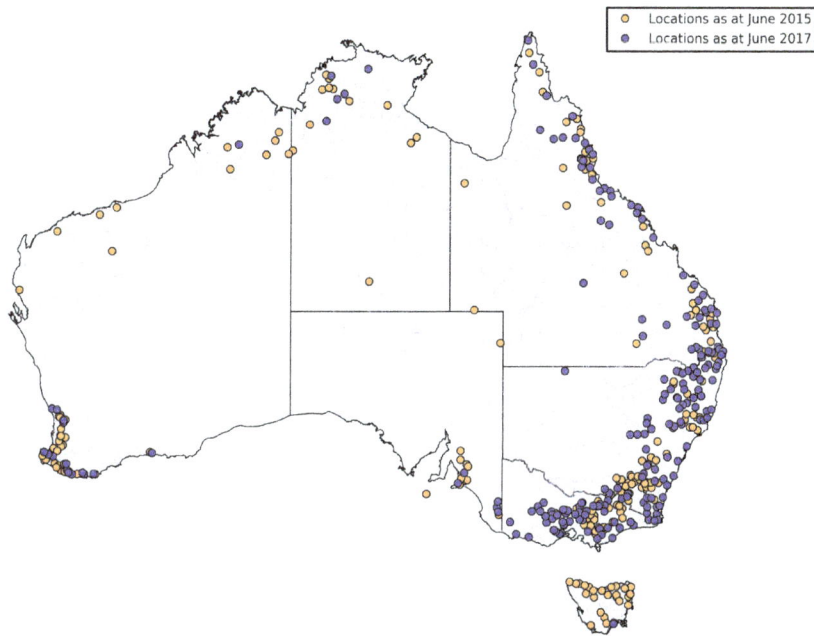

Fig. 12. Current (200; orange) and potential future (200, purple) locations in the seasonal streamflow forecasting service.

where the climate is highly variable. There will be challenges in modelling the hydrology of ephemeral and intermittent catchments to create more accurate and reliable seasonal streamflow forecasts.

The difficulty of creating accurate seasonal streamflow forecasts is compounded by the difficulty of accurately forecasting seasonal climate. The CBaM method makes the best use of outputs from current dynamic climate models, but better climate forecasts will depend on improvements in climate modelling.

An additional requirement is to make these forecasts reliable for hydrologic applications. Further consideration is given to the complexity of operational systems, and the need to keep these systems and service delivery only as complex as necessary. This includes rationalising the data harvesting methods from various water information systems of the Bureau.

Over the next 2–3 years, the Bureau will move from using the POAMA model for seasonal climate forecasting to using ACCESS-S. This

change will create challenges for the seasonal streamflow forecasting service, which currently relies on downscaled rainfall for generating dynamic forecasts. In contrast to the current situation, under ACCESS-S, ensemble members will be generated in staggered time, with hindcasts from ACCESS-S available at only a few dates in each month and with fewer ensemble members than operational forecasts.

Successful development and delivery of the current seasonal streamflow forecasting service across Australia depends on:

- the strong scientific foundation for the modelling and forecasting approach;
- the involvement of end users in the development of forecasting products; and,
- robust systems that allow for timely and efficient processing of large volumes of data.

These factors require the Bureau to engage in meaningful and sustained relationships both internally and with the stakeholders who are employing the forecasts to make informed decisions. Case studies provide a valuable way to show how seasonal streamflow forecasts can be a useful part of water resources planning and management.

Acknowledgments

We thank the many stakeholders who have, and continue to, engage with us to further improve the service. We gratefully acknowledge funding from CSIRO and the Bureau of Meteorology through the Water Information Research and Development Alliance to develop and deliver the Seasonal Streamflow Forecasting service. QJ Wang from CSIRO (now at the University of Melbourne) has led the research work on seasonal streamflow forecasting. Narendra Tuteja and Dasarath Jayasuriya have guided development of this service. Support from Graham Hawke and Rob Vertessy is greatly appreciated. We also acknowledge those who have and continue to contribute to the service, including Neil Plummer, Jeff Perkins, Richard Laugesen, Bat Le, Trudy Wilson, Urooj Khan, Christopher Pickett-Heaps, Andrew MacDonald, and David Kent (Bureau of Meteorology), Quanxi Shao, Ming Li, Yong Song, Enli Wang and Hongxing Zheng (CSIRO), George Kuczera (University of Newcastle), Mark Thyer, and Dmitri Kavetski (University of Adelaide). Neville Garland and David Dreverman (Murray-Darling Basin Authority) and many senior officers from water agencies across Australia have also provided critical support.

References

Bennett, J. C., Q. J. Wang, M. Li, D. E. Robertson, and A. Schepen, 2016: Reliable long-range ensemble streamflow forecasts: Combining calibrated climate forecasts with a conceptual runoff model and a staged error model. *Water Resour. Res.*, **52**, 8238–8259.

Charles, A., B. Timbal, E. Fernandez, and H. Hendon, 2013: Analog downscaling of seasonal rainfall forecasts in the Murray Darling Basin. *Mon. Weather Rev.*, **141**, 1099–1117.

Clark, M. P., S. Gangopadhyay, L. Hay, B. Rajagopalan, and R. Wilby, 2004: The Schaake shuffle: A method for reconstructing space-time variability in forecasted precipitation and temperature fields. *J. Hydrometeorol.*, **5**, 243–262.

Hawthorne, S., Q. J. Wang, A. Schepen, and D. E. Robertson, 2013: Effective use of GCM outputs for forecasting monthly rainfalls to long lead times. *Water Resour. Res.*, **49**, 5427–5436.

Hudson, D., A. G. Marshall, Y. Yin, O. Alves, and H. H. Hendon, 2013: Improving intraseasonal prediction with a new ensemble generation strategy. *Mon. Weather Rev.*, **141**, 4429–4449.

Kavetski, D., G. Kuczera, and S. W. Franks, 2006: Bayesian analysis of input uncertainty in hydrological modeling: 1. Theory. *Water Resour. Res.*, **42**, W03407.

Lerat, J. and Coauthors, 2015: Dynamic streamflow forecasts within an uncertainty framework for 100 catchments in Australia. *Hydrological and Water Resources Symposium*, Hobart, Australia, 7–10 December 2015.

Li, M., Q. J. Wang, and J. Bennett, 2013: Accounting for seasonal dependence in hydrological model errors and prediction uncertainty. *Water Resour. Res.*, **49**, 5913–5929.

Li, M., Q. J. Wang, J. C. Bennett, and D. E. Robertson, 2015: A strategy to overcome adverse effects of autoregressive updating of streamflow forecasts. *Hydrol. Earth Syst. Sc.*, **19**, 1–15.

Peng, Z., Q. J. Wang, J. C. Bennett, A. Schepen, F. Pappenberger, P. Pokhrel, and Z. Wang, 2014: Statistical calibration and bridging of ECMWF system4 outputs for forecasting seasonal precipitation over China. *J. Geophys. Res. Atmos.*, **119**, 7116–7135.

Perrin, C., C. Michel, and V. Andreassian, 2003: Improvement of a parsimonious model for streamflow simulation. *J. Hydrol. (Amst.)*, **279**, 275–289.

Plummer, N., N. K. Tuteja, Q. J. Wang, E. Wang, D. Robertson, and S. Zhou, 2009: A seasonal water availability prediction service: Opportunities and challenges. *International Congress on Modelling and Simulation*, Cairns, Australia, 13–17 July 2009, pp. 13–17.

Pokhrel, P., Q. J. Wang, and D. E. Robertson, 2013: The value of model averaging and dynamical climate model predictions for improving statistical seasonal streamflow forecasts over Australia. *Water Resour. Res.*, **49**, 6671–6687.

Robertson, D. E. and Q. J. Wang, 2012: A Bayesian approach to predictor selection for seasonal streamflow forecasting. *J. Hydrometeorol.*, **13**, 155–171.

——, 2013: Seasonal forecasts of unregulated inflows into the Murray River, Australia. *Water Resour. Manage.*, **27**, 2747–2769.

Robertson, D. E., P. Pokhrel, and Q. J. Wang, 2013: Improving statistical forecasts of seasonal streamflows using hydrological model output. *Hydrol. Earth Syst. Sc.*, **17**, 579–593.

Schepen, A. and Q. Wang, 2014: Ensemble forecasts of monthly catchment rainfall out to long lead times by post-processing coupled general circulation model output. *J. Hydrol. (Amst.)*, **519**, 2920–2931.

Schepen, A. and Q. Wang, 2015: Model averaging methods to merge operational statistical and dynamic seasonal streamflow forecasts in Australia. *Water Resour. Res.*, **51**, 1797–1812.

Schepen, A., Q. J. Wang, and D. E. Robertson, 2012: Combining the strengths of statistical and dynamical modeling approaches for forecasting Australian seasonal rainfall. *J. Geophys. Res.*, **117**, D20107.

Schepen, A., Q. J. Wang, and D. E. Robertson, 2014: Seasonal forecasts of Australian rainfall through calibration and bridging of coupled GCM outputs. *Mon. Weather Rev.*, **142**, 1758–1770.

Shao, Q. and M. Li, 2012: An improved statistical analogue downscaling procedure for seasonal precipitation forecast. *Stochastic Environmental Research and Risk Assessment*, doi:10.1007/s00477-012-0610-0.

Shin, D. and Coauthors, 2011: WAFARi: A new modelling system for seasonal streamflow forecasting service of the Bureau of Meteorology. *International Congress on Modelling and Simulation*, Perth, Australia, 12–16 December 2011, pp. 2374–2380.

Tuteja, N., 2015a: Reliable water availability forecasts for Australia. *Water J. Aust. Water Assoc.*, **42**, 70–72.

——, 2015b: The case for extended hydrologic prediction services for improved water resource management. *WMO Bulletin*, **64**.

van Dijk, A. I. J. M., H. E. Beck, R. S. Crosbie, R. A. M. de Jeu, Y. Y. Liu, G. M. Podger, B. Timbal, and N. R. Viney, 2013: The Millennium Drought in southeast Australia (2001–2009): Natural and human causes and implications for water resources, ecosystems, economy, and society. *Water Resour. Res.*, **49**, doi:10.1002/wrcr.20123.

Vertessy, R. A., 2013: Water information services for Australians. *Aust. J. Water Resour.*, **16**, 91–106.

Wang, G. and Coauthors, 2011a: POAMA-2 SST skill assessment and beyond. *CAWCR Res. Lett.*, **6**, 40–46.

Wang, Q. J., D. E. Robertson, and F. H. S. Chiew, 2009: A Bayesian joint probability modeling approach for seasonal forecasting of streamflows at multiple sites. *Water Resour. Res.*, **45**, W05407.

Wang, Q. J., A. Schepen, and D. E. Robertson, 2012a: Merging seasonal rainfall forecasts from multiple statistical models through Bayesian model averaging. *J. Climate*, **25**, 5524–5537.

Wang, Q. J., D. L. Shrestha, D. E. Robertson, and P. Pokhrel, 2012b: A log-sinh transformation for data normalization and variance stabilization. *Water Resour. Res.*, **48**, W05514.

Wang, Q. J., T. C. Pagano, S. L. Zhou, H. A. P. Hapuarachchi, L. Zhang, and D. E. Robertson, 2011b: Monthly versus daily water balance models in simulating monthly runoff. *J. Hydrol. (Amst.)*, **404**, 166–175.

Wang, Q. J., J. C. Bennett, Y. Song, M. Li, A. Schepen, and D. E. Robertson, 2014: Forecast guided stochastic scenarios of monthly streamflows out to 12 months. *HWRS2014: 35th Hydrological and Water Resources Symposium*, Perth, Australia, pp. 397–404.

Wilson, T., P. Feikema, and J. Ridout, 2014: *2013 User Feedback on the Seasonal Streamflow Forecasts Service. Internal report.* Bureau of Meteorology, 38 pp.

Chapter 11

A Holistic Framework to Assess Drought Preparedness

Ximing Cai*, Majid Shafiee-Jood, Yan Ge, Sylwia Kokoszka
and Tushar Apurv

Department of Civil and Environmental Engineering,
University of Illinois at Urbana-Champaign,
Urbana, IL 61801, USA
*xmcai@illinois.edu

Droughts continue to be a major natural hazard. Areas around the world face a pressing question: Are we prepared for another major drought given the likelihood of a remarkably drier future, where the frequency and intensity of droughts falls far outside our contemporary experience? This question is complex since it involves many risk management issues beyond traditional crisis management, including the impact of climate change, the status of infrastructures, the role of institutions and behaviors, and the limiting factors of the use of drought forecasts. These issues must be studied in a holistic framework considering the interdependence among the social, economic, technical, and environmental factors that jointly challenge drought preparedness. This chapter provides discussion of these issues and synthesis of relevant studies. We expect to shed light on an integrated research agenda for drought preparedness, which is urgently needed for our society.

1. Introduction

Droughts continue to be a major natural hazard around the world. For example, in early 2016 Southeast Asia endured one of the historic droughts and heat waves in the history of the region, with most of the countries struggling to cope with the severe conditions (hotter and drier than usual) caused by El Niño. In Australia, the Millennium Drought recorded the longest uninterrupted period (2001–2009) of below median rainfall in southeast Australia, severely impacting river systems and agriculture and imposing water restriction enforcement in major cities (van Dijk *et al.* 2013). In the United States, of the 178 weather-and-climate-related disasters that occurred during 1980–2014 (each reaching or exceeding $1 billion in damages), 22 were droughts, and six of these droughts were listed in the top 20. Among these, the 1988 drought ended with an estimated loss of $46 billion (NOAA 2015). More recently, the nation experienced the record-breaking Midwest drought in 2012; at present, the ongoing drought in California has led to unprecedented water use restrictions in the state. In the Fifth Assessment Report, IPCC admitted that the global increasing trend in drought suggested by their previous assessment report was probably an overstatement and concluded that there is not enough evidence available in favor of or against any global trend in drought with high confidence (IPCC 2013). However, there is high confidence for an increasing trend in drought in the Mediterranean and West Africa, Southern and Central Europe, the Alpine region, Central North America, the Mediterranean region, Australia, Southern Africa, Mexico, and Northeast Brazil (Calanca 2007; Planton *et al.* 2008; IPCC 2012, 2013; Dai 2013). This situation leaves us with a pressing question: Are we prepared for another major drought given the likelihood of a "remarkably drier future falling far outside our contemporary experience" (Cook *et al.* 2015)?

Due to the difficulty in mitigating the damages of past and ongoing droughts and the uncertainty in drought prediction, drought preparedness is complex and challenging, involving issues of many aspects:

- Drought and climate change: Has climate change already intensified the risk of drought in some places or will this happen in the future?
- Drought and institution and behavior: Does the current institution (organization and policy) support drought preparedness?
- Drought and forecast: How reliable is the state-of-the-art forecasts for drought mitigation and is the forecast information effectively used?
- Drought and infrastructure: Is current infrastructure under-prepared or over-prepared to mitigate the damage of future droughts?

This chapter provides a discussion of these issues and synthesis of relevant studies, highlighting challenges in drought preparedness. We then propose a holistic framework considering the interdependence among the social, economic, technical, and environmental factors that jointly challenge adaptations.

2. Drought Features

Drought occurs event by event, which can be characterized by multiple variables (Mishra and Singh 2010). Different drought events last a wide range of time periods and some can last over multiple years; thus, it is not appropriate to assess an inter-year drought event using the classic procedures applied to intra-year extreme events, which only consider annual maximum. The length of a drought event and the cumulative severity can be more important than the maximum intensity of the event for assessing its impact on natural and human systems (e.g., agriculture and water demand). Thus, it has been suggested to use multiple characteristics, including expected drought inter-arrival

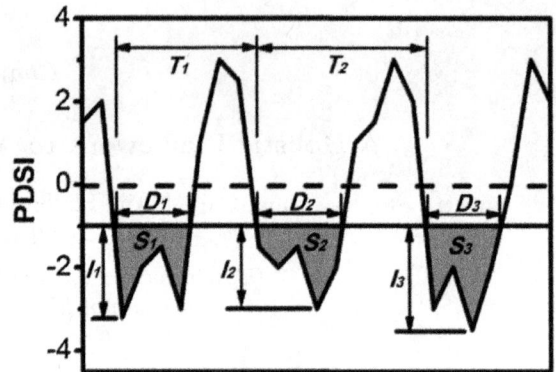

Fig. 1. Definition sketch of drought events; D_i is the duration, S_i is the severity, I_i is the peak intensity and T_i is the inter-arrival time (after Saghafian and Mehdikhani 2014).

time, severity, duration, and peak intensity for drought assessment (Shiau and Shen 2001; Shiau 2006). Using Palmer Drought Severity Index (PDSI) as an example, Fig. 1 defines these characteristics. The intensity (I) of a drought is the maximum absolute value of a drought index below the critical level. Duration (D) is the amount of time that a drought index is continuously below the critical level. Severity (S) is the cumulative deficiency of a drought event below the critical level. The inter-arrival time (T) is the time range between the initiation times of two consequent drought events. All of these features are correlated, and thus, drought should be referred to as a joint status of all of these aspects (Kao and Govindaraju 2010; Ganguli and Reddy 2013; Madadgar and Moradkhani 2013).

As an example to show the change of drought characteristics, Fig. 2 plots the trends and abrupt shifts in D, S and I for six locations in the United States from 1900 to 2012. Drought often lasts for a long period of time, so a 40-year moving window analysis was used to construct a continuous time series of D, S and I during 1990–2012. Changes in drought are displayed for three levels of drought events specified by the return period of 25-year, 50-year, and 100-year. As evidenced by Fig. 2a, Northern California has experienced a long-duration drought in the

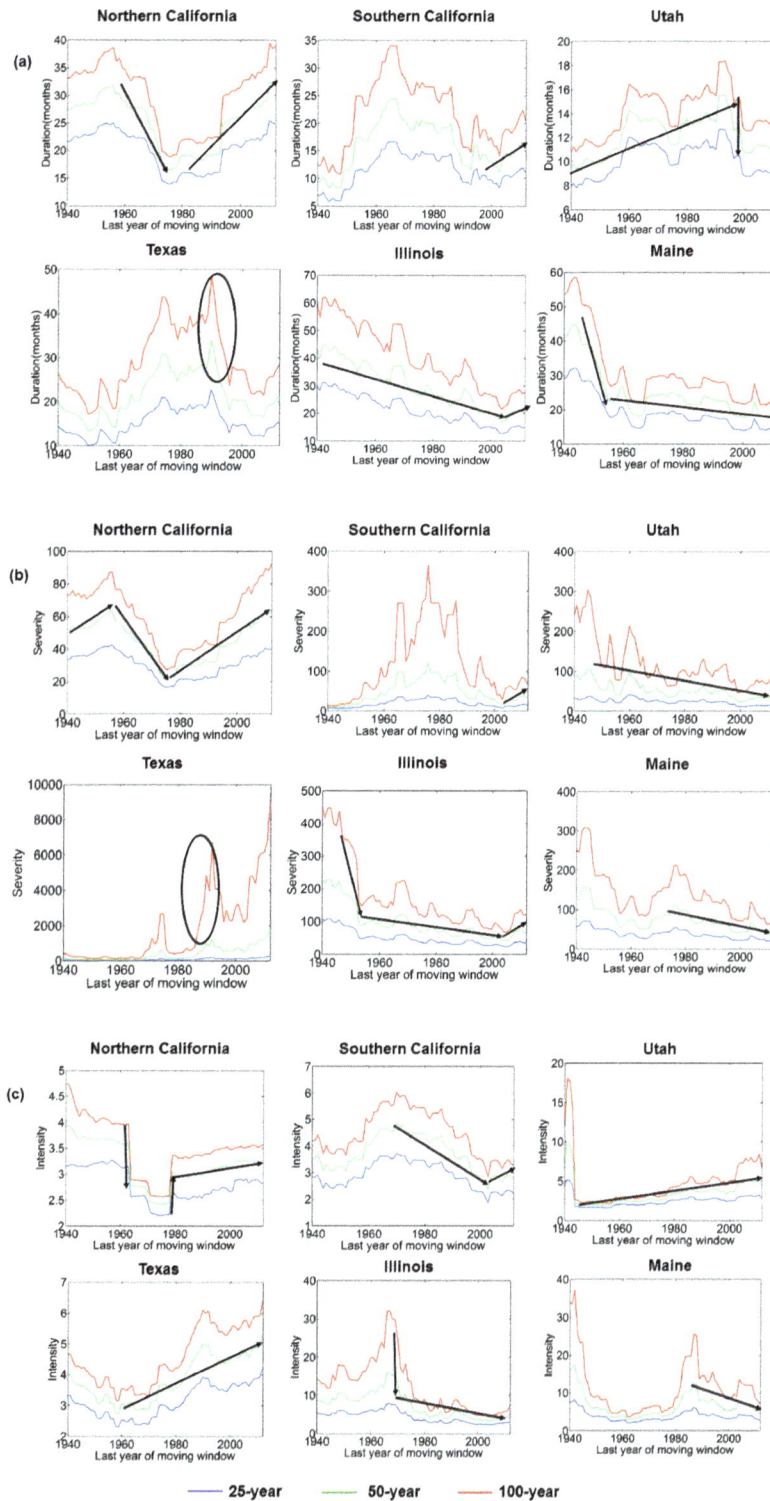

Fig. 2. (a) Duration, (b) severity, and (c) intensity of drought events, 1900–2012.

1960s, followed by a decrease in duration up to the 1980s. In recent decades, however, there has been an ongoing increase. In Illinois, drought duration has been steadily decreasing up to 2000 after which there has been an ongoing increase. As shown in Fig. 2b, Northern California's drought severity follows a similar pattern as for drought duration. In Illinois, severe drought occurred prior to the 1940s with an abrupt decrease in severity before the 1960s. After the 1960s, the drought severity in Illinois followed a similar pattern as for drought duration. Finally, Fig. 2c indicates high variability in drought intensity with a pattern different from those of duration and severity. There is a recent increase in drought intensity for Northern California, and a decrease in intensity for Illinois. There is also spatial variability over the six locations, with different trends and shifts at each location.

The natural (e.g. hydrology) and socio-economic (e.g., agricultural) impact of a drought event depends on a combined level of duration, severity and intensity. Under some cases, one of these characteristics is a dominating factor. Thus, understanding the combined impacts on multiple characteristics, as well as the dominating impact of one aspect of drought characteristics, is informative for more effective drought preparedness.

2.1. *Natural and socio-economic droughts*

Droughts propagate from meteorological systems to agricultural systems and can greatly impair water quantity, water quality, agricultural production, food security, and economy at local, regional and national scales (Calanca 2007; Schmidhuber and Tubiello 2007; Wang *et al.* 2011; Eitzinger *et al.* 2013). Droughts can then be classified into different categories: meteorological, agricultural, hydrological, and socio-economic/political. Meteorological drought results from precipitation deficits and/or temperature increase (heat).

Agricultural droughts can be identified by soil moisture deficits. Hydrological droughts are related to streamflow and water storage. Finally, socio-economic droughts can result from production loss, such as crop yield deficits. Meteorological droughts propagate impacts on other levels of drought (Wang *et al.* 2011). As demonstrated by an example in Fig. 3, an increase in drought duration decreases the emphasis on meteorological droughts (precipitation deficiency) and increases emphasis on water resource management. Increase in drought duration also increases the complexity of impacts and conflicts (also see Haslinger *et al.* 2014; Wang *et al.* 2011).

An example of nonlinear drought impact propagation in central Illinois is shown in Fig. 4 (Wang *et al.* 2011). Meteorological, agricultural, and hydrologic droughts are represented with standardized precipitation index (SPI), standardized soil water index (SSWI), and standardized runoff index (SRI), respectively. Regional climate models (RCM) and general circulation models (GCM) were used for two emission scenarios to assess changes in SPI, SSWI, and SRI in terms of number of droughts per year. The results in Fig. 4 may indicate concerns regarding the nonlinear propagation of meteorological drought to other types of drought (Wang *et al.* 2011). In particular, lower levels of meteorological drought can amplify the impacts on agricultural and hydrologic droughts (see also Schlenker and Roberts 2006).

Drought impact amplification is related to the increase of population and economy size and structure. In reference to the ongoing drought in California and the severe drought in the 1970s, "…the difference right now is the population is around 40 million while during the previous severe drought in 1970s it was 20 million" (Dimick 2015). A meteorological drought of similar size can have much larger socio-economic impacts in the future if engineering infrastructure and institution remain the same. Using the Midwest as another example, during a massive drought in 1995, approximately 55% of the

Decreasing emphasis on the natural event

Increasing emphasis on water/natural resources management

Increasing Complexity of impacts and conflicts

Hydrological

Agricultural

Meteorological

Socio-economic

Time/Duration of the event

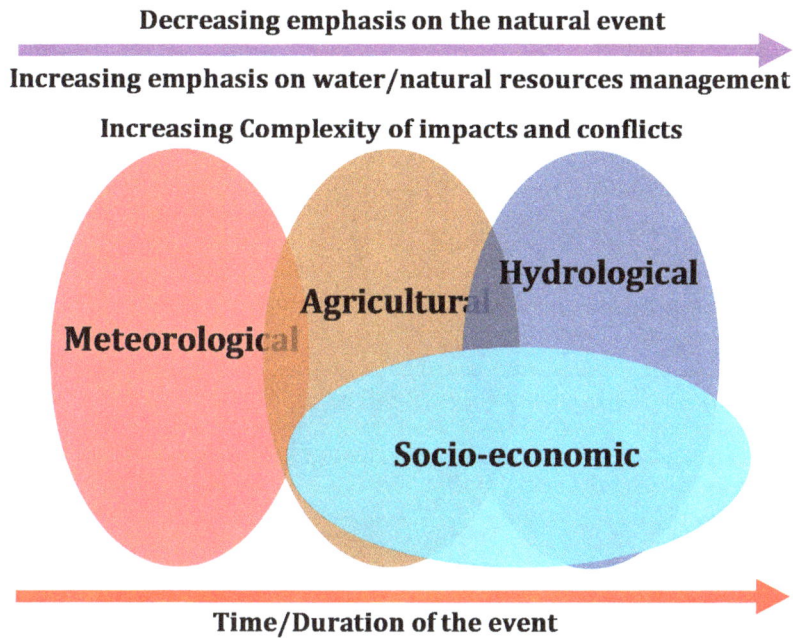

Fig. 3.   Natural and Social Drought (*Source*: National Drought Mitigation Center, University of Nebraska–Lincoln, USA).

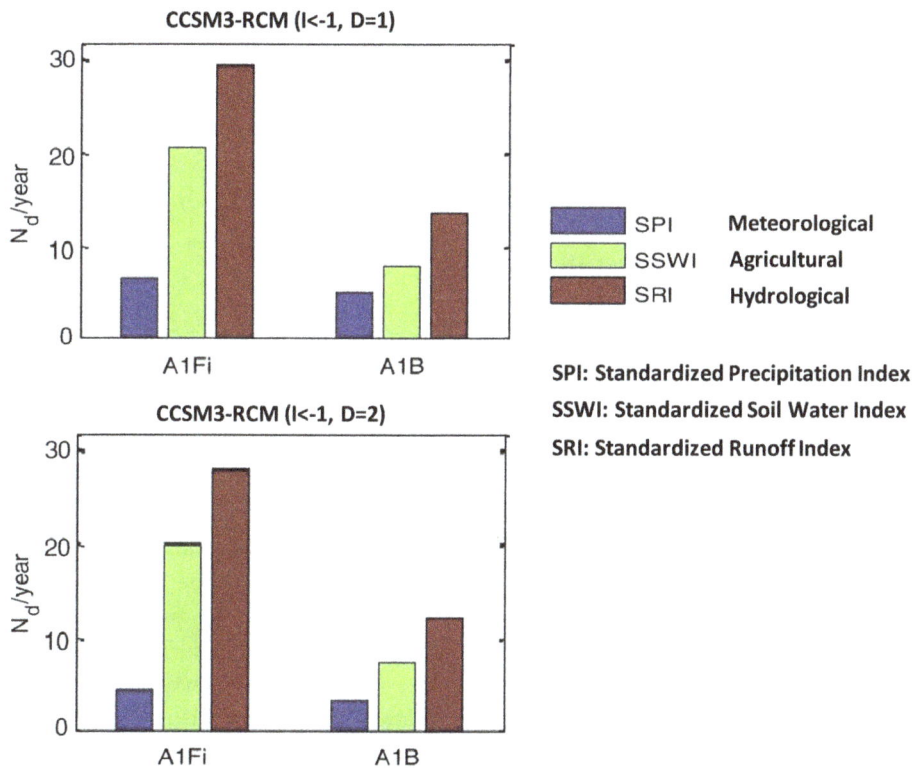

CCSM3-RCM (I<-1, D=1)

CCSM3-RCM (I<-1, D=2)

SPI   **Meteorological**

SSWI   **Agricultural**

SRI   **Hydrological**

SPI: Standardized Precipitation Index

SSWI: Standardized Soil Water Index

SRI: Standardized Runoff Index

Fig. 4.   Impacts of climate change on different levels of drought (*Source*: Wang *et al.* 2011).

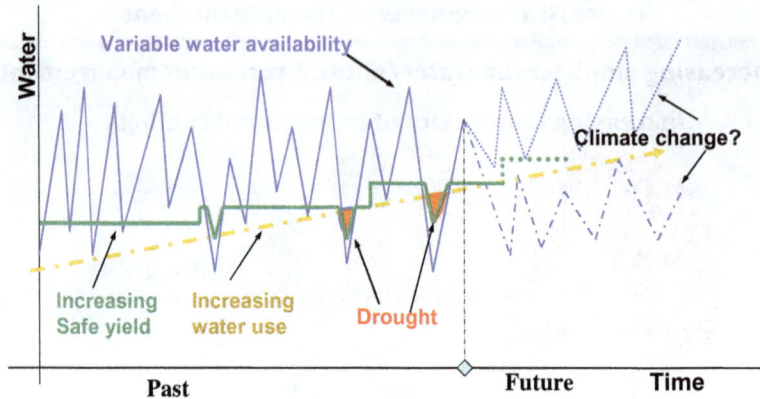

Fig. 5.   Drought as a consequence of a complex natural-social system.

corn was used for domestic feed grain. During the 2012 Midwest drought, more than 40% of the corn was used for domestic ethanol use (Adonizio *et al.* 2012), which implies different socio-economic consequences. Even if the meteorological droughts follow historical patterns, increase of population and economy size and shift in the economic structure can have a much larger effect on the socio-economic system.

Further, the impacts of droughts are affected by the level of socio-economic development in a region. In contrast to developed countries like the U.S. where the impacts are related to the decline in crop productivity farmers' profit, the impacts in less developed countries can be much more varied with more serious consequences. For example, droughts in Bangladesh have led to problems like food shortages and starvation as droughts affect rice cultivation, which is not only the major source of income for farmers but also the staple diet of the people (Paul 1995). More than 5 million people in Bangladesh are under a great threat due to severe impact of drought, which forces them to use arsenic contaminated ground water (Miyan 2015). In Cambodia, only 20% of the cultivated land has irrigation systems, making a large fraction of the crop land in the country vulnerable to droughts (WFP 2010).

Thus, drought can be viewed as a consequence of a complex natural-social system as shown in

Fig. 5. In the past, we have observed the variability in water availability due to changes in precipitation and temperature. Socio-economic development and increases in population have increased water consumption. At the same time, growing water supply is facilitated with an increase in engineering facilities, resulting in increases in safe yield. Considering the ongoing historic drought in Southeast Asia, Sanny Jegillos, the regional disaster reduction advisor for United Nations Development Programme Regional Centre for Asia and Pacific, commented:

> ... but the effects [of the drought] vary from country to another. Some countries have better water management therefore those are not as badly affected as others, but some have actually [like Cambodia] been impacted since 2014.

In fact, due to improvements in irrigation facilities and water management abilities, it appears that most countries in the region are better prepared, compared to the conditions during the previous major El Niño in 1997–98 (Southeast Asia Globe 2016).

Droughts occur when water availability drops drastically and as a result, safe yield drops below the level that is required for water use. In the future, there are numerous uncertainties, such as the effects of climate change, population growth and the status of water supply infrastructures, which make it very difficult to decide on necessary drought preparedness tactics. As can be seen, hydroclimatic, engineering, and socioeconomic factors are all linked to determine the level of a drought event and its impact; they must be considered together in dealing with drought preparedness.

To summarize, multiple correlated drought features must be considered when dealing with the risk of drought. Drought features have a lot of variability from region to region, and trends and abrupt shifts at each location can be identified. The impact of meteorological drought propagation to socio-economic systems is often nonlinear. Finally, in order to deal with drought preparedness, an interdisciplinary approach must be taken considering hydroclimatic, engineering, and socioeconomic factors.

3. The Role of Institution and Public Behavior

3.1. *Institutional development for drought preparedness in the U.S.*

Before the 1980s, states conducted crisis management, relying on the federal government for assistance during drought events. In the early 1980s, the U.S. experienced severe droughts, motivating 38 states to rapidly develop long-term drought response plans (Wilhite 2005). Emphasis on response planning continues today, stressing mitigation planning (i.e., risk management) rather than crisis management. Today, the federal government has already developed risk-based national policies to promote risk management, for example, the National Drought Preparedness Act has created a National Drought Council; the National Office of Drought Preparedness has also been developed, which consists of both federal and non-federal members. Further, U.S. Geological Survey (USGS) has developed the National Integrated Drought Information System (NIDIS) to collect data and monitor droughts. Recently, the President's Climate Action Plan, links information with drought preparedness in critical sectors to manage drought-related risks. In particular, this new policy considers the impact of climate change on drought (White House 2013). As evident by the above examples of risk-based national policies, the U.S. is institutionally heading in a promising direction in terms of drought preparedness, moving away from crisis management and instead promoting the risk-management approach.

3.2. *Institutional development for drought preparedness in Australia*

In Australia, the national government has long been involved in providing drought assistance, but a major policy reform occurred in 1989 which led to the National Drought Policy in 1992 (Botterill 2010). The new paradigm recognized that farming is a risky business, farmers are self-reliant, and risk management approach should be adopted to cope with and manage drought (Wilhite 2005). Moreover, the government has a limited direct role, except during an exceptional drought event, and is responsible to provide necessary skills and tools to farmers and help them manage climate variability (Wilhite 2005). Drought risk management strategies focus on either reducing water demands or enhancing water supplies, or both. Grant *et al.* (2013) point out contrasting drought management strategies in two drought affected regions: southeastern Australia and southwestern U.S. The drought management policies in southeastern Australia focus on the reduction of demand levels through reuse of water by storm-water harvesting and wastewater recycling; whereas the focus of drought management policies in southwestern U.S. has been enhancing water supplies through infrastructure enhancements and promotion of water markets permitting transfers to high valued users.

3.3. *The impact of institutional, market, and socio-economic conditions*

The various impacts of institutional and socioeconomic conditions on drought preparedness decisions have been observed with past and ongoing drought events. The 2012 Midwest drought and the ongoing California drought are used to discuss the impacts as follows. In response to the 2012 Midwest drought, President Obama commented:

> It is a historic drought and it is having a profound impact on farmers and ranchers all across many states.

Indeed, more than half the counties in the U.S., or approximately 1,584 counties across 32 states, were declared as disaster areas (Watson 2012). Despite such an extensive effect, the 2012 Midwest drought did not receive national attention, at least not as much as the ongoing California drought. As questioned by a news story from Forbes, why didn't this drought go social? (Watson 2012). Indeed, droughts are considered slow motion disasters. They have less of an impact on instant media unless they persist for a long time. This is evident by the California drought, which has been around for many years, yet did not receive attention in the media until April 2015 when the government started taking action. In addition, the general public perceives America as having a relatively stable food supply that is not threatened by drought (Watson 2012). Further, the negative impacts associated with drought disasters have been significantly mitigated by government agricultural policies, including insurance and subsidies, as discussed in details in the following.

In Dan Charles' article about the Midwest drought entitled "Secret side of the drought: Many corn farmers will benefit (Charles, 2012) he commented:

> But the next time you see a story about farmers suffering in the drought, just remember: There are winners and losers when food becomes scarce, and farmers are on both sides.

Many of the farmers' incomes were unaffected if not higher than during a regular crop season without drought. This was possible with the help of crop insurance and subsidies. In addition, the prices of crops increased later in the season, allowing farmers to sell their crops for more profit. On the other hand, some farmers made contracts with refineries ahead of the season to obtain a better price, but because of the drought, they could not provide the amount of crops that they had promised. These farmers were severely affected; they were forced to buy corn with higher price from the market to uphold their side of the contract.

Some existing market conditions affect undertaking drought mitigation measures. For the case of the ongoing drought in California, some people propose for California to copy Israel's institutional model in order to better combat the water shortage. With significant water shortages in 2009, Israel has implemented serious institutional change. Conditions in Israel have been greatly improved with about 40% of Israel's water coming from desalination and 85% of all household wastewater being purified. However, it would be difficult to duplicate Israel's model in California as Israel is a lot more prepared for institutional change and infrastructure development (Chabin 2015). For example,

> Israel's water policy and pricing are set for the whole country. In California, water prices vary from location to location and district to district... Los Angeles had over 100 different water suppliers (Chabin 2015).

In addition,

> Unlike Israel, where permits for desalination plants have been accelerated, California has a long process because of environmental concerns (Chabin 2015).

It is not so easy to duplicate an existing model combating drought because varying conditions may limit the model's feasibility in another region (Mishra and Singh 2010).

Drought management measures interact with water right systems that provide their functions under normal conditions. Farmers with "junior" rights in some California basins were ordered to stop using the river's depleting water in 2015, but only a fifth of such water holders confirmed that they were complying (Associated Press 2015). Some farmers even started groundwater pumping since it is not under regulation (Holthaus 2015). It is concerning to think that changes in institution can perhaps make drought conditions worse with the indirect encouragement of groundwater pumping (Famiglietti 2014).

It is also observed that some social conditions, such as income, interfere with drought management. In response to the worsening drought conditions, California's government has ordered a 25% cut on water usage. The cut actually ranges from 4% to 36% depending on the community's level of wealth. The wealthiest communities have the highest cuts, while the less wealthy communities have more modest cuts. According to Jeffrey Kightlinger, the general manager of the Metropolitan Water District of Southern California,

> We are finding it works with 90 percent of the public. You still have certain wealthy communities that won't bother. And the price penalty doesn't impact them.

Such policy causes an economic divide and further stresses the difficulty of implementing a given institution (Nagourney and Healy 2015).

4. The Use of Forecast Information for Decision-Making in Drought Preparedness

First of all, it is important to understand what aspects determine whether a given forecast is valuable for decision-making in drought preparedness. There are three types of forecasts for real-time (short), medium, and long-term decisions. Short-term forecasts are 1- or 2-week lead forecasts used for near real-time decision-making. This includes irrigation scheduling and some tactical measures (Hejazi et al. 2014). Seasonal forecasts typically have a 3- to 6-month lead and are used for seasonal planning. They are particularly important for rain-fed agriculture since little can be done to mitigate drought impacts after a crop is planted. Some decisions which are significantly affected by use of seasonal forecasts include land use, crop patterns, fertilizer use, insurance type, and forward contracts (Hill and Mjelde 2002; Shafiee-Jood et al. 2014). Finally, long-term forecasts are intra-annual or multi-decadal, which can be used for long-term planning, such as for infrastructure development and strategic measures (Cai et al. 2015).

The use of seasonal forecasts depends on multiple factors such as forecast accuracy, institutional intervention, and the relation between these factors. The historic 2012 drought in Midwestern U.S. is a good example to explore this issue. The 2012 Midwest drought had very limited long-lead predictability due to the internal variability of the atmosphere (Hoerling et al. 2014; Kumar et al. 2013). In other words, the seasonal forecast could not capture the drought and use of the forecasts could have made farmers worse off. For instance, Shafiee-Jood et al. (2014) compared the seasonal climate forecast based on a CWRF model (Liang et al. 2012) for the Midwest drought in 2012 with actual observations. The seasonal forecast was then improved by using available observation data for re-analysis. Comparing the forecast precipitation to observations revealed that the seasonal forecasts were not accurate and tended to overestimate the precipitation in crucial months (June and July). Therefore, they could have possibly made some farmers worse off if they were employed in their decision-making. On the other hand, the re-analysis improves the actual forecast significantly, suggesting that future forecasts could perhaps be improved to capture such droughts more accurately. Comparing improved and ideal forecasts with CWRF forecasts suggested that a better forecast is likely to have a positive value to the farmers, thereby increasing their profits. This study also showed that the ex-post value of forecasts to farmers with no forward contracts was not very different under the three varying forecast scenarios. This is due to the institutional intervention in the form of crop insurance: As crop insurance gets stronger, the skill of forecast is less important and the value of an improved forecast (e.g. the re-analysis) is then not significantly higher than the primary forecast (e.g., the actual forecast) (Shafiee-Jood *et al.* 2014).

The intervention of institutional setting to the use of a forecast was further illustrated by Craig (2015) in an analysis of forecast use by Uruguayan farmers. The Uruguayan government is working with Columbia's International Research Institute for Climate and Society (IRI) to develop sophisticated short-term and seasonal forecasts for Uruguayan farmers. Walter Beathgen, a Columbia agronomist and environmental scientist says,

> The reason you don't see systems like this operating in the US or Europe is because wealthy countries have government-subsidized insurance programs that bail out farmers whenever there is a bad harvest. As a result, there's not much demand within the agricultural industry for something like this. Developing countries, on the other hand, can't afford expensive farm subsidies. But they can afford to give their farmers the information they need to pursue efficient, climate-smart agriculture.

Whether or not a farmer uses forecasts can, therefore, be significantly affected by the aid that institution provides during drought. Thus, the use of forecasts can depend on the impact of economic conditions, which can vary greatly between developed and developing countries.

In general, the effective use of a forecast can be affected by three types of factors as shown in

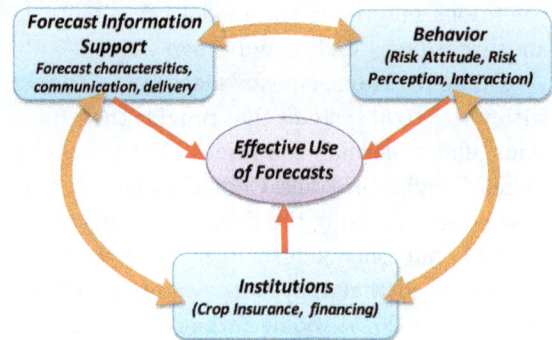

Fig. 6.   The various factors to the effective use of forecasts.

Fig. 6. The first type of factors involves forecast reliability and method of information delivery. The information must be delivered within a timeframe during which it will be useful to farmers. This is less of an issue today with an improvement in communication due to the significant increase in the use of mobile technology. The second component is users' behavior in response to forecast information and the heterogeneity of the users. Some farmers and institutions may be very sensitive to forecast information, while others may have no reaction. The third component involves the role of institution. Crop insurance and crop prices can significantly affect whether the forecast is used, as discussed above.

5. Decision Problems of Drought Preparedness

With the complexity of drought, decisions for effective drought preparedness can be very complicated. As previously mentioned, California's government has ordered a 25% cut on water usage in response to the ongoing drought. Governor Brown of California stressed,

> It's a different world" and "we have to act differently" (Megerian *et al.* 2015).

One interesting example regarding decision making for drought preparedness is the desalination plant in Santa Barbara. The desalination plant was constructed in the early

1990s when Santa Barbara faced what, at the time, was a record-breaking drought. The plant's construction was finished as the drought was ending. The Santa Barbara plant was too expensive to run and was shut down after three months of operation. The plant has been unused since the early 1990s, but with drought conditions worsening, they have considered placing it back online. Hesitation to re-start the plant revolved around environmental concerns and high costs. It can take up to $40 million to bring the desalination plant back online, plus $5 million per year to keep it running. Some environmental concerns include death of small marine life after getting sucked into the intake filters for ocean water pumping and detrimental effects on marine species resulting from the return of higher salt concentration water to the ocean (Roth 2015; Hiltzik 2015).

With the ongoing drought in California, Santa Barbara decided to re-start the desalination plant in 2015. At a time when drought within California is severe, decision-makers have placed the water shortage issue as a priority over costs and environmental concerns. Although it hasn't been discussed whether this plant will run after the drought is over, Mayor Helene Schneider of Santa Barbara commented,

> So instead of going through this panic mode of reauthorizing and restructuring a desalination plant, should we think differently: should desalination be a part of a regular water supply for the region, even at a small level, when we're not in a drought?" (Roth 2015).

The Mayor suggests that the decision to re-start the desalination plant in Santa Barbara is a crisis management measure. It may be beneficial to instead head towards a risk-management approach in order to be readily prepared for drought when it occurs. Such a decision could be expanded out to a global context: Should we continue to practice crisis management, or should we move to risk-management as preparation for future droughts?

The huge drought losses and great threats from climate change have intensified the need to move from emergency response-focused crisis management to risk management, which would place greater emphasis on preparedness planning and mitigation actions (Wilhite 2002). Decisions for drought preparedness include traditional, short-term tactical measures, (e.g., facility operation) and long term or in-advance strategic measures, which require capital investment. It is likely that current facilities are not enough to mitigate damage under future climate conditions, indicating the requirement for infrastructure investment (Cai *et al.* 2015). In other words, risk management, or long term strategic measures may be crucial.

Drought preparedness quickly becomes a messy decision problem due to tradeoffs involved in the decisions and the effects of uncertainty. Quantifying the unknown or deep risk in climate change projection and future economy can be very challenging. Further, disagreements in tradeoffs exist. Numerous groups have different priorities, which extends and complicates the decision process (Metlay and Sarewitz 2012). The Santa Barbara desalination plant, for example, had a huge tradeoff between environmental concerns and water supply, which lengthened the decision process. The city did not make a decision until a state of crisis was reached.

The decision for drought preparedness under "deep uncertainty" is a statistical decision problem that can be represented by Fig. 7 below. The question becomes: Should we take measures now? If the answer is yes and there is more serious drought in the future, we will have appropriate adaptations. However, if it turns out that the future droughts are no worse than historic droughts, the result will be over-preparedness. This case should be avoided since it is the objective of every engineer to minimize costs. If instead we delay infrastructure development, and more serious drought occurs in the future, we will be under-prepared.

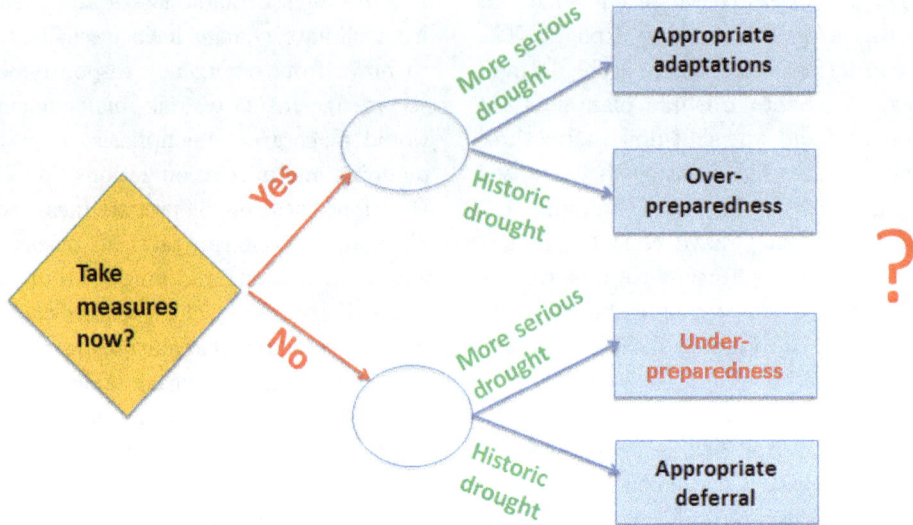

Fig. 7. Over-preparedness vs. under-preparedness under "deep uncertainty" (after personal communication with Richard Vogel).

6. The Need of a Holistic Framework to Assess Drought Preparedness

As discussed above, drought preparedness depends on natural, social, and technical conditions, which suggests an integrated framework for assessing drought preparedness as represented by Fig. 8. This figure proposes an approach based on risk management for major and especially unexpected droughts, although crisis management still plays a role in making a specific area prepared for some normal drought events. Risk analysis requires us to consider how climate change and variability affects both natural characteristics of drought (duration, severity, and intensity) and socio-economic drivers in order to have a more realistic assessment of drought risk. Our final decision and implementation is dependent on the drought risk assessment, information we obtain from monitoring and forecasts, and from public behavior, institution, and infrastructure. The decision is complicated by tradeoffs between different groups and between human society and the natural environment. Various uncertainties involved in forecasts, institutions and behaviors further complicate the decision.

The following questions must be considered in order to complete the holistic framework for the assessment of drought preparedness:

- Do climate forecasts/projections provide useful information?
- How will climate change affect policy and behavior (adaptation)?
- What happens if drought is unpredictable (deep uncertainty)?
- Will socio-economic drivers be a major source of vulnerability?
- How do we ensure forecasts are actually used?
- Does institution/infrastructure support decisions and implementations?
- Should we do something now or "wait and see" what will happen?

By addressing all of these questions in the holistic framework, we can identify the bottleneck in technology (especially on drought forecast and monitoring), infrastructure, and/or

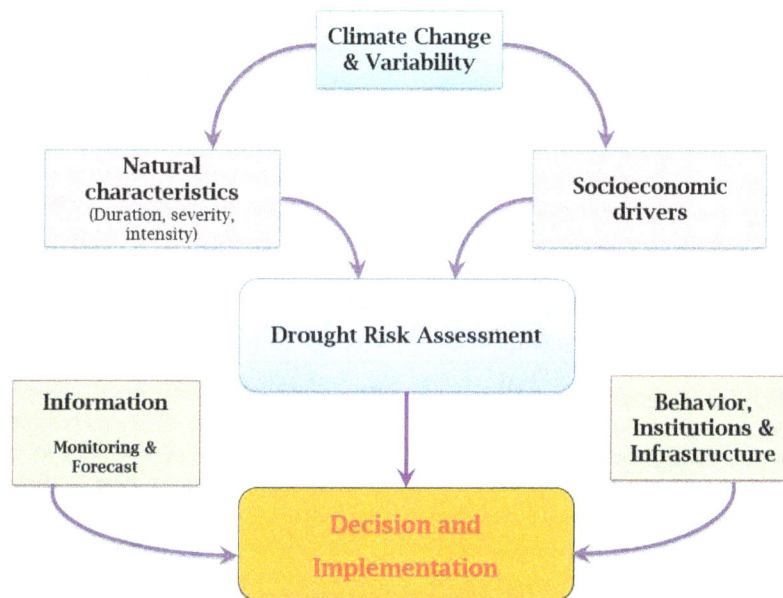

Fig. 8.   Integrated framework for drought preparedness.

institution that hinder effective drought preparedness, and we can also develop an approximate plan on the new development/refinement of these aspects to move forward to more effective drought management.

References

Adonizio, W., N. Kook, and S. Royales, 2012: Impact of the drought on corn exports: Paying the price. *U.S. Bureau of Labor Statistics*, Vol. 1, No. 17, 7 pp. Accessed 21 December 2015. [Available online at http://www.bls.gov/opub/btn/volume-1/pdf/impact-of-the-drought-on-corn-exports-pa ying-the-price.pdf]

Associated Press, 2015: California water cuts move to those with century-old rights. *CBS NEWS*. Accessed 21 December 2015. [Available online at http://www.cbsnews.com/news/california-droug ht-water-farmers-century-old-rights/]

Botterill, L., 2010: Risk management as policy: The experience of Australia's National Drought Policy. In: *Economics of Drought and Drought Preparedness in Climate Change Context*. Zaragoza: CIHEAM/FAO/ICARDA/GDAR/CEIGRAM/MARM, 241–248.

Cai, X., R. Zeng, W. H. Kang, J. Song, and A. J. Valocchi, 2015: Strategic planning for drought mitigation under climate change. *J. Water Res. Plan. Man.*, **141**, 04015004. doi:10.1061/(ASCE)WR.1943-5452.0000510.

Calanca, P., 2007: Climate change and drought occurrence in the Alpine region: How severe are becoming the extremes? *Global Planet. Change*, **57**, 151–160. doi:10.1016/j.gloplacha.2006.11.001.

Chabin, M., 2015: Israel to California: Here's how to save water. *USA Today*. Accessed 21 December 2015. [Available online at http://www.usatoday.com/story/news/world/2015/05/07/israel-droug ht-california-desalination/26923503/]

Charles, D., 2012: Secret of the drought: Many corn farmers will benefit. *National Public Radio*. Accessed 21 December 2015. [Available online at http://www.npr.org/sections/thesalt/2012/08/14/158752153/secret-side-of-the-drought-corn-fa rmers-will-benefit]

Cook, B. I., T. R. Ault, and J. E. Smerdon, 2015: Unprecedented 21st century drought risk in the American Southwest and Central Plains. *Science Advances*, **1**, e1400082. doi:10.1126/sciadv.1400082.

Craig, D. J., 2015: Seeds of Hope. *Columbia Magazine*. Accessed 21 December 2015. [Available online at http://magazine.columbia.edu/featur es/spring-2015/seeds-hope]

Dai, A., 2013: Increasing drought under global warming in observations and models. *Nat. Clim. Change*, **3**, 52–58. doi:10.1038/nclimate1633.

Dimick, D., 2015: 5 things you should know about California's water crisis. *Natl. Geogr.* Accessed 21 December 2015. [Available online at http://news.nationalgeographic.com/2015/04/15 0406-california-drought-snowpack-map-water-sci ence/]

Eitzinger, J. and Coauthors, 201: Regional climate change impacts on agricultural crop production in Central and Eastern Europe–hotspots, regional differences and common trends. *J. Agr. Sci.*, **151**, 787–812. doi:10.1017/S002185961200 0767.

Famiglietti, J. S., 2014: The global groundwater crisis. *Nat. Clim. Change*, **4**, 945–948. doi:10. 1038/nclimate2425.

Ganguli P. and M. J. Reddy, 2013: Spatio-temporal analysis and derivation of copula-based intensity-area-frequency curves for droughts in western Rajasthan (India). *Stoch. Env. Res. Risk A.*, **27**, 1975–1989. doi:10.1007/s00477-013-0732-z.

Grant, S. B., T. D. Fletcher, D. Feldman, J. D. Saphores, P. L. M. Cook, M. Stewardson, K. Low, K. Burry, and A. J. Hamilton, 2013: Adapting urban water systems to a changing climate: Lessons from the millennium drought in Southeast Australia. *Environ. Sci. Tech.* **47**, 10727–10734. doi:10.1021/es400618z.

Haslinger, K., D. Koffler, W. Schöner, and G. Laaha, 2014: Exploring the link between meteorological drought and streamflow: Effects of climate-catchment interaction. *Water Resour. Res.*, **50**, 2468–2487. doi:10.1002/2013WR01505.

Hejazi, M. I., X. Cai, X. Yuan, X.-Z. Liang, and P. Kumar, 2014: Incorporating reanalysis-based short-term forecasts from a regional climate model in an irrigation scheduling optimization problem. *J. Water Res. Plan. Man.*, **140**, 699–713. doi:10.1061/(ASCE)WR.1943-54 52.0000365.

Hill, H. S. J. and J. W. Mjelde, 2002: Challenges and opportunities provided by seasonal climate forecasts: A literature review. *J. Agr. Appl. Econ.*, **34**, 603–632.

Hiltzik, M., 2015: Desalination plants aren't a good solution for California drought. *Los Angeles Times.* Accessed 21 December 2015. [Available online at http://www.latimes.com/business/hilt zik/la-fi-hiltzik-20150426-column.html]

Hoerling, M., J. Eisceid, A. Kumar, R. Leung, A. Mariotti, K. Mo, S. Schubert, and R. Seager, 2014: Causes and predictability of the 2012 Great Plains drought. *Bull. Am. Meteor. Soc.*, **95**, 269–282. doi:10.1175/BAMS-D-13-00055.1.

Holthaus, E., 2015: It's gotten this bad: California moves to restrict farmers' oldest water rights. *The Slatest.* Accessed 21 December 2015. [Available online at http://www.slate.com/blogs/the_sl atest/2015/05/21/california_drought_water_restric tions_are_coming_for_farmers_with_century.html]

IPCC, 2012: *Managing the Risks of Extreme Events and Disasters to Advance Climate Change Adaptation. A Special Report of Working Groups I and II of the Intergovernmental Panel on Climate Change.* Cambridge University Press, 582p.

IPCC, 2013: Climate Change 2013: *The Physical Science Basis. Contribution of Working Group I to the Fifth Assessment Report of the Intergovernmental Panel on Climate Change.* Cambridge University Press, 1535 pp.

Kao, S. C. and R. S. Govindaraju, 2010: A copula-based joint deficit index for droughts. *J. Hydrol.*, **380**, 121–134. doi:10.1016/j.jhydrol.2009.10.029.

Kumar, A., M. Chen, M. Hoerling, and J. Eischeid, 2013: Do extreme climate events require extreme forcings? *Geophys. Res. Lett.*, **40**, 4900–4905. doi:10.1002/grl.50657.

Liang, X.-Z. and Coauthors, 2012: Regional climate-weather research and forecasting model. *Bull. Am. Meteorol. Soc.*, **93**, 1363–1387. doi:10.1175/ BAMS-D-11-00180.1.

Madadgar, S. and H. Moradkhani, 2013: Drought analysis under climate change using copula. *J. Hydrol. Eng.*, **18**, 746–759. doi:10.1061/ (ASCE) HE.1943-5584.0000532.

Megerian, C., M. Stevens, and B. Boxall, 2015: Brown order California's first mandatory water restrictions: 'It's a different world'. *Los Angeles Times.* Accessed 21 December 2015. [Available online at http://www.latimes.com/local/lanow/ la-me-ln-snowpack-20150331-story.html]

Metlay, D. and D. Sarewitz, 2012: Decision Strategies for Addressing Complex, 'Messy' Problems. *The Bridge on Social Sciences and Engineering*, **42**, 6–16. [Available online at https://www. nae.edu/Publications/Bridge/62556/62558.aspx]

Mishra, A. K. and V. P. Singh, 2010: A review on drought concepts. *J. Hydrol.*, **391**(1–2), 202–216. doi:10.1016/j.jhydrol.2010.07.012.

Mishra, A. K. and V. P. Singh, 2011: Drought modeling — A review. *J. Hydrol.*, **403**(1–2), 157–175. doi:10.1016/j.jhydrol.2011.03.049.

Miyan, A., 2015: Drought in Asian least developed countries: Vulnerability and sustainability. *Weather and Climate Extreme*, **7**, 8–23. doi:10.1016/j.wace.2014.06.003.

Nagourney, A. and J. Healy, 2015: Drought frames economic divide of Californians. *The New York Times*. Accessed 21 December 2015. [Available online at http://www.nytimes.com/2015/04/27/us/drought-widens-economic-divide-for-californians.html?_r=2]

NOAA., 2015: Billion-dollar weather and climate disasters: Table of events. Accessed 21 December 2015. [Available online at https://www.ncdc.noaa.gov/billions/events]

Paul, B. K., 1995: Farmers' and Public Responses to the 1994–1995 Drought in Bangladesh: A case study, Quick Response Report #76, Department of Geography, Kansas State University.

Planton, S., M. Déqué, F. Chauvin, and L. Terray, 2008: Expected impacts of climate change on extreme climate events. *C. R. Geosci.*, **340**, 564–574. doi:10.1016/j.crte.2008.07.009.

Ross, T. and N. Lott, 2003: A climatology of 1980–2003 extreme weather and climate events. NCAR Tech. Report No. 2003-01, 15 pp. Accessed 21 December 2015. [Available online at https://www.ncdc.noaa.gov/billions/docs/lott-and-ross-2003.pdf]

Roth, S., 2015: California's last resort: Drink the Pacific. *The Desert Sun*. Accessed 21 December 2015. [Available online at http://www.desertsun.com/story/news/environment/2015/04/20/californias-last-resort-drink-pacific/26081355/]

Saghafian, B. and H. Mehdikhani. 2013: Drought characterization using a new copula-based trivariate approach. *Nat. Hazards*, **72**, 1391–1407. doi:10.1007/s11069-013-0921-6.

Schlenker, W. and M. J. Roberts, 2006: Nonlinear effects of weather on corn yields. *Appl. Econ. Perspect. P.*, **28**, 391–398. doi:10.1111/j.1467-9353.2006.00304.x.

Schmidhuber, J. and F. Tubiello, 2007: Global food security under climate change. *P. Natl. A. Sci. USA*, **104**, 19703–19708. doi:10.1073/pnas.0701976104.

Shafiee-Jood, M., X. Cai, L. Chen, X.-Z. Liang, and P. Kumar, 2014: Assessing the value of seasonal climate forecast information through an end-to-end forecasting framework: Application to U.S. 2012 drought in central Illinois. *Water Resour. Res.*, **50**, 6592–6609. doi:10.1002/2014WR015822.

Shiau, J. T., 2006: Fitting drought duration and severity with two-dimensional copulas. *Water Resour. Manag.*, **20**, 795–815. doi:10.1007/s11269-005-9008-9.

Shiau, J. T. and H. W. Shen, 2001: Recurrence analysis of hydrologic droughts of differing severity. *J. Water Res. Plan. Man.*, **127**, 30–40. doi:10.1061/(ASCE)0733-9496(2001)127:1(30).

South Asia Globe, 2016: Southeast Asia facing severe drought. Accessed 24 June 2016. [Available online at http://sea-globe.com/southeast-asia-facing-severe-drought/]

van Dijk, A. I. J. M., H. E. Beck, R. S. Crosbie, R. A. M. de Jeu, Y. Y. Liu, G. M. Podger, B. Timbal, and N. R. Viney, 2013: The Millennium Drought in southeast Australia (2001–2009): Natural and human causes and implications for water resources, ecosystems, economy, and society. *Water Resour. Res.*, **49**, 1040–1057. doi:10.1002/wrcr.20123.

Wang, D., M. Hejazi, X. Cai, and A. Valocchi, 2011: Climate change impact on meteorological, agricultural, and hydrological drought in central Illinois. *Water Resour. Res.*, **47**, W09527. doi:10.1029/2010WR009845.

WFP, 2010: Lao PDR Country Strategy 2011–2015. *World Food Program* [Available online at http://www.wfp.org/sites/default/files/WFP%20Lao%20PDR%20Country%20Strategy_ENG.pdf]

Watson, T., 2012: Beneath a parched surface, the great Midwest drought of 2012 doesn't go social. *Forbes*. Accessed 21 December 2015. [Available online at http://www.forbes.com/sites/tomwatson/2012/08/10/beneath-a-parched-surface-the-great-midwest-drought-of-2012-doesnt-go-social/]

Wilhite, D. A., 2002: Combating drought through preparedness. *Nat. Resour. Forum*, **26**, 275–285. doi:10.1111/1477-8947.00030.

Wilhite, D. A., 2005: Australian and U.S. Drought Policy Experiences: Are Lessons Learned Transferable to Africa? *First African Drought Adaptation Forum*. Nairobi, Kenya. [Available online at http://web.undp.org/drylands/docs/drought/ADAF1/ADAF2005_1.4_Wilhite.pdf]

White House, 2013: The President's Climate Action Plan. Executive Office of the President. Accessed 21 December 2015. [Available online at https://www.whitehouse.gov/sites/default/files/image/president27sclimateactionplan.pdf]

Chapter 12

Priorities of the WMO Commission for Hydrology in the Context of Water, Climate and Risk Management

Liu Zhiyu

Vice President of the WMO Commission for Hydrology,
Director, Hydrological Forecast Center,
Ministry of Water Resources of China
2 Lane 2, Baiguang Road, Beijing 100053, China
liuzy@mwr.gov.cn
www.wmo.int/pages/prog/hwrp/index_en.php,
www.mwr.gov.cn/english/

This paper introduces the priorities of the WMO Commission for Hydrology. The programme of work adopted by the Commission for the period 2013–2016 focuses on five thematic areas, namely Quality Management Framework — Hydrology; Data Operations and Management; Water Resources Assessment; Hydrological Forecasting and Prediction; and Water, Climate and Risk Management. In addition to the five thematic areas, the WMO Congress has tasked the Commission with contributing to several WMO priorities, including the Global Framework on Climate Services, the WMO Integrated Global Observing System, Disaster Risk Reduction and Capacity Development. The paper elaborates major activities within the theme area of Water, Climate and Risk Management. Chinese practices of hydrological forecasting and prediction for decision-making support to flood management is also presented. Thanks to the WMO's efforts, collaboration between the water and climate communities has expanded in recent years. Climate products and services through the incorporation of hydrological observations have been improved. These advances, coupled with recent improvements in monthly to seasonal climate and hydroclimatic outlooks are beginning to yield improvements in hydrological services delivery, particular with respect to increased use of short-term climate forecasts and outlooks in water resources systems operations and seasonal planning. The paper will describe these issues and future direction associated with them.

1. Introduction

The issues of water stress, scarcity and uneven distribution in time and space are still with us. Superimposed on these are increasing populations, increasing water demand (particularly for potable water and agriculture) and uncertainties of climate change. This calls for better water resources management practices. And, it is very important for climate data, services and products to sustainable water resources management, especially those related to droughts and floods.

To cope with the water, climate and disaster issues, the World Meteorological Organization (WMO) coordinates the activities of National Meteorological and Hydrological Services (NMHSs), and makes great efforts to contribute to the safety and well-being of people throughout the world and to the economic benefit of all. The WMO works through the NMHSs to ensure that, among other things, effective 24-hour operational early warning systems for weather-, climate- and water-related hazards are made available in a timely manner and with longer lead times across political boundaries to all affected populations. The organization has a tremendous reputation in international cooperation in the world and has done a great contribution in the meteorological and hydrological

---

field. WMO accomplishments in the past six decades have made a difference in a number of areas, such as promotion of free and unrestricted international exchange of data and products, facilitation of coordinated observation system and standardization of instrument and data, sponsoring research and promotion of operational application of scientific and technological achievements, improving capacity-building with a focus on technical cooperation, education and training, providing world leadership in expertise and international cooperation in weather, climate, water and related environmental issues.

This paper presents briefly the priorities of the Commission for Hydrology, which is one of eight technical commissions of the WMO. Major activities within the theme area of Water, Climate and Risk Management will be elaborated with a focus on hydrological forecasting and prediction, water resources assessment, and contribution to the Global Framework for Climate Services. The paper will also describe the issues and key challenges associated with them.

2. WMO Programmes and WMO Commission for Hydrology

The WMO carries out its work through 10 international, scientific and technical programmes (Fig. 1), namely World Weather Watch Programme, WMO Space Programme, Natural Disaster Prevention and Mitigation

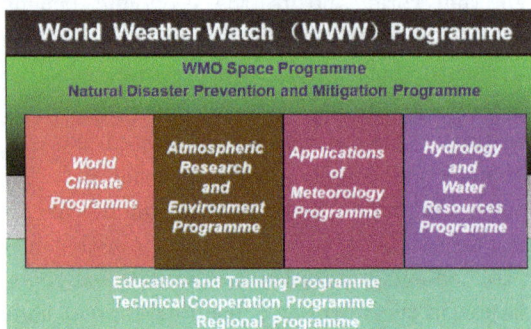

Fig. 1. The WMO programmes sketch map.

Programme, World Climate Programme, Atmospheric Research and Environment Programme, Application of Meteorology Programme, Hydrology and Water Resources Programme, Education and Training Programme, Technical Cooperation Programme, Regional Programme. Full WMO programme description can be found at http://www.wmo.int/pages/summary/progs_struct_en.html.

These programmes are designed to assist all Members to provide, and benefit from, a wide range of meteorological and hydrological services and to address present and emerging problems.

A brief review of the Hydrology and Water Resources Programme in the context of water and climate is presented below.

2.1. Hydrology and water resource programme

Hydrology and Water Resources Programme, short as HWRP, is one of the four core WMO professional programmes (the other three are World Climate Programme, Atmospheric Research and Environment Programme, Application of Meteorology Programme).

HWRP fulfils one of the major purposes of WMO, namely to promote application of hydrology and to further close cooperation between Meteorological and Hydrological Services. The Programme is concerned with the assessment of the quantity and quality of water resources, the mitigation of water-related hazards like floods and droughts, efforts to facilitate horizontal cooperation and transfer of technologies to strengthen the capacities of the hydrological services across the world.

The Programme effectively contributes to the implementation of all the Expected Results (ERs) in the WMO Strategy Plan 2012–2015 (http://www.wmo.int/pages/about/spla_en.html) and particularly to ER 3 along with ER2 and ER4. Many of the activities under HWRP are closely linked with other Programmes, in particular the World Weather Watch Programme, the World Climate Programme, the Tropical Cyclone Programme, the Education and Training Programme, and contributes to the overall objectives of the Disaster Risk Reduction Programme, the Regional Programme and the Least Developed Countries Programme.

The HWRP contributes to the implementation of the Global Framework on Climate Services (GFCS). This programme will act as an interlocutor and conduit to provide climate services under GFCS for use in the water sector. At the same time HWRP also contributes to GFCS through hydrological communities.

The HWRP is implemented through three mutually supporting components: (a) Basic Systems in Hydrology; (b) Forecasting and Applications in Hydrology; (c) Capacity-building in Hydrology and Water Resources Management.

More details about HWPR and publications such as technical regulations, guide to hydrological practices, manuals, technical reports and series, statements and other documents can be downloaded free of charge from the HWRP website (http://www.wmo.int/pages/prog/hwrp/index_en.php)

2.2. *WMO commission for hydrology*

The WMO Commission for Hydrology (CHy) shapes the water related activities of the WMO and addresses issues related to the basic hydrological observation network, water resources assessment, flood forecasting and management, adaptability to climate variability and change and promotes exchange of technology and capacity building. In particular the outcomes of its deliberations provide guidance to WMO Member countries and WMO Secretariat for the implementation of the Hydrology and Water Resources Programme of WMO.

Areas of activity include integrated hydrological networks, hydrological forecasting systems, hydrological aspects of natural disasters and environmental issues, assessment of water

use, latest advances in hydrological technology and practical implementation of IWRM principles with particular attention to integrated flood management. This information is incorporated into technical standards, reports or guidelines that can be used by NHSs in their work.

The Commission for Hydrology organizes activities under the framework of the HWRP according to Thematic Areas and specific topics, in order to align its work plan with the WMO Strategy Plan. The five thematic areas for the inter-sessional period 2012–2016 are: *Quality Management Framework — Hydrology (QMF-H)*; *Water Resources Assessment*; *Hydrological Forecasting and Prediction*; *Water, Climate and Risk Management*; *and Data Operations and Management* (WMO 2012).

The commission stressed to continue to work on capacity development, measurement of uncertainty, assessment of water availability, hydrological forecasting, data operation and management, including data exchange and in the development and implementation of demonstration projects which show the application of the outputs from the commission activities. Please visit http://www.wmo.int/pages/prog/hwrp/chy/ for more details.

3. Current Activities of CHy in the Theme Area of Water, Climate and Risk Management

The overall objective of HWRP and the mission of the WMO-CHy is to apply hydrology in meeting the needs for sustainable management through integration of hydrological, meteorological and climatological information and forecasts for use in water resources management; prevention and mitigation of water-related disasters; and in climate change adaptation in the water sector at national, regional and international levels.

From a hydrological perspective, the major disaster risk reduction activities are undertaken through the WMO Flood Forecasting

Initiative (FFI), which was established by WMO in 2003. The Components of the WMO — FFI include: Strategy and Action Plan, Action Plan and Activities Plan, Publications, Flash Flood Guidance Systems, Projects, Demonstration Projects, Associated Program on Flood Management and International Flood Initiative. The WMO and the Global Water Partnership are also working on a similar activity based on the concept of Integrated Drought Management.

3.1. *Hydrological forecasting and prediction in support of flood and drought management*

Hydrological forecasting and prediction has been a core priority of CHy for the past five decades. During the past years, the WMO Flood Forecasting Initiative has been the basis for improving the capacity of NMHSs to detect flood-critical situations and to improve the capacity of NHSs to use meteorological forecasting information in the provision of accurate and timely flood forecasting services for Emergency and Disaster Risk management.

Flash flood guidance system

The WMO Flood Forecasting Initiative has incorporated the Flash Flood Guidance System (FFGS) for global level application, which has close linkages to the Severe Weather Forecasting Demonstration Project (SWFDP) activities, to assist National Meteorological and Hydrologic Services in alerting their populations in data sparse areas of potentially hazardous near-term hydrologic conditions. The implementation of Flash Flood Guidance Systems globally is a WMO program to provide a tool for members to develop flash flood warnings.

Associated programme on flood management

In conjunction with GWP, WMO has expanded this effort to focus increasingly on flood management, particularly in terms of providing policy guidance, technical tools, and capacity building documents. The CHy Associated Programme on Flood Management has operationalized its HelpDesk function to include demand-based outreach capabilities, and it is currently finding success in documenting the implementation of flood management practices in developing countries. The concept of this HelpDesk is founded on a decentralized support base, among which UNESCO-IHE is one of 25 support base partners. The APFM continues to compile and produce guidance documents and tools in support of Integrated Flood Management. The most recent tools include Floodplain Mapping, flood forecasting and early warning, transboundary aspects of flood management and coastal flood management (APFM tools series could be found at http://www.apfm.info/?page_id=696). Training activities in support of countries wishing to adopt Integrated Flood Management Strategies are continuing.

Integrated drought management programme

In addition, significant progress has been made during the past three years, through collaboration with CAgM and CCl, in establishing operational support for drought prediction capabilities within NHMSs and drought management capabilities through the Integrated Drought Management Programme (IDMP). The IDMP, also in partnership with GWP, is now operational with a Technical Support Unit in the WMO Secretariat, to which GWP also contributes through a seconded expert working in WMO. A number of country and regional workshops have been held, also in close linkage with GWP initiatives.

3.2. *Water resources assessment under climate change background*

Water resources assessment (WRA) is an inveterate functional theme within CHy and, for many years, was focused primarily on estimation of regional water balance components and their implications for water supply. During the past six years, however, CHy activities associated with WRA have shifted substantially toward a broader evaluation of water availability and use, with an increasing emphasis being placed on dynamic assessments of various water resources conditions at weekly, monthly, and seasonal time steps. In selected assessments, it is also beginning to incorporate predictions and outlooks for future time periods, using access to improved climate services provided under the implementation of the GFCS.

With improved data availability, water resources assessment will go beyond providing information about the quantity and quality of the resources in time and space to value-addition to such data and information by developing hydrological products, tools and decision support systems for decision making.

3.3. *Contribution of the HWRP to the implementation of GFCS*

The GFCS (http://gfcs-climate.org/) has been identified as one of the five key priority areas for the WMO by the 16th Congress. Its primary goal is to ensure the greater availability of access to, and use of climate services for all countries and a wide range of users, of whom the water sector including hydrologists is one major key group. It aims to enhance the availability of needed climate information and services to plan ahead of time and make sustainable management decisions. It is concerned particularly with communities which are most vulnerable to climate variability and change.

Water is one of the five priority areas of the GFCS, the other three are Agriculture and Food Security, Disaster Risk Reduction, Energy, Health. The GFCS structure is based on five pillars (Fig. 2): a user interface platform; a climate services information system; observation and

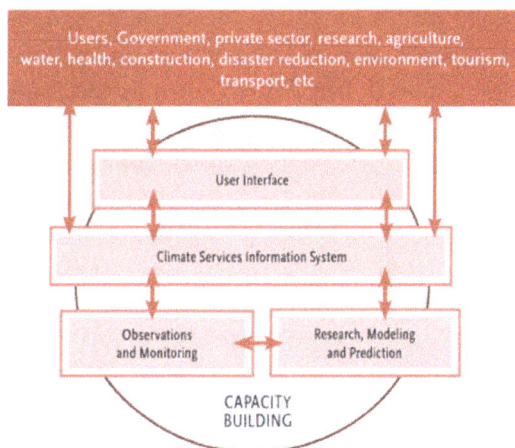

Fig. 2.   The components of the GFCS.

monitoring; research, modeling and application, and capacity building.

- *User Interface Platform* — provide ways for climate service users and providers to interact and improve the effectiveness of the Framework and its climate services.
- *Climate Services Information System* — produce and distribute climate data and information according to the needs of users and to agreed standards.
- *Observations and Monitoring* — develop agreements and standards for generating necessary climate data.
- *Research, Modeling and Prediction* — harness science capabilities and results to meet the needs of climate services.
- *Capacity Building* — support the systematic development of the institutions, infrastructure and human resources needed for effective climate services.

The GFCS will be implemented over 2-, 6-, and 10-year timeframes. The first two years are a start-up phase for establishing the Framework's infrastructure and for initiating and facilitating demonstration projects in the priority areas. During the second phase of implementation (the development phase), demonstration projects will be replicated in other parts of the world, and there will be worldwide improvements in climate services for the priority areas. After ten years the GFCS should have facilitated access to improved climate services worldwide and across all climate-sensitive sectors.

The HWRP contributes to the GFCS implementation in an indispensable manner. For example, the User Interface Platform is seen as a key part of the GFCS structure. The opinion of hydrologists is likely to be sought in terms of functionality and design. In addition, aspects of hydrological understanding and/or data are to be routinely included in atmospheric circulation models in the anticipation of improved predictive performance. Linked climatologic and hydrological modeling is increasingly offering potential in the exploration of water management options. Hydrologists also provide a useful link between climatologic information and others influenced by water regimes, notably the agricultural sector.

The CHy has a number of activities closely linked with the GFCS pillars. For example, Observations and monitoring, Research, Modeling and Applications, Climate service information system, User interface platform, and Capacity development.

Observations and monitoring

CHy contributes to the GFCS in this pillar through WHOS (WMO Hydrological Observation System), WHYCOS (World Hydrological Cycle Observation System), and GHSF (Global Hydrometry Support Facility). WHOS is a portal to the online holdings of National Hydrological Services (NHS) around the world that publish their data without restrictions or cost. It represents the hydrological component of the WMO Integrated Global Observing System (WIGOS). Access to the data comprising WHOS can be obtained via map-based links.

Research, modeling and prediction

There is an obvious link to the UNESCO-IHP and IAHS (especially with the new Panta Rhei "Change in hydrology and society" programme). CHy contributes through WMO FFGS, EHP(Extended Hydrological Prediction) activities, and GHSF. The other contribution could be development and implementation of tools in collaboration with climate community for assessing, coping with, and adapting to climate variability and change.

Climate services information system

CHy contributes through WHOS, APFM, FFI, EHP activities and GHSF. There is also a possibility to somehow make the existing data centers (e.g. WHYCOS, GRDC, HYDROLARE, IGRAC) more visible for GFCS and easily accessible for users by a common web directory. The other contribution could be the link to WMO-RCCs, i.e. which are the products hydrologists could contribute and or use/need from RCCs at a regional or sub-regional level.

User interface platform

CHy contributes to this pillar through RCOF (Regional Climate Outlook Forum) participation and feedback on a regional basis, Activities of JEG-CFW (Joint CCl/CAgM/CHy Expert Group on Climate, Food, and Water), EHP activities, and Helpdesk of APFM and IDMP. Operational hydrologists should be considered full partners in defining the specific climate products.

Capacity development

CHy contribution in this pillar includes WHYCOS, GHSF, FFI, EHP, APFM, IDMP, QMF (Quality Management Framework) — Hydrology (http://www.wmo.int/pages/prog/hwrp/index_en.php). The other contribution can be: (1) the WMO-CHy Guides, Manuals, Guidelines, Technical reports, etc. (2) WMO-RTCs, Training courses including e-learning, roving seminars, workshops (like EHP for South America together with IRI). (3) Cooperation with UNESCO-ICHARM, UNESCO-IHE, United Nations University, etc. (4) Technology transfer and knowledge management, e.g., Working groups on theme areas, open sources and community of practice solutions for issues of high relevance to NHSs and GRCS program.

4. Chinese Practices of Hydrological Forecasting and Prediction for Decision-making Support to Flood Management

China is one of those countries suffering from severe impacts of frequent floods and droughts. Through decades of efforts in preventing and controlling water-related natural disasters, substantial progress has been achieved in such aspects as flood forecasting, hydrological monitoring and hydrologic networking, and abundant experience has been accumulated in respect of hydrological and water resources management. China also participates and contributes to WMO hydrological activities through hydrological data sharing, expert exchange and consultancy, on-job training, operational flood forecasting systems development, etc. These efforts have been contributed to the national, regional and international hydrometeorological services.

With the deepening understanding of the disaster risk change, as well as socio-economic development, flood control strategies and measures in China have been continuously improved in the process of coping with floods over the years. Modern flood risk management in China has evolved from the early "build and protect" flood control approach that relied largely on structural interventions to a broader flood risk management approach. The development of hydrological forecasting and prediction methods in terms of how these methods have informed decision-making as flood risk management has evolved over the years.

4.1. Evolution process of flood management in China

Since the founding of the P. R. China in 1949, the government has attached more attention to the prevention and control of floods. Continuous efforts have been exerted to summarize experiences and lessons from major flood events, innovative work mechanisms, enhance structural development, and improve non-structural measures, leading to remarkable progress in capacity building for flood prevention and management. The accomplishments China has achieved during the past six decades run through the following three phases of flood control and management.

Start-up phase: before the mid 1970s

The 1950s and 1960s saw increasing efforts to harness major rivers, control and prevent floods. In 1950, the State Flood Control Headquarters and the Yellow River Flood control Headquarters were formally established, followed by the setting up of the Yangtze River Flood Control Headquarters and some key provincial flood control headquarters in succession. During this period, based on the principles of attaching equal importance to enhancing flood-water storage and discharge capacity, with the emphasis on the latter, the comprehensive management planning for the Huaihe River, the Yellow River, the Yangtze river, and the Haihe River basin have been successively carried out, which laid a solid foundation for rivers harnessing afterwards. Moreover, an engineering system for flood control has been preliminarily constructed, including the construction of trunk embankments for protecting major rivers, building of large- and medium-sized reservoirs, completing flood storage and retention areas on major rivers, and progressively draining major rivers.

*Development phase: from the mid
1970s to 1998*

In August 1975, the Great Flood of the Huaihe River caused by the heavy rainstorm and dam-failures, devastated Henan province, killing 26,000 people. A subsequent profound reflection on the "75.8" flood led to the transition towards a broader, comprehensive flood management approach where the emphasis shifted from engineering measures to a broad portfolio of structural and non-structural measures for managing flood risk. From this time onwards, the national specification on design flood estimation was revised, and the standard of flood control projects has been significantly enhanced. Besides, investment on the water conservancy has been progressively increased, with the emphasis on dyke reinforcement and restoration on major rivers, safety control and reinforcement of dangerous reservoirs by stages and in groups, construction of flood storage and retention areas with a view to guaranteeing the safety of people living there. By 1998, continuous efforts have been made to improve the nonstructural measures including development of early warning and forecasting systems, formulating flood risk management plans, improvement of the legal framework, establishing the administrative leader responsibility system, strengthening the organizational and institutional system, and setting up flood relief rescue teams.

Improvement phase: from 1998 to the present day

In 1998, the most devastating floods occurred in some major rivers such as the Yangtze, Songhuajiang, Pearl and Min Rivers, which left over 4,000 people dead and caused material damage of US$30 billion. In the aftermath of the Great Flood of 1998, the central government launched the strategic plan encompassing post-disaster reconstruction, restoration of rivers and lakes, start construction of water conservancy,

and developed the water governance policy, namely overall planning and comprehensive treatment, promoting what is beneficial and abolish what is harmful, attaching equal importance to increasing income and reducing expenditure, and putting equal emphasis on flood and drought management. Thus, flood control work entered a new period of development.

Flood risk concept has been introduced and gradually applied in flood management in China. In 2003, the China Ministry of Water Resources proposed to re-direct flood prevention from flood control to flood management, which aims to enhance understanding about interconnecting systemic issues and risk awareness. The flood prevention efforts thus shifted from attempts to eliminate floods forward to building capacity to endure floods of certain risk degree. In flooding areas, spatial planning is introduced to regulate human activities. Appropriate and feasible flood control standards were established together with flood prevention schemes and flood regulation plans, where a variety of measures are taken to ensure safety under the established flood control standard and minimize loss caused by exceeding-standard flood. Meanwhile, efforts are also being taken to utilize storm water as a complementary resource for water supply.

4.2. *Development of operational hydrological forecasting and prediction in China*

In response to growing demands and development of forecasting theory as well as computer science, China has built up the hydrological forecasting and prediction technology and capacity.

Hydrological forecasting method

The constraints of hydrological data availability and computer science in the past pin down China hydrological forecasting method on an experience-based level. In the 1960s, hydrologic models started to be developed and applied following introduction of mathematical simulation into hydrology field. China focused on bringing in suitable hydrologic models and developing those applicable to local hydrological conditions. Currently there are basically two types of hydrological forecasting methods: empirical hydrological forecasting method and conceptual hydrologic model.

The empirical hydrological forecasting method comes from the experiences that hydrological practitioners gathered over a long time, and this proves to be very effective in operational hydrological forecasting. At present, the seven major river basin authorities in China have compiled rather complete empirical hydrological forecasting schemes, and the 600 plus hydrological forecasting stations in China have nearly 1,000 schemes. In general, the empirical hydrological forecasting schemes apply such methods as antecedent precipitation index (API), corresponding stage (discharge) method, resultant discharge method, stage (discharge) fluctuating rate method, multi-factors combined axes correlation method, rainfall runoff correlation method, and so on.

China has extensive territory, with varied underlying surface conditions and different climates such as humid, dry, semi-humid and semi-dry climate. For decades, Chinese hydrologists have taken efforts to learn from foreign hydrologic models and succeeded in developing series of river basin hydrologic models applicable to local conditions. The Xin'anjiang model proposed by Prof. Zhao Renjun *et al.* in 1973 is the representative fruit. The main feature of the model is the concept of runoff formation on repletion of storage, which means that runoff is not produced until the soil moisture content of the aeration zone reaches the field capacity, and thereafter runoff equals the rain fall excess without further loss. The model can be used for real time flood forecasting, investigations for gaining understanding of hydrological processes

and study of the impacts of land use change on runoff, etc. Nowadays, hydrologic models used in operational hydrological forecasting in China comprise three types: those developed by Chinese professional, foreign ones and adapted foreign models. Local developed model include examples of Xin'anjiang model (Zhao *et al.* 1980; Zhao 1992; Zhao and Liu 1995), double excess runoff yield model, Hebei storm flood model, Jiangwan Bay runoff model, and double attenuation curve model. Tank, Sacramento, NAM and SMAR are major imported models, while continuous API model and SCLS model were introduced and modified to adapt to local conditions (WMO 2011).

In recent years, environmental changes brought challenges to river basin hydrological forecasting. Distributed hydrological models based on underlying terrain geographic information were developed and started to be used in flood forecasting and warning for flash floods as well as medium-small river flooding.

Hydrological forecasting system

The computer science and network technology are dispensable for the development and application of China hydrological forecasting systems. In early 1980s, China developed the real-time flood forecasting system run on the VAX machines and on a single computer. As the consequence of the limited database management and software development technology at that time, data were mostly managed in a file system and on-line operational forecasting was at a very low level with very simple functions, mainly programming existing forecasting scheme and model. From 1990 onwards, hydrologists in China learned from peers in developed countries with more mature and advanced technologies, and achieved progress in developing forecasting system with functions of database management, model parameter calibration and optimized calculation, operation forecasting, real-time forecast updating, and visualized data

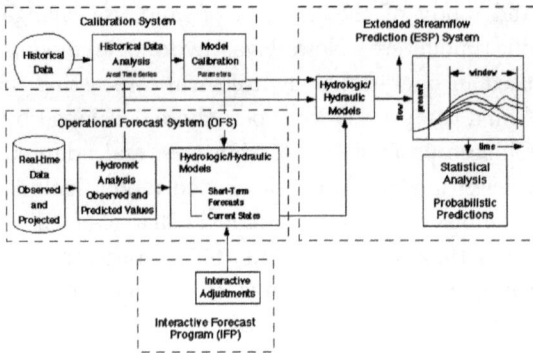

Fig. 3.   Overall structure chart of China National Flood Forecasting System.

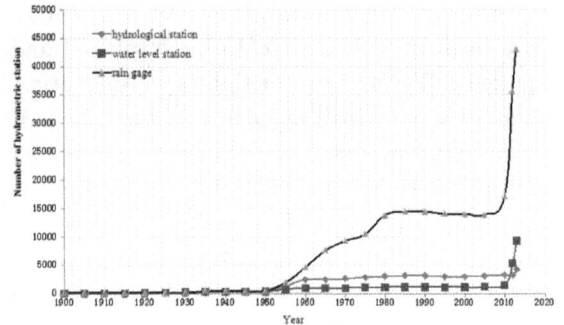

Fig. 4.   Development of hydrometric station network in China.

and forecasting results. After the 1998 Great Flood, China hydrological sector took efforts to enhance hydrological monitoring capacity and develop the universal flood forecasting system software platform: China National Flood Forecasting System (CNFFS) (Zhang and Liu 2007). The CNFFS consists of three major components of calibration system, operational forecast system and extended stream flood prediction system, which are shown in Fig. 3. The CNFFS contains common forecasting models and a method library allowing prompt building of forecasting schemes. Many different flood forecasting models are employed in the system, including the Xin'anjiang model, and other models such as API, Sacramento, Tank, SMAR, and the synthetic constrained linear system (SCLS).

At present, there are over 70,000 hydrometric monitoring stations, including 3,200 hydrological stations and 35,000 rain gauges (see Fig. 4), and more than 1,700 hydrological forecasting sites along rivers all over China. Most large reservoirs and key medium-sized reservoirs built for flood prevention purpose have established telemetry system, flood forecasting and regulation systems. The CNFFS has been successfully used in 25 national and provincial hydrology departments in support of flood management at different levels in the country.

4.3. Application of new technology in hydrological forecasting and prediction

The scientific and technological development during recent 10 years enables China to apply new and advanced technology to the hydrological forecasting, which improves the forecasting software and operational forecasting accuracy. The following are three examples.

Rainfall and flood joint forecast

In real-time operational forecasting, the rainfall and flood joint forecast technology applies satellite and radar images for rainfall quantitative analysis and estimation. In the first step, meteorological department provides data needed by hydrological department for rainfall forecast and prediction in terms of location, quantity and time. The next step is flood forecast. The results of the two steps are then combined to improve the lead time and prediction accuracy of hydrological forecasting and prediction.

Interactive forecast program

Since introducing the NWSRFS from NOAA, USA and Interactive Forecast Program (IFP), interactive river forecast proto-system has been established for river basins in China. IFP combines operational forecast system (OFS) and

GUI, providing hydrological forecasters with information desired for judging data or simulating results, and verifying interactively to improve forecast accuracy.

Application of DEM, RS and GIS technologies

More recently, deterministic spatially distributed hydrologic models have been gradually used in China. Based upon equations representing the storage and the movement of water in the soil and on the surface, distributed models are potentially easier to calibrate by relating the values for their physically meaningful parameters to additional information provided in the form of Digital Elevation Maps, Soil Maps and Land Use Maps. These models account for the spatial distribution of rainfall (now available at pixels of between 1×1 km and 10×10 km from RADAR images or in terms of QPF from NWP models) and are better equipped to address the problem of extrapolating the parameter values to the ungauged sub-catchments, a problem that has not yet found a proper solution. All these potential properties, combined to the increased efficiency of computers, has given rise to a number of fully distributed rainfall-runoff models, some of which, for instance, the TOPKAPI model (Liu *et al.* 2005; Liu *et al.* 2008), have been incorporated into real-time flood forecasting systems.

5. Conclusions and Outlook

5.1. *Conclusions*

Hydrometeorological disasters, arising from floods and droughts in the past decades have become more frequent and devastating in many countries in the world. The WMO through its Hydrology and Water Resources Programme and the WMO Commission for Hydrology assists and supports National Meteorological and Hydrological Services of all Members to provide timely and accurate weather, climate and water forecast and prediction, early warning and outlooks, which contribute to world peace and socio-economic development.

The activities within the CHy thematic area of Water, Climate and Risk management have gained considerable focus, among others, hydrological forecasting and prediction in support of flood and drought management, water resources assessment under climate change and variability background, and assistance in the implementation of water-related initiatives within the Global Framework for Climate Services (GFCS). The Chinese best practices of hydrological forecasting and prediction demonstrated a good case of decision-making support for flood management in disaster reduction.

Improved forecasts and predictions call for more collaboration between hydrologists and meteorologists and increased utilization of integrated data. Also, NMHSs need to make use of new technologies, tools and products that have become available to improve outputs of models, predictions and forecasting. Among the cutting edge technologies and tools available for improved hydrological modeling and forecasting are the use of radar and satellite-based rainfall estimation, quantitative precipitation forecasts, numerical weather prediction and climate outlooks for water resources management, among others.

5.2. *Suggestions for future direction in hydro-climate services*

(1) All observations should be used for generation of integrated and quality assured climate products

The advantages of systematic observations of hydrological conditions worldwide is receiving wider recognition. However, data availability and accessibility in many areas continues to be a significant problem, and new approaches to improving capabilities of least advanced NHSs to produce quality-assured data and to induce reluctant NHSs to share data, are required.

The WMO-CHy established a Water and Climate Services Exchange (WACSE) as a consultative, coordination, and capacity building mechanism to identify, implement and evaluate a range of climate and related water information services in support of improved water resources management. This mechanism has already improved selected climate products and services through the incorporation of hydrological observations.

(2) Seasonal forecasting should continue with high priority under the Extended Hydrological Prediction activities

Seasonal forecasting and the development of hydrological services based on these forecasts was an area in which the water sector had a significant role to play. Countries are seeking hydrological products based on seasonal forecasts. There was a need for better information on the quality and limitations of seasonal forecasts and also their potential benefits to hydrological services, especially in terms of seasonal prediction of river flows. The focus in the area of seasonal forecasting should continue with high priority under the Extended Hydrological Prediction activities. Linkages should continue to be established with the Commission for Climatology (CCl), the GFCS and NMHSs also undertaking activities in this area.

Recent improvements in monthly to seasonal climate and hydroclimatic outlooks are beginning to yield improvements in hydrological services delivery, particular with respect to increased use of short-term climate forecasts and outlooks in water resources systems operations and seasonal planning.

(3) Regional Climate Centers and the Regional Climate Outlook Forums need to be looking more seriously into water issues than they currently are

The GFCS has been identified as one of the key priority areas for the WMO. Its primary goal is to ensure the greater availability of access to, and use of climate services for all countries and a wide range of users, of whom the water sector including hydrologists is one major key group. It aims to enhance the availability of needed climate information and services to plan ahead of time and make sustainable management decisions. We call for strengthened cooperation in the future between WMO and its Members in implementation of GFCS.

WMO Regional Climate Centers create regional products including long-range forecasts that support regional and national activities and thereby strengthen the capacity of WMO Members in a given region to deliver the best climate services to national users. While WMO Regional Climate Outlook Forums (RCOFs), active in several parts of the world, routinely provide real-time regional climate outlook products. RCOFs need to be looking more seriously into water issues than they currently are. It is expected for regional climate centers, regional climate outlook forums (monitoring, prediction and assessment) to play a larger role in GFCS. And, it is also important to build a link between climate and user (hydrological) communities in the field of seasonal forecasts in GFCS.

References

Arduino, G., P. Reggiani, and E. Todini (eds.), 2005: Special issue: Advances in flood forecasting. *Hydrol. Earth Syst. Sci.*, **9**(4), 280–284.

Linnerooth-Bayer, J. and A. Amendola, 2003: Introduction to special issue on flood risks in Europe. *Risk Anal.*, **23**(3), 537–543.

Liu, Z., M. Martina, and E. Todini, 2005: Flood forecasting using a fully distributed model: Application to the upper Xixian catchment. *Hydrol. Earth Syst. Sc. (HESS)*, **9**(4), 347–365.

Liu, Z., B. Tan, X. Tao, and Z. Xie, 2008: Application of a distributed hydrologic model to the catchments of different characters for flood forecasting. *J. Hydrol. Eng.*, **13**(5), 378–384.

WMO, 2011: Manual on Flood Forecasting and Warning. WMO Publication, No. 1072, pp. 48–49.

WMO, 2012: Report of the 14th Session of WMO-CHy. WMO Publication, No. 1105, pp. 1–87.

Wu, Z. G. Lu, Z. Liu, J. Wang, and H. Xiao, 2013: Trends of extreme flood events in the Pearl river basin during 1951–2010. *Advances in Climate Change Research*, **4**(2), 110–116.

Zhang, J. and Z. Liu, 2006: Hydrological monitoring and flood management in China. In: *Frontiers in Flood Research* (eds. I. Tchiguirinskaia, K. Thein and P. Hubert), *IAHS Publ.*, No. 305, pp. 93–102.

Zhang, J. and Z. Liu, 2007: Hydro-meteorological monitoring and operational hydro-systems for flood management in China, in *Proceedings of the International Symposium on Methodology in Hydrology*, Nanjing, China. *IAHS Publ.*, No. 311, pp. 3–9.

Zhao, R. J., 1992: The Xinanjiang model applied in China. *J. Hydrol.*, **135**, 371–381.

Zhao, R. J. and X. R. Liu, 1995: The Xinanjiang model. In: *Computer Models of Watershed Hydrology* (ed. V. P. Singh), Water Resources Publications.

Chapter 13

From Prediction to Scenario Analysis: A Brief Review and Commentary

Bryson C. Bates

CSIRO Oceans and Atmosphere,
Wembley, Western Australia, Australia
bryson.bates@csiro.au; bryson.bates@iinet.net.au

This chapter explores the issue of handling climate uncertainty in strategic water planning with a focus on drought risk. It builds on the author's work experience at the interface between climate science and water planning practice in Australia. This experience suggests that there is an ongoing disconnect between the academic literature and practice due to: differences in terminology owing to the different disciplines involved; an emphasis on accurate and precise prediction versus a preparedness and resilience paradigm that evolves to meet emerging threats; the perceived need to accurately characterise uncertainty versus the disempowering impacts of large levels of uncertainty and complexity on practical decision-making (which can lead to overly conservative decisions); and a lack of synthesis of climate adaptation research findings in terms of their implications for practical decision-making.

1. Introduction

The freshwater resources chapter of the Fifth Assessment Report of the Intergovernmental Panel on Climate Change's Working Group 2 (Jiménez Cisneros *et al.* 2014) concluded that for each degree of global warming, approximately 7% of the global population is projected to be exposed to a decrease of renewable water resources of at least 20%. For most subtropical regions, marked reductions in renewable water resources are possible whereas increases are possible at high latitudes. Projected changes in the frequency of multi-annual droughts were deemed to be uncertain because they depend on accumulated precipitation over long periods. It also concluded that there is no evidence that surface water and groundwater drought frequency has changed over the last few decades, although impacts of drought have increased mostly due to increased water demand.

Global climate models (GCMs), driven by several greenhouse gas and aerosol emissions scenarios, provide the basic input information for climate change impact assessments. This information can be used to define an envelope of uncertainty for projections of future climate. The width of this envelope expands over time, resulting in increasingly divergent trajectories that may need to be accounted for by planners (Stafford-Smith *et al.* 2011). Jiménez Cisneros *et al.* (2014) note two basic approaches to the handling of this uncertainty: the use of a small set of scenarios to characterise the range of impacts on water resources and systems, and the use of very large numbers of scenarios to generate likelihood distributions of indicators of impact for use in risk assessments. While there has been much debate on this issue in the literature, there is yet no consensus on the most suitable approach. This presents a key challenge for water planners who make adaptation decisions that have long lead times or that have consequences through the decades ahead when there will be an interplay between increasing scientific knowledge and persistent if not growing uncertainty and complexity.

Bridging Science and Policy Implication for Managing Climate Extremes
Edited by Hong-Sang Jung and Bin Wang
© 2017 by World Scientific Publishing Co. and APEC Climate Center

This chapter explores the issue of handling climate uncertainty in strategic water planning with a focus on drought risk. It draws on the author's work experience at the interface between climate science and water planning practice in Australia. This experience suggests that there is an ongoing disconnect between the academic literature and practice due to: differences in terminology owing to the different disciplines involved; an emphasis on accurate and precise prediction versus a preparedness and resilience paradigm that evolves to meet emerging threats; the perceived need to accurately characterise uncertainty versus the disempowering impacts of large levels of uncertainty and complexity on practical decision-making (which can lead to overly conservative decisions); and a lack of synthesis of climate adaptation research findings in terms of their implications for practical decision-making. This article is likely to be of most relevance to climate services personnel who provide science-based and user-specific information relating to projected climate conditions and water planners who are charged with the responsibility of developing and implementing transitional or transformational adaptation measures. It will also be relevant to members of the climate modelling community who seek a deeper understanding of the needs and the perspective of the water sector. The intent is to assist these parties in developing a common understanding of some of the well and not-so-well known aspects of the development and dissemination of climate change projections and their incorporation into water planning, and to stimulate further discussion.

2. Predictions, Forecasts, Projections and Scenarios

The terms prediction, forecast, projection and scenario appear throughout the climate change literature. MacCracken (2001) notes that they are often used interchangeably during public discussions about weather and climate. Personal experience and that of others (see, e.g., Pielke Sr. and Wilby 2012) suggests that this is not an uncommon occurrence in academic circles too. In succinct terms, a prediction can be defined as a specific statement about the future, a forecast as the most likely picture of the future, a projection as what could happen if certain assumed conditions prevail in the future, and a scenario as a series of events that could lead from the present to a plausible but not assured future. The terms prediction and forecast are appropriate in the context of numerical weather prediction (one to a few days ahead). Prediction is primarily an initial value problem as memory of initial conditions will persist over the timescale at which the prediction can be regarded as skilful. In the context of anthropogenic climate change, the assumed conditions for a projection are benchmark emissions scenarios. The latest scenarios are called representative concentration pathways (RCPs). The RCPs are designated as RCP2.6, RCP4.5, RCP6.0 and RCP8.5, and are named according to radiative forcing values in the year 2100 relative to preindustrial values (2.6, 4.5, 6.0 and 8.5 W m^{-2}). They have no assigned likelihood of occurrence, but they are considered plausible. While a projection is primarily a solution to a boundary value problem, it can to some extent depend on the initial state of the climate system due to the decadal timescales involved in ocean, cryosphere and biosphere processes (Collins 2002). Scenarios are useful in situations involving high risk, considerable complexity, deep or poorly characterised uncertainty and a low degree of control over possible future events (Schwartz 1996; Peterson et al. 2003; Mietzner and Reger 2005; Duinker and Greig 2007).

Bammer and Smithson (2008) regard uncertainty as not just a problem to be overcome or managed but also as an essential source of opportunity, discovery and creativity. Scenario planning focuses attention on what are the consequences and most appropriate adaptation responses under different circumstances rather than the estimation of a "best guess". Scenarios focus on describing images of quite different but plausible futures that challenge current assumptions, gain or broaden perspectives and spark creativity. When used as intended, they:

(1) Enable "what if" questions to explore the consequences of uncertainty.
(2) Allow for and warn of sharp discontinuities.
(3) Enable anticipation of future threats and opportunities.
(4) Lead planners and decision-makers to question their assumptions and to acquire knowledge.

Postma and Liebl (2005) point out that the conventional approach to scenario planning overlooks the existence of "unknowables" (or total surprises), which stem from our ignorance of the system(s) of interest. In a climate change context, this term reflects our inability to predict changes in natural forcing (e.g., the radiant power emitted by the Sun, large volcanic emissions and variations in the Earth's orbit), socio-economic conditions and human behaviour in a deterministic manner over long time frames (New and Hulme, 2000).

While total surprises cannot be "forecasted" they can become very relevant to decision-making in a much shorter time frame than might have been anticipated.

A personal observation is that the above perspectives have not been widely adopted by water planners. Climate change projections are often viewed as predictions and their range as a definitive guide to the extent of future climate risk. There is also a propensity to treat the median projection across all GCMs and RCPs considered as the "best guess". Such an approach ignores many of the sources of uncertainty that exist in climate change impact assessments and adaptation planning and practice. Some of these sources and the limitations of conventional planning approaches are explored below, and possible ways forward discussed.

3. Sources of Uncertainty

Ascough II et al. (2008) describe four major sources of uncertainty in environmental decision-making:

(1) *Knowledge.* This source includes incomplete process understanding, model structural error, measurement error, parameter estimation error, errors in initial and boundary conditions, and software and hardware errors. All of the mathematical models used to quantify climate change impacts suffer from this source of uncertainty (Giorgi 2005; Wilby 2005; Steele-Dunne et al. 2008; Dessai et al. 2009).
(2) *Variability.* This includes natural and technological factors and the extent to which: practice is based on historical preferences and use; adaptation plans interact with other; and there are situations where values, interests, or institutions constrain societal responses to change (Stafford-Smith et al. 2011; Wise et al. 2015).
(3) *Linguistic.* This includes vagueness (e.g., when a precise definition of a term is not available), ambiguity (e.g., cases where words have more than one meaning and their precise meanings are not given), and under-specificity (e.g., unwanted generality in data or a less of detail during the process of transcription).
(4) *Decision-making.* This arises from ambiguity or controversy about how to quantify and compare social values and objectives, the interpretation and communication of climate change projections or scenarios, and performance measure selection.

It is often asserted that accurate and precise climate projections or knowledge about their levels of uncertainty are essential to informed decision-making (Giorgi 2005; Dessai et al. 2009 and references cited therein). However, climate change projections have an intrinsic level of uncertainty because of the: limitations in our knowledge about the climate system; imperfections in our models and data, stochastic aspects of the climate system (e.g., noise, turbulence) and future natural and anthropogenic forcings (e.g., changes in solar and volcanic activity, land-use). The emissions scenarios/RCPs are ephemeral in nature as they are always likely to be updated over time (Hulme and Dessai 2008). Similarly, the development of climate models, growth in our knowledge of the climate system, changing instrumentation and observational networks are ongoing processes that will influence each other across numerous applications. Thus, our knowledge of relevant processes and deficiencies in our data, methods and models is likely to remain imperfect for the foreseeable future (Jones 2000; Visser *et al.* 2000). This situation necessitates the use of scenario-based planning together with a commitment to continuous monitoring of developments in climate science, revisiting impact assessments and, when required, amendment of adaptation measures in genuine consultation with the community and internal as well as external stakeholders.

4. Approaches to Systems Planning

Peterson *et al.* (2003) suggest that a response to uncertainty should depend on the level of uncertainty and the degree to which the system of interest can be controlled (Fig. 1). Water management in Australia has traditionally been carried out on the "stationarity" assumption that the instrumental record is representative of current and future climatic conditions. What has not been widely appreciated by practitioners in Australia and elsewhere is that the determination of a baseline (the state against which change is measured) is highly dependent on the origin and period of the instrumental record and that, in the presence of low-frequency climate fluctuations or climate change, the baseline will change with time. In cases where the latter is recognised, the trend in the changing baseline is often characterised as linear or consisting of a set of

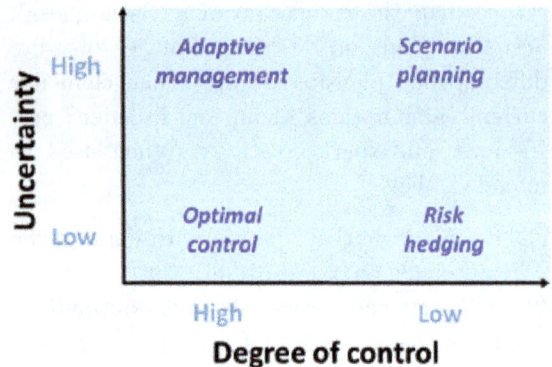

Fig. 1. Approaches to planning under uncertainty. (Adapted from Peterson *et al.* 2003.)

step functions even when a more robust representation is possible (see, e.g., Bates *et al.* 2010).

Several decades ago, the most popular technique for water resources planning in Australia was hedging (i.e. the elimination or mitigation of risk). More recently, water planners have adopted an adaptive management approach in which water plans spanning several decades ahead are updated every five years or so, or sooner if unforeseen circumstances arise. Adaptive management removes the focus from a once-off incremental process to continuous risk evaluation, and recognises that anthropogenic climate change will be an enduring phenomenon for the foreseeable future (Stafford-Smith *et al.* 2011, Haasnoot *et al.* 2013; Wise *et al.* 2015). It has proven to be a particularly useful approach in southwest Australia where the hydroclimatic baseline has changed markedly over the last 40 years or so (see, e.g., Bates *et al.* 2008a; 2009; 2010).

While climate change projections for water availability are gaining traction with Australian water planners, there is currently little evidence of the adoption of a scenario-based approach that blends climate change projections with existing knowledge of the intensity, frequency and duration of severe droughts in instrumental, historical and prehistorical records. This is an important point as the impacts of the so-called Millennium Drought (1997–2009) on streamflows in the northwest of the State of Victoria in Australia at that time exceeded the worst-case climate change projection for 2070 (Rae Moran, pers. comm.). Moreover, reconstructions of the eastern Australian climate over the last millennium have shown that severe droughts are not unusual, with some droughts lasting up to 40 years (Vance *et al.* 2015). The existence of such information suggests the need to consider scenarios that incorporate the characteristics of past multi-decadal droughts as well as climate change projections.

In scenario planning, the use of two to five scenarios is usually considered optimal, but the use of three often leads to the interpretation that the middle scenario is the "most likely". Cornish (2004) suggests five generic themes: (1) continuation (surprise-free); (2) pessimistic; (3) disastrous; (4) optimistic; and (5) transformational (miraculous). The exploration of such themes enables managers and stakeholders to make informed adaptation and mitigation decisions that are robust (work well) under all or most scenarios, prepares decision and policymakers to respond in an appropriate and timely manner if their expectations prove to be false and provides insight into events that could indicate which, if any, scenario is being played out (Duinker and Greig 2007). It also allows planners and managers to understand where vulnerabilities may lie (Dessai *et al.* 2009) and to hedge against risk.

5. Climate Change Impact Assessments

There is a general sequence of steps undertaken in climate change impact assessments for water resources and systems. They are: coupled atmosphere-ocean GCM runs driven by climate change forcings; physical (dynamical) and/or statistical models to "downscale" GCM outputs to the smaller spatial scales relevant for hydrologic processes and to obtain a better match with observations; hydrologic models to simulate potential changes in surface water availability and groundwater recharge; and the simulation of water resource systems to assess potential changes to systems performance. This sequence can also be used to assess the effectiveness of policy responses for adaptation and mitigation, and to incorporate natural and human feedbacks (e.g., changes in societal, ecological and environmental water demand). Each step in the sequence is subject to uncertainty which propagates to the next leading to the so-called "uncertainty cascade". Uncertainty accumulates in a nonlinear fashion through the cascade leading to large levels of uncertainty in climate change impacts (see, e.g., Mearns *et al.* 2001; Wilby and Harris 2006; CSIRO 2008). For the mid-21st century and beyond, the largest sources of uncertainty in the cascade are the emissions scenarios/RCPs and cross-GCM responses to external forcing. While there is no consistently superior downscaling method, the use of downscaling is regarded as providing "added value" (Fowler *et al.* 2007). However, the combined use of dynamical and statistical methods does not necessarily lead to a marked improvement in performance (see, e.g., Charles *et al.* 2007). Also, even when the use of a suite of downscaling methods leads to a degree of consistency for present conditions, the application of the same methods to changed climate simulations can lead to very different projections. Some form of statistical downscaling is usually required to create "application-ready" datasets for use in impact assessments.

To date, research on water-related impacts has focused strongly on potential changes in mean annual or monthly runoff (e.g., CSIRO 2008). In relative terms, there is a paucity of information about impacts on groundwater recharge, surface water-groundwater interactions, saltwater intrusion into aquifers and wetlands due to rising sea levels and storm surge, and water quality issues (e.g., salinity, nutrients, turbidity, algal blooms, and biological oxygen demand).

It has long been argued that improved characterisation of uncertainty could help water planners and managers to adapt to future climatic conditions and that the incorporation of such information in a risk management approach would be useful (see, e.g., Kundzewicz *et al.* 2008). Efforts have been made to obtain multi-model assessments of impacts on catchment water yield. These works typically use multiple emissions scenarios/RCPs, and corresponding output from multiple GCMs and several downscaling techniques (e.g., Bates *et al.* 2008b; CSIRO 2008; Jiménez Cisneros *et al.* 2014). The importance of using a large suite of models and ensembles of model runs is often stressed and estimates of uncertainty are inferred directly from the model simulations obtained (e.g., Jiménez Cisneros *et al.* 2014 and references cited therein). This approach often leads to high estimates of cumulative uncertainty that may not allow effective conclusions to be drawn about projected changes in water availability and immobilise decision-making. This problem is compounded by other sources of uncertainty that are not routinely considered or addressed in most impact assessments:

(1) The available number of climate models and emissions scenario/RCP combinations is but a small sample drawn from the population of all possible models and all possible emissions/RCP trajectories. Moreover, the models are not physically (and hence statistically) independent owing to the sharing of computer code and computational schemes. Thus, only a part of the "uncertainty space" is explored, the projections could be biased and the total level of uncertainty understated. This could lead to maladaptation if large suites of projections are accepted without question, misinterpreted or used incorrectly (Hall 2007; Dessai *et al.* 2009).

(2) GCMs and downscaling techniques contain parameters that are tuned to fit observational data during the model development process. In the case of dynamical models, the tuning is often performed by adjusting uncertain or non-observable parameters related to processes that occur at a spatial scale smaller than that of the model's computational grid. In comparison, model tuning is a critical step in the development of statistical downscaling techniques and the use of hydrological models. Despite these facts, the sensitivity of climate change projections to model tuning is not widely investigated.

(3) Observational data are subject to both systematic and random errors and structural uncertainty. This is an important consideration when observations are used to derive climatic variables such as potential evaporation (see, e.g., Kay and Davies 2008).

(4) Changes in the water use efficiency of plants, vegetation composition and structure that may impact water availability and irrigation requirements. The water use efficiency of vegetation with the C3 photosynthetic pathway is likely to increase due to CO_2 fertilisation. This may be a short or long term phenomenon dependent on changes in rainfall and temperature regimes and water and nutrient availability. It is likely to be a significant factor in future water balances (see, e.g., Alo and Wang 2008; Kruijt *et al.* 2008).

(5) Changes in the meteorological risk of bushfire (wildfire) which may have implications for surface water quality and increased water use due to vegetation regrowth.

(6) Changes in hydrologic regime (e.g., spatial and temporal changes in the timing, magnitude and frequency of low and high streamflows). Such changes may pose threats to ecosystem health and alter the characteristics of flow duration curves used for water supply, water quality and irrigation planning and management (see, e.g., Viney *et al.* 2007).

(7) The potential sensitivity of hydrologic model parameters to climate variability and change. Most studies on the impact of climate change on surface water availability

have used "lumped" rainfall-runoff models that require calibration and it is assumed that the tuned parameter values hold for changed climate conditions. It is not *a priori* obvious that this should be the case.

(8) Hydrologic variability at decadal to multi-decadal timescales is of critical importance in water resources planning. Climate change impact assessments are often done by comparing the statistics for a 20- or 30-year time slice centred on a particular year (e.g. 2050) against those for the reference ("baseline") period (e.g., 1961–1990). Current GCMs do reproduce not observed low-frequency variability in the hydrologic cycle (Hornberger *et al.* 2012) and this, coupled with the use of short time slices, compromises reliable assessments of safe reservoir yield under projected extreme drought conditions in places such as Australia.

(9) The potential impacts of compound extreme weather events (Leonard *et al.* 2014). Infrastructure systems have interdependencies that are often under-estimated in risk assessments (e.g., extreme rainfall/wind/lightning/fire combined with poor drainage maintenance, power failure, pump failure, sewage/ash/debris contamination of waterways, soil or shoreline erosion, and inundation.

Consideration of the above factors suggests that the application of the traditional prediction paradigm to impact assessment is perhaps unwarranted and potentially misleading. A better approach is to simply but effectively characterise the range of plausible future climates, assess their potential impacts on the systems of interest and seek adaptation measures that are insensitive to the resolution of uncertainty.

6. Climate Futures Web Tool

In Australia, the approach of using a small set of climate model results to characterise the range of impacts has been facilitated by the development of the Climate Futures web tool (CSIRO and Bureau of Meteorology 2015). Climate Futures is a flexible, multi-purpose decision-support tool designed to assist the use of climate change projections in impact assessments and adaptation planning. It provides a unique way of exploring regional climate projections by allowing users to view the projected changes in two climatic variables simultaneously. Climate change projections are focussed on Natural Resource Management (NRM) "clusters" or regions for which information, data and reports are available (see http://www.climatechange inaustralia.gov.au/en/impacts-and-adaptation/nrm-regions/). Projected changes from the latest (Coupled Model Intercomparison Project Phase 5, CMIP5) GCMs can be explored for 14 20-year periods centred on 2025, 2030, ..., 2090; and the four RCPs that were used to drive the GCMs. Data are available from up to 40 GCMs, 6 dynamical downscaling and 22 statistical downscaling simulations. The selection of models is based on the most extensive, independently peer-reviewed climate model evaluation ever undertaken in Australia.

Users of Climate Futures can explore, and obtain data for, projected monthly, 3-monthly, 6-monthly and annual changes in up to 14 climate variables. The data are usually displayed in an easy-to-understand colour-coded matrix (Fig. 2). It subdivides the projected changes in two climate variables (such as annual temperature and annual rainfall) from the full suite of GCMs into several classes, e.g., warmer-wetter, hotter-drier, much hotter-much drier. The changes are relative to the 20-year (1986–2005) baseline adopted by the Intergovernmental Panel on Climate Change (IPCC) for its Fifth Assessment report. The resultant classification displays the spread and clustering of the projected changes. This provides model consensus information for each classification and assists the selection of the classifications that are of

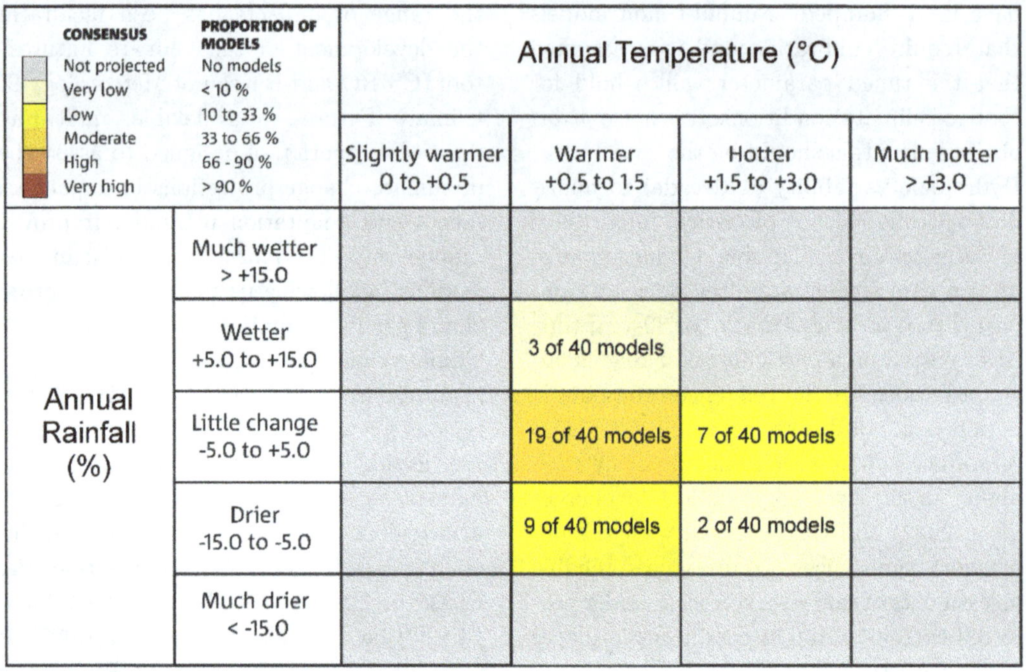

CONSENSUS	PROPORTION OF MODELS	Annual Temperature (°C)			
Not projected	No models				
Very low	< 10 %				
Low	10 to 33 %				
Moderate	33 to 66 %	Slightly warmer 0 to +0.5	Warmer +0.5 to 1.5	Hotter +1.5 to +3.0	Much hotter > +3.0
High	66 - 90 %				
Very high	> 90 %				
Annual Rainfall (%)	Much wetter > +15.0				
	Wetter +5.0 to +15.0		3 of 40 models		
	Little change -5.0 to +5.0		19 of 40 models	7 of 40 models	
	Drier -15.0 to -5.0		9 of 40 models	2 of 40 models	
	Much drier < -15.0				

Fig. 2. Schematic of output from Climate Futures web tool for Southern Slopes NRM Cluster, RCP4.5 and 20-year time slice centred on 2060. Changes are relative to the 1986–2005 baseline.

most importance for a given impact assessment. Users can then identify the "best" and "worst" cases in terms of likely impacts. It is also possible to identify a "maximum model consensus" case. For the classification displayed in Fig. 2, this is the warmer annual temperatures — little change in annual rainfall case.

A key feature of Climate Futures is the Representative Model Wizard. This identifies a suitable subset of models that can be used to represent the selected projections, e.g. one model to represent a "worst" case, one to represent a "best" case and one to represent the "maximum consensus" case. It provides a pragmatic solution to the issue of dealing with "too many models" and presents climate change projections as a range of possibilities. Thus the information obtained can be used to inform a scenario analysis. Further information can be found at (http://www. climatechangeinaus tralia.gov.au/ en/climate-projections/climate-futures-tool/introduction-climate-futures/).

7. Water Planning and Management under Climate Uncertainty

As noted above, surface and groundwater water management systems have customarily been designed and operated under the assumption of statistical stationarity of inflow series. Milly *et al.* (2008) argue that stationarity should no longer serve as the basis for water resources risk assessment and planning given the growing evidence of hydroclimatic change. Recent studies reporting such change include: Mauget (2003); Kahya and Kalayci (2004); Dixon *et al.* (2006); Zheng *et al.* (2007); and Bates *et al.* (2010). These changes may mean that water systems are ill-conceived, or under- or over-designed, resulting in inadequate performance or excessive costs (Kundzewicz *et al.* 2008).

The growing interest in the incorporation of climate change projections in Australian water planning is largely due to ongoing hydrological drought in the southwest since the mid-1970s

(Bates *et al.* 2008a, 2010) and the Millennium Drought in southeast Australia (Marsden Jacob Associates, 2006; van Dijk *et al.* 2013). These droughts have led to extensive policy and management change in both rural and urban water systems (Reisinger *et al.* 2014). Climate change projections for these regions indicate declines of 0 to 40% in the far southeast and 20 to 70% in the far southwest for a 2°C warming due to reductions in winter precipitation.

Water planners have two options for maintaining the balance between water demand and supply: to encourage water conservation and impose restrictions; and/or to increase supply (Marsden Jacobs Associates 2006). Ongoing adaptive responses to the drought in southwest Australia has enabled water planners to (Bates and Hughes 2009; http://www.watercorporati on.com.au/water-supply-and-services/solutions-to-perths-water-supply?pid=res-wss-np-spw):

(1) Undertake a gradual de-rating of surface water resources.
(2) Rapidly develop, review and amend source development programs in response to observed as well as projected climate risk and population growth in the Greater Perth region. This process has changed the supply mix from surface water — shallow groundwater sources in the 1980s to one consisting of deep groundwater extraction, aquifer replenishment, seawater desalination, recycling of treated wastewater, and water demand reduction. In 2014–15 financial year, 17% of potable supply was sourced from surface water (including a proportion of water that originated from groundwater or desalination); 42% from groundwater; and 41% from desalinated seawater. It has also involved the construction of an extensive interbasin water transfer system.
(3) Sensitize customers to a drying climate and the need for water conservation.

In southeast Australia, additional adaptation strategies include replacing open/leaky irrigation channels with pipes, rainwater capture and storage. Other options such as the construction of new dams is often considered to be too expensive or contentious.

Community attitudes to water restrictions in the southwest have been canvassed over several years (Bates and Hughes 2009). Survey results have generally shown that while there is general support for water conservation and regular restrictions, there is a noticeable reduction in support for outright sprinkler bans. There is also resistance to the use of reclaimed wastewater for potable purposes, and this is an issue nationwide. Thus, recycled water is typically used for groundwater replenishment, industrial purposes and the irrigation of parkland and sport grounds.

8. Robust Adaptation Strategies

As noted above, the level of uncertainty in climate change projections can be so large that it renders many traditional water planning methods inadequate. It also precludes reliance on a single climate model and RCP combination in system performance assessments. Lempert and Schlesinger (2000), Pittock *et al.* (2001), Dessai *et al.* (2009), Hallegatte (2009) and others suggest the use of a planning approach based on the search for strategies that are robust against a wide range of plausible climate futures, rather than "optimal" in the sense of minimisation of costs and maximisation of benefits. Hallegatte (2009) lists five approaches to risk mitigation:

(1) "No-regret" strategies that provide benefits in the absence of climate change.
(2) Reversible options, e.g., adoption of cheap-to-retrofit designs.
(3) "Safety margins", i.e., over-design at null or low cost.
(4) Soft adaptation strategies. Examples include the use of institutional tools to create a long-term prospective (such as state or city water plans), and demand control and water reuse.

(5) Reduction of decision time horizons, e.g., avoiding long-term commitments and choosing short-lived decisions.

The five strategies listed above provide a reduction in sensitivity to climate model/RCP selection, model error and (probably irreducible) uncertainty. While these strategies may preclude optimal performance, they are more likely to succeed than optimal decision- and policy-making based on unrealistic expectations for predictive accuracy. It is essential to consider both the adverse and beneficial side-effects of adaptation measures, adaptation-mitigation interactions and social acceptability. The desktop study by Viney *et al.* (2007) provides an example of a "no regret" strategy. They investigated the management of algal blooms at Maude Weir on the Murrumbidgee River in New South Wales, Australia. Thirty-year time slices for present and projected conditions at 2050 were derived from a statistically-downscaled transient run of the CSIRO Mark 3 GCM driven by the IPCC A2 (high) emissions scenario (IPCC 2000). They found that a simple, resource neutral, adaptive management strategy (using flushing flows as and when required to break thermal stratification in the weir pool) could substantially reduce the frequency and duration of algal blooms for both sets of conditions. This was even though 45% of the projected 52% decrease in discharges at Maude Weir were for flows above the historical median. Such results support the notion that planning can potentially lead to beneficial outcomes in the present.

More recently, Stafford-Smith *et al.* (2011) have presented a more general framework for reducing complexity and uncertainty that is aimed at helping decision-makers to arrive at better adaptation solutions. It subsumes the five approaches articulated by Hallegatte (2009) and involves a systematic approach to categorizing the interactions between decision lifetime (the sum of lead time and consequence time), the type of driver for uncertainty (monotonic (with an

uncertain rate of change) or indeterminate) and the nature (type and extent) of adaptation response options. These three factors combine to require different approaches to risk management. In these contexts, the framework provides useful risk mitigation options and describes some available options for reducing decision risk. For these reasons, it has been incorporated into a new set of climate change adaptation guidelines for the Australian urban water industry (WSAA 2016).

9. Concluding Remarks

While the number of climate change impact assessments continues to grow, it has been challenging for practitioners to make informed and robust decisions on adaptation strategies under considerable uncertainty. This chapter has emphasised the importance of a full recognition of climate uncertainty in the management of drought risk provided effort is made to avoid "decision paralysis" due to complexity. There are at least seven key ingredients for successful management of this risk:

(1) Climate is only one of many stressors that influence water planning and management decisions and outcomes. Climate-driven hydrological changes will combine with, and possibly amplify, the impacts of other sources of stress, including population growth, economic priorities, infrastructure maintenance/replacement, loss of biodiversity and hence ecosystem services and amenity, invasive exotic species, social behaviour and preferences, and land-use change. These factors could compromise water security if they are inappropriately monitored and managed. Thus, a systematic examination of potential adaptation responses over a wide range of plausible futures will need to consider the interplay between climatic, economic, environmental and cultural factors. This will lead to improvements in water security.

(2) The primary focus of adaptation planning should be preparedness and resilience rather than prediction. The state of preparedness needs to evolve with emerging threats. The emerging threats are likely to involve unexpected incremental changes in the factors listed above as well as "total surprises".

(3) Consideration of a small number but wide range of scenarios to facilitate decisiveness and action rather than "paralysis by analysis" (Peterson *et al.*, 2003). Low probability, high impact events (such as past multi-decadal droughts) should be included in the assessment and notions of the "most likely" projection or scenario or the "best guess" abandoned.

(4) Use of reliability, resilience and vulnerability criteria to assess asset performance under selected projections or scenarios and adaptation responses (e.g. water demand management). Also, the benefits and costs of immediate adaptation and retrofitting should also be assessed.

(5) Adoption of a continuous cycle of planning and evaluation. This will lead to the development of contingency plans with identified trigger points for alternative courses of action (the adaptation pathways approach, see Wise *et al.* 2015). Such a process may involve "transformational" change rather than small, once-off adjustments to existing practices. Planning, resourcing and construction lead times should be identified, as well as any means for reducing them. Also, maintenance of regional observational networks with long-term, high-quality records is needed to facilitate: detection of climate-related trends; evaluation of predictive models under non-stationary hydroclimatic conditions; and assessment of potential adaptation strategies.

(6) Diagnosis of the psychological, social and institutional limits and barriers to adaptation (Jones and Boyd 2011; Stafford-Smith *et al.* 2011; Wise *et al.* 2015). The removal of these barriers may require a preparedness to consider and plan for radical change.

(7) Adoption of adaptive governance arrangements and markets (e.g. flexible water pricing).

Finally, it is important to acknowledge that adaptive responses will be shaped by the geographical, hydrological, socio-economic and political settings confronting planners and the financial resources available. Effective communication and knowledge brokering will require ongoing and close contact between scenario developers and stakeholders, and the distribution of science-based information that is tailored to meet their needs (Jacob and van den Hurk 2009). Thus, future water planning will require deeper collaboration between practitioners and the social, economic, hydrologic and climate sciences.

Acknowledgments

I thank Kevin Hennessy (CSIRO Oceans and Atmosphere) for providing comments on the content of an earlier draft of this chapter. Much of the material presented here is based on discussions over the last 25 years with researchers and practitioners from several countries. These discussions have taken place in the context of: the Indian Ocean Climate Initiative (http://www.ioci.org.au/); the South Eastern Australian Climate Initiative (http://www.seaci.org/); IPCC Technical Paper IV — Climate Change and Water (http://ipcc.ch/pdf/technical-papers/climate-change-water-en.pdf); the Second to Fourth IPCC Assessment Reports; and the Australian Climate Change Science Program (https://www.environment.gov.au/climate-change/climate-science/australian-climate-change-science-program).

References

Alo, C. A. and G. Wang, 2008: Hydrological impact of the potential future vegetation response to climate changes projected by 8 GCMs. *J. Geophys. Res.*, **113**, G03011. doi:10.1029/2007JG000598.

Ascough II, J. C., H. R. Maier, J. K. Ravalico, and M. W. Strudley, 2008: Future research challenges for incorporation of uncertainty in environmental and ecological decisionmaking. *Ecol. Model.*, **219**, 383–399. doi:10.1016/j.ecolmodel.2008.07.015.

Bammer, G. and M. Smithson, 2008: *Uncertainty and Risk: Multidisciplinary Perspectives*, Earthscan, Cromwell Press, Trowbridge, U.K., 320pp.

Bates, B., P. Hope, B. Ryan, I. Smith, and S. Charles, 2008a: Key findings from the Indian Ocean Climate Initiative and their impact on policy development in Australia. *Clim. Change*, **89**, 339–354. doi:10.1007/s10584-007-9390-9.

Bates, B. C., Z. W. Kundzewicz, S. Wu, and J. P. Palutikof, 2008b: *Climate Change and Water, Technical Paper of the Intergovernmental Panel on Climate Change*, IPCC Secretariat, Geneva, 210 pp.

Bates, B. C. and G. Hughes, 2009: Adaptation measures for metropolitan water supply for Perth, Western Australia, In: *Climate Change Adaptation in the Water Sector*, F. Ludwig, P. Droogers, M. van der Valk, H. van Schaik, and P. Kabat, (eds.), Earthscan, pp. 187–204.

Bates, B. C., R. E. Chandler, S. P. Charles, and E. P. Campbell, 2010: Assessment of apparent non-stationarity in time series of annual inflow, daily precipitation and atmospheric circulation indices: A case study from southwest Western Australia. *Water Resour. Res.*, **46**, W00H02. doi:10.1029/2010WR009509.

Charles, S. P., M. A. Bari, A. Kitsios, and B. C. Bates, 2007: Effect of GCM bias on downscaled precipitation and runoff projections for the Serpentine catchment, Western Australia. *Int. J. Climatol.*, **27**, 1673–1690. doi:10.1002/joc.1508.

Collins, M., 2002: Climate predictability on interannual to decadal time scales: The initial value problem. *Clim. Dynam.*, **19**, 671–692.

Cornish, E., 2004: *Futuring: The Exploration of the Future*, World Future Society, Bethesda, MD, 313 pp.

CSIRO, 2008: *Water Availability in the Murray-Darling Basin, A report to the Australian Government from the CSIRO Murray-Darling Basin Sustainable Yields Project*, Commonwealth Scientific and Industrial Organisation, Australia, 67 pp.

CSIRO and Bureau of Meteorology, 2015: *Climate Change in Australia, Projections for Australia's NRM Regions*. Technical Report, CSIRO and Bureau of Meteorology, Australia. Retrieved from www.climatechangeinaustralia.gov.au/en.

Dessai, S., M. Hulme, R. Lempert, and R. Pielke, Jr., 2009: Climate prediction: a limit to adaptation? Chapter 5 in *Living with Climate Change: Are there Limits to Adaptation?* W. N. Adger, I. Lorenzoni and K. O'Brien (eds.), Cambridge University Press, Cambridge, U.K., pp. 64–78.

Dixon, H., D. M. Lawler, and A. Y. Shamseldin, 2006: Streamflow trends in western Britain. *Geophys. Res. Lett.*, **33**, L19406. doi:10.1029/2006GL027325.

Duinker, P. N. and L. A. Greig, 2007: Scenario analysis in environmental impact assessment: Improving explorations of the future. *Environ. Impact Assess. Rev.*, **27**, 206–219. doi:10.1016/j.eiar.2006.11.001.

Fowler, H. J., S. Blenkinsop, and C. Tebaldi, 2007: Linking climate change modelling to impact studies: Recent advances in downscaling techniques for hydrological modelling. *Int. J. Climatol.*, **27**, 1547–1578. doi:10.1002/joc.1556.

Giorgi, F., 2005: Climate change prediction. *Clim. Change*, **73**, 239–265. doi:10.1007/s10584-005-6857-4.

Haasnoot, M., J. H. Kwakkel, W. E. Walker, and J. ter Maat, 2013: Dynamic adaptive policy pathways: A method for crafting robust decisions for a deeply uncertain world. *Global Environ. Change*, **23**, 485–498. http://dx.doi.org/10.1016/j.gloenvcha.2012.12.006.

Hall, J., 2007: Probabilistic climate scenarios may misrepresent uncertainty and lead to bad adaptation decisions. *Hydrol. Process.*, **21**, 1127–1129. doi:10.1002/hyp.6573.

Hallegatte, S., 2009: Strategies to adapt to an uncertain climate change. *Global Environ. Change*, **19**, 240–247. doi:10.1016/j.gloenvcha.2008.12.003.

Hornberger, G. M., E. Bernhardt, W. E. Dietrich, D. Entekhabi, G. E. Fogg, E. F.-Georgiou, W. J. Gutowski Jr., W. B. Lyons, K. W. Potter, S. W. Tyler, H. J. Vaux Jr., C. J. Vorosmarty, C. Welty, C. A. Woodhouse, and C. Zheng, 2012: *Challenges and Opportunities in the Hydrologic Sciences*. National Research Council of the National Academies, The National Academies Press, Washington, D. C., 188 pp.

Hulme, M. and S. Dessai, 2008: Predicting, deciding, learning: Can one evaluate the 'success' of national climate scenarios? *Environ. Res. Lett.*, **3**, 045013. doi:10.1088/1748-9326/3/4/045013.

IPCC, 2000: Emission Scenarios. *Special Report of Working Group III of the Intergovernmental Panel on Climate Change*, N. Nakićenović, Coordinating Lead Author, Cambridge University Press, Cambridge, U.K., 599 pp.

Jacob, D. and B. van den Hurk, 2009: Climate change scenarios at the global and local scales. In: *Climate Change Adaptation in the Water Sector*, F. Ludwig, P. Droogers, M. van der Valk, H. van Schaik, and P. Kabat (eds.), Earthscan, pp. 23–33.

Jiménez Cisneros, B. E., T. Oki, N. W. Arnell, G. Benito, J. G. Cogley, P. Döll, T. Jiang, and S. S. Mwakalila, 2014: Freshwater resources. In: *Climate Change 2014: Impacts, Adaptation, and Vulnerability. Part A: Global and Sectoral Aspects. Contribution of Working Group II to the Fifth Assessment Report of the Intergovernmental Panel on Climate Change*, Field, C. B., V. R. Barros, D. J. Dokken, K. J. Mach, M. D. Mastrandrea, T. E. Bilir, M. Chatterjee, K. L. Ebi, Y. O. Estrada, R. C. Genova, B. Girma, E. S. Kissel, A. N. Levy, S. MacCracken, P. R. Mastrandrea, and L. L. White (eds.). Cambridge University Press, Cambridge, UK and New York, USA, pp. 229–269.

Jones, L. and E. Boyd, 2011: Exploring social barriers to adaptation: Insights from Western Nepal. *Global Environ. Change*, **21**, 1262–1274. doi:10.1016/j.gloenvcha.2011.06.002

Jones, R. N., 2000: Managing uncertainty in climate change projections — issues for impact assessment. *Clim. Change*, **45**, 403–419.

Kahya, E. and S. Kalayci, 2004: Trend analysis of streamflow in Turkey. *J. Hydrol.*, **289**, 128–144.

Kay, A. L. and Davies, H. N., 2008: Calculating potential evaporation from climate model data: A source of uncertainty for hydrological climate change impacts. *J. Hydrol.*, **358**, 221–239. doi:10.1016/j.jhydrol.2008.06.005.

Kruijt, B., J. P. M. Witte, C. M. J. Jacobs, and T. Kroon, 2008: Effects of rising atmospheric CO_2 on evaporation and soil moisture: A practical approach for the Netherlands. *J. Hydrol.*, **349**, 257–267. doi:10.1016/j.jhydrol.2007.10.052.

Kundzewicz, Z. W., L. J. Mata, N. W. Arnell, P. Döll, B. Jimenez, K. Miller, T. Oki, Z. Şen, and I Shiklomanov, 2008: The implications of projected climate change for freshwater resources and their management. *Hydrol. Sci. J.*, **53**, 3–10.

Lempert, R. J. and M. E. Schlesinger, 2000: Robust strategies for abating climate change. *Clim. Change*, **45**, 387–401.

Leonard, M. S. Westra, A. Phatak, M. Lambert, B. van den Hurk, K. McInnes, J. Risbey, S. Schuster, D. Jakob, and M. Stafford-Smith, 2014: A compound event framework for understanding extreme impacts. *WIREs Clim. Change*, **5**, 113–128. doi:10.1002/wcc.252.

MacCracken, M., 2001: Prediction versus projection — forecast versus possibility. *Weather Zine*, *Number 26*, Guest Editorial, February 2001.

Marsden Jacobs Associates, 2006: *Securing Australia's Urban Water Supplies: Opportunities and Impediments*, discussion paper prepared for the Department of the Prime Minister and Cabinet, 58 pp.

Mauget, S. A., 2003: Multidecadal regime shifts in U.S. streamflow, precipitation, and temperature at the end of the twentieth century. *J. Climate*, **16**, 3905–3916.

Mearns, L. O., M. Hulme, T. R. Carter, R. Leemans, M. Lal, and P. H. Whetton, 2001: Climate scenario development. In: *Climate Change 2001: The Scientific Basis, Chapter 13, Contribution of Working Group I to the Third Assessment Report of the Intergovernmental Panel on Climate Change*, Houghton J. T., Y. Ding, D. J. Griggs, M. Noguer, P. J. van der Linden, and D. Xiaoxu (eds.), Cambridge University Press, Cambridge, U.K. pp. 739–768.

Mietzner, D. and G. Reger, 2005: Advantages and disadvantages of scenario approaches for strategic foresight. *Int. J. Technol. Intell. Plann.*, **1**, 220–239.

Milly, P. C. D., J. Betancourt, M. Falkenmark, R. M. Hirsch, Z. W. Kundzewicz, D. P. Lettenmaier, and R. J. Stouffer, 2008: Stationarity is dead: Whither water management. *Science*, **319**, 573–574. doi:10.1126/science.1151915.

New, M. and M. Hulme, 2000: Representing uncertainty in climate change scenarios: A Monte Carlo approach. *Integrated Assess.*, **1**, 203–213.

Peterson, G. D., G. S. Cumming, and S. R. Carpenter, 2003: Scenario planning: A tool for conservation in an uncertain world. *Conserv. Biol.*, **17**, 358–366.

Pielke Sr., R. A. and R. L. Wilby, 2012: Regional climate downscaling: What's the point? *Eos*, **93**, 52–53.

Pittock, A. B., R. N. Jones, and C. D. Mitchell, 2001: Probabilities will help us to plan for climate change — without estimates, engineers and planners will have to delay decisions or take a gamble. *Nature*, **413**, 249.

Postma, T. J. B. M. and F. Liebl, 2005: How to improve scenario analysis as a strategic management tool?. *Technol. Forecast. Soc. Change*, **72**, 161–173. doi:10.1016/j.techfore.2003.11.005.

Reisinger, A., R. L. Kitching, F. Chiew, L. Hughes, P. C. D. Newton, S. S. Schuster, A. Tait, and P. Whetton, 2014: Australasia. In: *Climate Change 2014: Impacts, Adaptation, and Vulnerability. Part B: Regional Aspects, Contribution of Working Group II to the Fifth Assessment Report of the Intergovernmental Panel on Climate Change*, Barros, V. R., C. B. Field, D.J. Dokken, M.D. Mastrandrea, K.J. Mach, T.E. Bilir, M. Chatterjee, K.L. Ebi, Y. O. Estrada, R. C. Genova, B. Girma, E. S. Kissel, A. N. Levy, S. MacCracken, P. R. Mastrandrea, and L. L. White (eds.). Cambridge University Press, Cambridge, U.K. and New York, USA, pp. 1371–1438.

Schwartz, P., 1996: *The Art of the Long View: Planning for the Future in an Uncertain World*, Currency Doubleday, New York, N.Y., 258 pp.

Stafford-Smith, M., L. Horrocks, A. Harvey, and C. Hamilton, 2011: Rethinking adaptation for a 4°C world. *Philos. T. R. Soc. A*, **369**, 196–216. doi:10.1098/rsta.2010.0277.

Steele-Dunne, S., P. Lynch, R. McGrath, T. Semmler, S. Wang, J. Hanafin, and P. Nolan, 2008: The impacts of climate change on hydrology in Ireland. *J. Hydrol.*, **356**, 28–45. doi:10.1016/j.jhydrol.2008.03.025.

van Dijk A. I. J. M., H. E. Beck, R. S Crosbie, R. A. M. de Jeu, Y. Y. Liu, G. M. Podger, B. Timbal, and N. R. Viney, 2013: The Millennium Drought in southeast Australia (2001–2009): Natural and human causes and implications for water resources, ecosystems, economy, and society. *Water Resour. Res.*, **49**, 1040–1057. doi:10.1002/wrcr.20123.

Vance, T. R., J. L. Roberts, C. T. Plummer, A. S. Kiem, and T. D. van Ommen, 2015: Interdecadal Pacific variability and eastern Australian megadroughts over the last millennium. *Geophys. Res. Lett.*, **42**, 129–137. doi:10.1002/2014GL062447.

Viney, N. R., B. C. Bates, S. P. Charles, I. T. Webster, and M. Bormans, 2007: Modelling adaptive management strategies for coping with the impacts of climate variability and change on riverine algal blooms. *Global Change Biol.*, **13**, 2453–2465. doi:10.1111/j.13652486.2007.01443.x.

Visser, H., R. J. M. Folkert, J. Hoekstra, and J. J. deWolff, 2000: Identifying key sources of uncertainty in climate change projections. *Clim. Change*, **45**, 421–457.

WSAA, 2016: *Climate Change Adaptation Guidelines*, Project Report WSA 303–2016-v1.2, Water Services Association of Australia Ltd., February 2016, 89 pp.

Wilby, R. L., 2005: Uncertainty in water resource model parameters used for climate change impact assessment. *Hydrol. Process.*, **19**, 3201–3219. doi:10.1002/hyp.5819.

Wilby, R. L. and I. Harris, 2006: A framework for assessing uncertainties in climate change impacts: Low-flow scenarios for the River Thames, UK. *Water Resour. Res.*, **42**. doi:10.1029/2005WR004065.

Wise R.M., I. Fazey, M. Stafford-Smith, S. E. Park, H. C. Eakin, E. R. M. Archer Van Garderen, B. Campbell, 2015: Reconceptualising adaptation to climate change as part of pathways of change and response. *Global Environ. Chang.*, **28**, 325–336, http://dx.doi.org/10.1016/j.gloenvcha.2013.12.002.

Zheng, H., L. Zhang, C. Liu, Q. Shao, and Y. Fukushima, 2007: Changes in stream flow regime in headwater catchments of the Yellow River basin since the 1950s. *Hydrol. Process.*, **21**, 886–893. doi:10.1002/hyp.6280.